The Social Context of Technological Change

Egypt and the Near East, 1650–1150 BC

Proceedings of a conference held at St Edmund Hall, Oxford
12–14 September 2000

Edited by
Andrew J. Shortland

Oxbow Books

First published in the United Kingdom in 2001. Reprinted in 2017 by
OXBOW BOOKS
The Old Music Hall, 106–108 Cowley Road, Oxford OX4 1JE

and in the United States by
OXBOW BOOKS
1950 Lawrence Road, Havertown, PA 19083

Paperback Edition: ISBN 978-1-84217-050-2
Digital Edition: ISBN 978-1-78570-564-9

A CIP record for this book is available from the British Library

For a complete list of Oxbow titles, please contact:

UNITED KINGDOM
Oxbow Books
Telephone (01865) 241249, Fax (01865) 794449
Email: oxbow@oxbowbooks.com
www.oxbowbooks.com

UNITED STATES OF AMERICA
Oxbow Books
Telephone (800) 791-9354, Fax (610) 853-9146
Email: queries@casemateacademic.com
www.casemateacademic.com/oxbow

Oxbow Books is part of the Casemate Group

Cover image: Detail of the reliefs, decorating the exterior of the Temple of Ramesses II at Abydos, showing a Hittite chariot (photograph by Ian Shaw).

To my wife Anja,
on our wedding day
30th June 2001

Contents

List of Contributors

JANINE BOURRIAU
McDonald Institute for Archaeological Research,
Downing Street,
Cambridge CB2 3ER

PETER HAARER
Institute of Archaeology
University of Oxford
Beaumont Street
Oxford OX1 2PG

ANJA HEROLD
Goldbekweg 3
D-22303 Hamburg
Germany

CAROLINE JACKSON
Department of Archaeology and Prehistory,
University of Sheffield,
Northgate House,
West Street,
Sheffield S1 4ET,

P. R. S. MOOREY
Department of Antiquities
Ashmolean Museum
Oxford OX1 2PH

PAUL NICHOLSON
The School of History and Archaeology
Cardiff University
PO BOX 909
Cardiff CF1 3X

SARAH PAYNTER
Centre for Archaeology
English Heritage
Fort Cumberland
Fort Cumberland Road
Eastney
Portsmouth PO4 9LD

JACKE PHILLIPS
McDonald Institute for Archaeological Research,
Downing Street,
Cambridge CB2 3ER

THEA A. POLITIS
Dept of Archaeology
Faculty of Letters & Social Sciences
The University of Reading
Whiteknights
PO Box 218
Reading RG6 6AA

EDGAR B. PUSCH
Grabung Qantir
Pelizaeus-Museum
Am Steine 1-2
D-31134 Hildesheim
Germany

THILO REHREN
Institute of Archaeology
University College London
31–34 Gordon Square
London WC1H 0PY

ELEANOR ROBSON
All Souls' College
Oxford OX1 4AL

MARGARET SERPICO
Institute of Archaeology
University College London
31–34 Gordon Square
London WC1H 0PY

IAN SHAW
Department of Archaeology (SACOS)
University of Liverpool
14 Abercromby Square
Liverpool L69 3BX

Andrew and Susan Sherratt
Institute of Archaeology
University of Oxford
Beaumont Street
Oxford OX1 2PG

Andrew Shortland
Research Laboratory for Archaeology and the History of Art
6 Keble Road
Oxford OX1 3QJ

Laurence Smith
McDonald Institute for Archaeological Research,
Downing Street,
Cambridge CB2 3ER

Rachael Sparks
Institute of Archaeology
University College London
31-34 Gordon Square
London WC1H 0PY

Zophia Stos-Gale
Research Laboratory for Archaeology and the History of Art
6 Keble Road
Oxford OX1 3QJ

Mike Tite
Research Laboratory for Archaeology and the History of Art
6 Keble Road
Oxford OX1 3QJ

Preface and acknowledgements

The technologically capabilities of the ancient world have long fascinated scholars and the general public alike. Much scholarly debate and no little public interest has gone into the study of such topics as how the pyramids and Stonehenge were built, when the first tools were made and the rise of early agriculture. W. M. F. Petrie did some of the most important early work on Egyptian technology and industries, recognising that the archaeological record contained not only palaces and temples, but evidence of workshops and the habitations of ordinary people. Using his wide ethnographic knowledge, Petrie was able to put together some of the first *chaine opératoire* for the manufacture of ancient objects. Petrie's work was built upon and greatly extended by Alfred Lucas, who produced the first systematic description of Egyptian materials and their associated industries. However, apart from Petrie, Lucas and one or two others, in the past most Near Eastern archaeologists have treated material culture not in a technological or industrial way, but as a tool for defining chronology and delineating the level of interactions between neighbouring societies. In this way material culture has fed into the wider study of social, political and religious processes which has dominated the Near Eastern archaeology over the last twenty five years or so. In addition, Near Eastern studies have become increasingly specialised over this time, each archaeologist concentrating on a narrower and narrower area of study. These areas are often limited to one geographical region (e.g. to Egypt, Mesopotamia or Crete, etc.) or to one field (language, archaeology, material culture, religion). Often these specialists have found themselves split into different university departments and only meeting those working in related fields on an occasional basis. This fragmentation of archaeological study into a myriad of specialisms and sub-specialisms has probably been detrimental to the study of areas that require the input of many different specialisms to make adequate progress. The social implications of ancient technology and technological change is just such an area, and this fragmentation has probably delayed the realisation that technology also plays a vital role in past socio-economic systems.

The meeting from which this volume has developed was held in September 2000 at St Edmund Hall, Oxford. The idea for the meeting was inspired by a conference organised by Bill Sillar and Brian Boyd at Lampeter University in September 1999, entitled "Embedded technologies: reworking technological studies in archaeology". Much of the format of the Lampeter conference was copied in the Oxford meeting. The aim of both was to bring together specialists from many different fields to exchange ideas in a relatively informal atmosphere, with the emphasis on discussion and exchange of ideas. In order to provide a setting in which the discussion would be of greatest relevance to the participants, the Oxford meeting (unlike the Lampeter conference) greatly restricted the geographical area and time period discussed. The geographical area chosen was the Near East and associated areas, including Greece, Crete, Cyprus, Anatolia, the Levant, Mesopotamia and Egypt. Since technological change was to be the main subject of the meeting, 1650–1150 BC – the end of the Middle Bronze age through to the Late Bronze Age – was chosen, a period when many technological innovations appear for the first time.

Over the three days of the meeting, seventeen papers were given, fourteen of which appear in this volume. The other three presentations were by Sariel Shalev (Middle Bronze Age metal production), Don Evley (Minoan crafts) and Valérie Matoïan (glass from Ugarit). A total of 30 people attended the meeting at sometime over the three days. In response to the Oxford meeting, a further conference "Social context of technological change II" was proposed for 2002, to take place in Cambridge.

The editor would like to thank Professor Mike Tite and Mrs Jane Simcox for their help and encouragement throughout the planning and organisation of the meeting. Thanks shoul also go to the Principal, fellows and staff of St Edmund Hall, Oxford for their help and flexibility which greatly assisted in making the meeting run so smoothly.

A.J.S.
Oxford
1st June 2001

Chapter 1

The Mobility of Artisans and Opportunities for Technology Transfer between Western Asia and Egypt in the Late Bronze Age

P. R. S. Moorey

Abstract

In the Near East and Egypt, the second half of the second millennium B.C. is the first period in their history when an unprecedented combination of information from objects and written documents allows for an investigation of technology transfers across the region. The Interaction of the great powers in Egypt, Turkey, Syria and Mesopotamia at this time, in war and peace, facilitated an unprecedented mobility of craftsmen and women across the whole area. This has an obvious potential for the transfer of new or evolving specialist technologies at court level. The surviving evidence is intermittent and scattered, often complex, and always difficult to integrate. This paper will examine particularly the role of military activities and international diplomacy as contexts for facilitating communication and change in élite technologies.

INTRODUCTION

In a famous passage in the Old Testament, to which I shall return, there is a description of Hiram of Tyre, the Phoenician craftsman sent by the City's ruler, of the same name, to assist King Solomon with the construction and furnishing of the Temple in Jerusalem in the tenth century B.C. (Iron Age II). Of this passage the distinguished art historian Ernst Gombrich (1993, 69) has remarked 'who can doubt that the account reflects the situation, so often repeated in history, that in the absence of a local tradition a master was called from a centre of acknowledged excellence... I frankly do not know if any social historian has set himself the task of investigating this phenomenon as such; maybe he would find that there are too many variables involved to make a profitable field of study... availability of materials... tools and equipment... what matters most is obviously a kind of skill... what we call "know-how".' This brief essay (and it is simply that) on technology transfer, a complex topic indeed, at a particular point in ancient history, will undoubtedly bear out Gombrich's caution; but it is a factor too central to the chosen subject of this workshop to be overlooked.

It was Homer, in Book VII (382–6) of the *Odyssey*, who epitomized more vividly than any other ancient author known to me the age-old significance of the mobility of craftsmen and other specialists in society at large rather than just at court level. 'For who of his own accord ever approaches and summons a stranger from elsewhere unless it be one of those who practise a craft of public benefit: a seer, or a healer of illnesses, or a master at building with timbers, or inspired by song, who gives pleasure with his singing? For these men are summoned throughout the boundless earth'.

Although Homer's words reflect the circumstances of the Iron Age in the Greek World in the earlier first millennium B.C., when free movement of labour existed in a way not evident in the Late Bronze Age during the second half of the second millennium in the Near East, which is our concern here, the motivation he epitomises here is timeless. Manufactured goods had travelled widely in the Near East and Egypt since time immemorial, whether as gifts, booty or traded items; but to what extent had they ever made craftsmen conversant with new techniques of manufacture and production? This way fresh designs, novel decoration and previously unknown raw materials might well have been transmitted, but not the means of manufacture and the usually well-guarded secrets of craftsmanship. In an influential pioneer study of the relationships between the arts of Egypt, the Aegean and Western Asia in the Late Bronze Age written over a generation ago, William Stevenson Smith (1965, XXV-VI) noted that:

> 'Sometimes we can detect contacts only through the appearance of foreign raw products or unfamiliar animals. In other cases the effect of the introduction of new features is more striking, as with the rapid spread of the use of the horse and chariot. Foreign manufactures should provide the most useful kind of evidence, but it is not always possible to follow them through various intermediaries to their source. Tracing out the interplay of decorative motives is even more difficult. Most important of all would be the identification of ideas which had been exchanged... however we can do little more than examine instances in which narrative representation might have been affected by a new point of view from abroad.'

Trace elements of rather a different kind are the business of materials science; but their significance for tracking the migration of craft skills and technologies unsupported by other evidence, is often no less elusive than the clues provided by art and archaeology. It is the ideas in the brains of men and women, and the skills in their hands, as Gombrich emphasized, that have made them the vital carriers in all ages and places. To bring this into our own time we have only to recall the reported words of an American University President in the 1930s. 'Hitler shook the trees and I picked up the apples'; or, then again, a decade later the activities of the victorious allies, competing one against the other, to capture the 'brains' behind Hitler's precocious rocket technology.

ROYAL COURTS: POWER, POLITICS AND PATRONAGE

In the Late Bronze Age (c. 1650–1150 B.C.) there was a network of international contacts and military conflicts between the major powers in the Near East and Egypt, previously unparalleled on such a scale in this region. The main players were the Egyptian pharaohs of the New Kingdom, the kings of the Hittites in Turkey, of the Kassites in Babylonia in southern Iraq and of the Mittani in northern Iraq and Syria, where they were later superseded politically by the Assyrians (see Fig. 1.1). In his seminal studies of Near Eastern technologies, notably glass and glazed materials, the philologist Leo Oppenheim (1978) always emphasised the role of royal courts in the record and transmission of technological innovations. He was working with texts compiled by families of experts ('scholars'), long associated with the local courts, who claimed their 'wisdom' was rooted in earlier, even mythical, periods. The texts they used had been arranged in standard form towards the end of the second millennium B.C., though many had precursors dating hundreds of years earlier.

He, and a number of other modern scholars, have seen the Hurrian-speaking peoples of the Kingdom of Mittani as particularly instrumental during the Late Bronze Age in a series of technological innovations: the combined introduction of the light horse-drawn war-chariot, the composite bow and scale-armour (cf. de Vaux 1967); the pioneering of iron technology (cf. Schaeffer 1939; Stech-Wheeler *et al.* 1981); and the development of core-formed glass vessels and related innovations in glaze technology (cf. Peltenburg 1987, 22). If the crucial period for many of these innovations had been in the first, rather than in the second quarter, of the second millennium B.C., as I

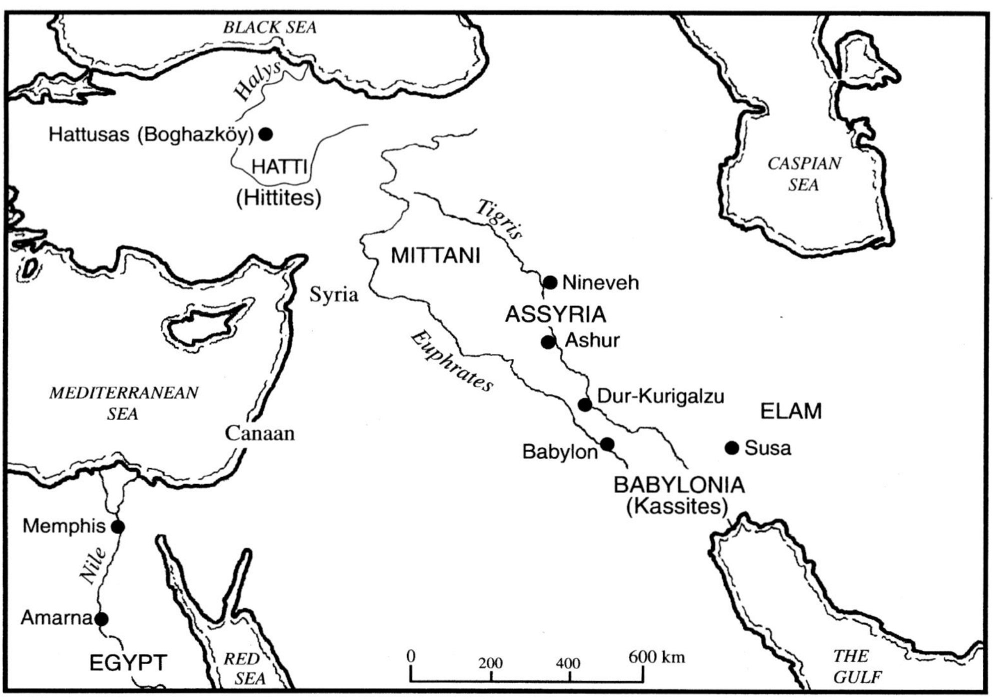

Figure 1.1: Map of the Near East showing places mentioned in the text.

believe may have been the case (cf. Moorey 1989), then the matter is not so simple. What the Mittani may have facilitated was the scholarly tradition at the heart of Oppenheim's written evidence and the diffusion of a number of influential innovations.

In all times and places royal patronage has particularly stimulated luxury crafts. Cyril Smith (1981, 195) always emphasised in his studies of the history of technology: a particularly 'courtly' motivator for innovation: 'over and over again scientifically important properties of matter and technological important ways of making and using them have been discovered or developed in an environment which suggests the dominance of aesthetic motivation'. Also the military were an integral part of ancient Near Eastern and Egyptian courts and warfare has always been a powerful driving force in technological innovation. More is consequently known about the technology of warfare than of any other ancient activity save for farming.

It was not accidental that courts were central to technological developments. Only in such places were the necessary resources in terms of labour and materials, tools and workshops readily available on an appropriate scale. It was, moreover, within this environment that vital subsidiary factors fostering technology transfer came into play: craftsmen of all types would have been working closely together in associated workshops, many of them foreigners. Many individual luxury artefacts, from chariots to royal tableware, required the skills of a diversity of specialist craftsmen interacting with one another to manufacture them. It is also clear, as so often in the pre-industrial world of technology, remembering only Benvenuto Cellini, that leading craftsmen were men of many practical talents. The famous biblical description at 2 *Chronicles* 2: 14–16 of the craftsman Hiram lent by the Phoenician king to Solomon to assist in creating the Temple in Jerusalem illustrates this albeit in an exaggerated way: 'he is an experienced worker in gold and silver, copper and iron, stone and wood, as well as in purple, violet and crimson yarn, and in fine linen; he is also a trained engraver who will be able to work with your own skilled craftsmen and those of my lord David your father, to any design submitted to him'. Despite current controversies over the time at which this was written, whether in the tenth century B.C. or centuries later, it epitomizes two of the factors vital to technological transfer amongst the royal courts of the Near East and Egypt in the Late Bronze Age: the exchange of skilled craftsmen between rulers as a matter of diplomacy in times of peace and the versatility of the best of them.

There is no clear evidence for completely independent specialist craftsmen trading on their own account in the Late Bronze Age; but, if they existed, they might be expected to have fallen outside the range of the surviving documentation. Throughout history, however, such artisans have generally produced utilitarian goods in response to a social demand outside the *élite* community, where neither technological innovation nor precious materials are to be expected. Within villages a high level of self-sufficiency was normal, both in domestic economies and in basic craft technologies, using a restricted range of locally available raw materials. In such circumstances, as an English village craftsman recorded early in the last century: 'There was nothing for it but practice and experience of every difficulty... Every detail stood by itself, and had to be learnt either by trial and error or by tradition' (Sturt 1923, 19–20).

The advantage of royal courts for my purposes is that they have left relevant documentation. Even in Egypt, where pen, ink and papyrus were the customary and perishable means of written record, diplomatic correspondence in the New Kingdom (c.1550–1070 B.C.) was written on clay tablets in the cuneiform script and the Babylonian (Akkadian) language. Such tablets, once baked, are almost as durable as potsherds. It is now more than a century since a chance find by local peasants revealed the diplomatic archives of Tell el-Amarna in middle Egypt. This was the new capital, established in the middle of the fourteenth century B.C. on a virgin site, by Amenophis IV (Akhenaten), the so-called heretic pharaoh and probably father of Tutankhamen. It was abandoned as a royal city within a generation. Akhenaten had taken with him from the old capital Thebes the diplomatic archives not only of his early years as pharaoh, but also those of his father who had reigned for nearly forty years (c. 1391–53 B.C.). Many of the tablets found by peasants disappeared before some 350 were rescued for the Berlin and British Museums. These, the renowned 'Amarna Letters', were published definitively in German within twenty years (Knudtzon 1907–1915) and in an English translation by Moran in 1992. Regrettably few have so far been found subsequently in continuing controlled excavations at Tell el-Amarna.

Just over a decade after Petrie's pioneering excavations at Amarna in 1891–2, Winckler and Macridy Bey, excavating between 1906 and 1912 close to Boghazköy, a modern village some one hundred miles east of Ankara in Turkey, revealed the site of the ancient Hittite capital. Some 10,000 clay tablet fragments, many inscribed in Akkadian, others in the local language, were then recovered. They were rapidly recognized as the royal archives of the Hittite Kingdom of Hattusa in the fourteenth and thirteenth centuries B.C. By 1915, the year in which the 'Amarna Letters' were fully published in German, the Czech scholar Hrozny had deciphered the previously unknown Hittite language in which most of the Boghazköy texts are written. This archive revealed that the Hittites had recurrently contested overlordship of Syria with the New Kingdom pharaohs of Egypt. In the fourteenth century B.C., when Egypt was ruled by pharaohs without marked military ambitions in Western Asia, the Hittites controlled this strategic area between Canaan and Mesopotamia. When later the ambitious and long-lived Ramesses II (c. 1290–24 B.C.) ruled Egypt, he vigorously challenged them.

The documents from Tell el-Amarna and Boghazköy indicate that international royal correspondence was carried by couriers, though the exact means of transport is not usually evident. It has been calculated on internal evidence (cf. Kühne 1973, 105 ff.) that overland contacts between Egypt and the Hittite capital took an average of between 23 and 31 days, depending on whether the courier had started from Memphis, Amarna or Thebes in Egypt. Royal retinues passed backwards and forwards; rich gifts were exchanged; and strict protocols governed such exchanges. The 'Great Kings' addressed one another as 'brother', beginning their letters: 'I and my house, horses, chariots, officials and my country are well indeed. May everything be likewise well with my brother and his house, his horses, chariots, officials and his country.' The significance of the priority given to horses and chariots will be evident later in this paper.

LANGUAGE AS AN INDICATOR OF TECHNOLOGY TRANSFER

Before passing on to more thorough examination of the evidence provided by these written sources for technology transfer, brief attention needs to be given to the role of lexicography in tracking it. The impact of the technology of one region on that of another is likely to be most vividly indicated by the technical terms of foreign origin in the host language, though they do not, of course, necessarily indicate the origin of the innovation. The lexicon of ancient Egypt has been intensively researched in this respect (cf. Redford 1992: 236–7). The relative influence of southwest Asia as a source of new materials, techniques and manufactures may be gauged from such linguistic evidence. Technical expressions describing the chariot, its parts and accoutrements account for half the relevant terms; others refer to types of weapons and military activities. Food and its preparation account for 8% of foreign vocabulary, whilst 7% embrace architectural features, construction techniques and materials (cf. Redford 1992, 236–7).

Lexicography was central to Leo Oppenheim's outstanding contributions to the study of Mesopotamian texts relating to the early manufacture of glazed materials and glass. It was he (Oppenheim 1973, 263), who cautiously pointed out that words for glasses are either isolated in Akkadian or clearly 'West Semitic'. He consequently attributed the 'invention' of glass to North Syria at a time when this formed part of the kingdom of Mittani in the sixteenth century B.C. (Oppenheim 1973, 263). The reign of Tuthmosis III (c.1479–25 B.C.) is commonly taken to mark the beginning of intentional glass making in Egypt. Lilyquist and Brill (1993, 43) have argued on art-historical and analytical grounds that Asiatic craftsmen were present in Egypt from an initial stage. Eyre (1987, 195), working from the evidence of Egyptian texts, is in no doubt that 'the influx of foreign craftsmen into temple and royal service was considerable.' However, his deductions are more cautious: 'they may have been of critical importance in the technological advances seen, particularly in areas such as metal working, arms manufacture, glass making and glazing, and may even have influenced artistic styles, although the putative foreign origins of individual workmen is meagre evidence for stylistic or technical borrowing'. This is a perfect cue, indeed, for the entry of archaeological scientists and other students of material culture to provide further evidence to weigh in the balance of probabilities in any particular case.

WARFARE AND THE MOBILITY OF SPECIALISTS

It is no surprise to discover in such an imperialistic age that the documentation of War and Peace: Conflict and Diplomacy provide the poles round which the best evidence relevant to the mobility of artisans and the first-hand transfer of specialist craft knowledge is most conveniently assembled. This information primarily relates to male artisans. However, the role of women in technology transfer should be mentioned here, not least since the organic nature of the specialized industry most relevant to them, textile manufacture, normally excludes it from the surviving archaeological record (cf. Eyre 1987, 200–1; van de Mieroop 1997, 185-7). In New Kingdom Egypt, for example, within the orbit of the 'royal harem', an organization of women, dependent on the royal household, had a significant role in the textile production. Texts indicate that this establishment included foreigners.

In the rare conditions provided by the tomb of Tutankhamum (cf. Vogelsang-Eastwood 1999, 80) two groups of foreign (Western Asiatic) garments had survived. Those definitely not of Egyptian origin, presumably royal diplomatic gifts, include a full-length tunic, two 'hip-wraps', a pair of leather sandals and a headrope, like the modern Arab aqal, which had actually been placed on the pharaoh's head underneath the famous gold mask. The other group, of non-Egyptian forms, but probably made in Egypt, include several pairs of socks, gloves and gauntlets, which illustrate a migration of Asiatic fashions and possibly foreign workmanship to Egypt.

Returning to expatriate labour, account has to be taken of two types of transfer in conditions of war: those men captured in battle, who were predominantly young and those systematically deported from the towns and villages of specific conquered regions, often with women and children, who should never be overlooked as active members in many ancient industrial enterprises. Specialist artisans travelled routinely with armies to provide basic repair and support services. Such workmen, by definition, would have had the particular skills essential to the tasks at hand, predominantly in working wood and metal for military purposes. Deportees, on the other hand, would have embraced a much wider range of skills. Most conspicuous amongst them would have been those required to manufacture luxury goods: gold and silverwork; glazed materials and glass and fine work in base metals, amongst which iron was the novelty at this time. The sack of towns also provided the victor with a wide range of raw materials, whereas armies are likely to have moved routinely only with the raw materials required to service their essential needs on campaign.

As it is recorded in 2 *Kings* 25, Nebuchadnezzar's sack of Jerusalem in 598/7 B.C., is perhaps the best known example of a deportation. He 'deported the rest of the people left in the city, those who had deserted to the King of Babylon and any remaining artisans. He left only the poorest people to be vine-dressers and labourers'. What happened then, at least to the *élite* young male exiles, is made explicit at *Daniel* 1:4: 'Then the king ordered Ashpenaz, his chief eunuch, to take certain of the Israelite exiles, of the blood royal and of the nobility... at home in all branches of knowledge, well-informed, intelligent, and fit for service in the royal court; and he was to instruct them in the literature and language of the Chaldaeans (i.e. Babylonians).'

The antiquity of such practices is illustrated some two thousand years earlier by a Sumerian epic poem, not necessarily historical, describing the stormy relationship of the King of Uruk in southern Iraq and the ruler of Aratta in Eastern Iran. It relates how the victorious Sumerians took away not only raw materials, such as gold and precious stones, but also jewellers and goldsmiths, as well as moulds for casting metal objects (Wilcke 1969, lines 409–12; cf. Gelb 1973).

It was, however, the Assyrians in the first half of the first millennium B.C., as their monumental palace relief sculptures and numerous inscriptions make clear, who used deportation systematically as an instrument of imperial exploitation. Indeed, Tiglath-Pileser III (c.745-27 B.C.), who instituted the unprecedented policy of mass deportations, changed the face of the Near East for generations by such a policy (cf.

Oded 1979). In the Late Bronze Age one of his most famous predecessors, Tukulti-Ninurta I (possibly Nimrod of the Old Testament) (c.1244–08 B.C.), built himself a new capital across the Tigris from the traditional capital at Ashur in Assyria, using artisans and labourers captured from Babylonia and Syria, who lived on site and were engaged in a wide range of crafts (cf. Freydank 1976–80) to judge from surviving inscriptions.

Chariots and chariotry are the most evident instance of military technology transfer in the Late Bronze Age from southwest Asia into Egypt. Not only did the horse and chariot come from the east, but, as already noticed, so did the relevant technical vocabulary (cf. Schulman 1963, 87–8). The fortunate survival of a number of actual chariots in Egyptian tombs allows for more detailed examination of this transfer (cf. Littauer and Crouwel 1985). Although there has never been a systematic expert study of the woods of which they were made, sufficient identifications are available for our purposes. Wheels were made of elm and tamarisk (cf. Western 1973), whilst birch bark was used to cover parts of the sides, the wheel naves and the spokes. The tamarisk wood could be of local Egyptian origin; but elm and birch are likely to have been imported from Turkey. Elm, used in modern wooden wheels, is not only susceptible of steam-bending, but also, being hard and tough, it is an ideal timber for chariot bodywork and wheels. The late Professor Pye, a well-known designer and worker in wood, observed to me many years ago, with respect to an Egyptian wheel-hub in the Ashmolean (Western 1973) that the suitability of wood for certain purposes varies not only with the sub-species, but also with the part of the tree from which it is taken, and even with the particular soil or climate in which the tree has grown. He was of the opinion that the bent wood obvious in the construction of this hub might have been trained to grow at source in the required shape rather than steam-bent later. This would imply some kind of 'do-it-yourself kit' from Turkey or thereabouts, transported for assembly and finishing in Egypt. It is not true, as has sometimes been claimed, that birch-bark can only be used fresh, requiring that its use demonstrates manufacture close to source. In fact, it remains pliable for a 'considerable time, and even after drying it can be softened by moisture' (Mcleod 1970, 36).

Chariots at this time were not, as is often popularly assumed, used *en masse* like a tank formation; they were pre-eminently mobile firing platforms for archers often equipped with composite bows, a weapon invented in Western Asia. These were commonly made of ash, birch and possibly cherry (McLeod 1970, 36); woods again native to Western Asia. Egyptian tomb reliefs of the 18th Dynasty show angular bows being manufactured in Egypt, so foreign woods and initially the skills of foreign craftsmen were certainly made available to local workshops, even if captured or traded bows of Western Asiatic manufacture were also current in Egypt.

New instruments of war, like composite bows fired from fast moving light chariots, generate new defensive measures. In Western Asia, with the development of light chariotry as a war machine in the second quarter of the second millennium B.C. (Cf. Moorey 1986) leather or metal scale-armour emerged as the response for protecting men, horses and chariot bodies. By the beginning of the New Kingdom such armour was being made in Egypt, where it has been found on a number of sites, not least a suit of scale-armour placed in the tomb of Tutankhamen (1333–23 B.C.). Howard Carter described this as: 'made up of scales of thick tinted leather [perhaps red and yellow] worked on a linen basis or lining, in the form of a close-fitting bodice without sleeves' (Vogelsang-Eastwood 1999, 109–111). Such armour was always regarded as a particularly acceptable gift for pharaoh from a foreign ruler and came also as booty. Tuthmosis III (1479–25 B.C.) in his *Annals* made special mention of the inlaid bronze armour belonging to the chiefs of Kadesh in Syria and Megiddo in Palestine (Aldred 1970, 107) captured in the 23rd year of his reign, and distinguishes it from other suits belonging to their chariot warriors (*maryannu*). On the Asiatic side, excavations at Nuzi in eastern Iraq seventy or so years ago, yielded examples of the component parts of scale-armour (Starr 1937, pl.126 A-L; 1938, 475 ff.), whilst texts from the same site provided the local terminology for such armour, which may then be traced westwards when it was adopted in international correspondence as at Amarna (Knudtzon 1915, 22 iii 37–40) and Boghazköy (Speiser 1950, 48).

Personal names of craftsmen may be no less indicative of transfer. In the later 18th Dynasty, a father and son team of boat-builders, bearing the title 'chief craftsmen of the king', have foreign names of Syrian origin. It is not, however, known whether the family had arrived as captives or as free immigrants. The Craftsman with the title 'chief draughtsman' of the god Amun of Thebes may be traced through at least seven generations of a family whose founder bore the 'evidently foreign name Pt-b^{c}r' (Eyre 1987, 195). This highlights another very important point: the way in which particular crafts and skills ran in families for generations. This is a recurrent factor in the history of applied technology before the Industrial Revolution, reminding us how specialist transfers might have been sustained once established.

ROYAL GIFT-EXCHANGE AS A MECHANISM OF TECHNOLOGY TRANSFER

My quotation from the Old Testament referring to Hiram of Tyre called attention to the mobility of craftsmen at the highest level through diplomacy rather than war. Gift exchange, including human experts, between rulers is a much discussed aspect of the diplomatic archives of this period. Oppenheim (1970, 14) took the gifts to represent: 'the pride of the king in the quality of the products of his country, in the techniques of the craftsmen of his palace or in the range of his personal international contacts'. This is obviously true, but only up-to-a-point. As Liverani (1979) subsequently pointed out, such transactions were in reality trade, in the sense that exchange, like any activity, is a total social phenomenon 'dictated concurrently by economic and non-economic factors (political, religious, psychological, etc) and is conditioned by custom-bound ideological schemes which over shadow the economic aspect.' Amongst the commodities sent from one court to another by the 'Great Kings' were specialised personnel, viewed as prestige goods (cf. Zaccagnini 1983, 250–1). Here there is an interesting distinction, in so far as surviving records go, between the artisans who generally moved through war and deportation and the physicians, soothsayers, and other specialists who were sent and returned through diplomacy; in modern terminology the tradesmen on the one hand, the professionals on the other, with the rare 'sculptor or mason' grouped with the professions. It was primarily the rulers of Egypt and Babylonia who provided their foreign 'brothers', above all the King of the Hittites, with the latter. Sometimes requests were ingeniously refused. For instance,

Ramesses II of Egypt, in replying to a King of the Hittites who needed medicines for his sister in order that she might bear a child, argued that since the Hittite Princess in question was already fifty or sixty years of age it was not the job of a doctor to help her bear a child. If the gods had recommended favourably, he would immediately have sent a 'conjuror' and a doctor who would have prepared medications for the lady (Zaccagnini 1983, 252).

Such 'specialists' did not, it would appear, contribute to the transfer of their knowledge, they went as agents for a particular purpose and generally returned. Indeed, this is confirmed by an exceptional case. Amenophis III of Egypt, a year or so before his death, requested the King of Mittani to send a statue of the goddess Ishtar of Nineveh to Egypt as an ultimate remedy for his illness. In responding the King of Mittani writes (Zaccagnini 1983, 254–5): 'Thus (says) Ishtar of Nineveh, the Lady of all lands: 'To Egypt, the country which I love, I will go, and from there I will come back'. And now I sent her and she is gone. Behold, at the time of my father, the Lady went to that land, and in the same way she was honoured when she formerly stayed (there), so may my brother honour her and may he send her back with joy, and may she come back... Ishtar is a goddess of mine; she is not a goddess of my brother.'

There were clearly also protocols about how such guest specialists were to be treated; the best example relates to a doctor from Babylon sent to Turkey. When the doctor had fallen ill and died the Hittite king stressed how he had looked after him, sending a tablet to Babylon listing the presents he had given him. Of another doctor the King of the Hittites writes: 'the lady whom he has taken (as wife) was of my own family, and he has a nice house (here). [Should he say:] 'I want to go back to my country!!,' he may leave and go back' (Zaccagnini 1983, 254). These were, of course, not free artisans; the autocratic tone of the royal correspondents emphasises that, if emphasis is needed.

Occasionally, in gift exchanges, technology emerges rather more specifically, as when an envoy witnesses the casting of a gold statue at the Egyptian Court (cf. Moran 1992, AE 27). This would appear to have been to impress rather than to instruct. More relevant to technology transfer is the movement of sculptors to Turkey from Babylon or of stone masons from Turkey to Egypt (cf. Zaccagnini 1983, 252). When Hattusilis III of the Hittites (c.1275/64–1245/39 B.C.) asked the Kassite King of Babylonia to lend him a sculptor, he promised to return him as he had done already when borrowing a sculptor previously from the Kassite king's father. Unfortunately, as Henri Frankfort (1958, 122) remarked, 'There is no Kassite sculpture in the round for comparison, but in one respect Hittite carving has closer analogies to Babylonian than to Egyptian or Assyrian work; it was a very heavy, thoroughly plastic relief, while Egypt and Assyria use flat relief with a subtly modulated surface.'

THE PROBLEM OF 'TRAVELLING' CRAFTSMEN:

Up to this point I have restricted my discussion to sources directly relevant to the period with which this conference is primarily concerned. In closing, I would like to draw attention to a palatial archive of the nineteenth century B.C., from Mari in Syria, comparable in significance for our purposes to the later Amarna and Boghazköy archives and the earliest of its kind known at present.

The mobility of artisans, within the orbit of royal courts, locally and internationally, is already apparent (cf. Durand 1992). Records of military activity and mass deportations, as well as royal gift exchanges, embrace numerous opportunities for technology transfer through artisans on the move. At Mari the archives make very apparent a real problem at the time over the availability of experts, be they diviners, musicians, scribes or technicians and their consequent mobility when in demand. Musicians represent a particularly mobile group, with practitioners listed at Mari from such distant cities as Aleppo and Carchemish, Hazor and Qatna to the west and northwest, Assur to the northeast. Doctors, valued at Mari for a variety of skills rather than just a single one, as well as architects and builders, appear moving from court to court as they do in the Late Bronze Age archives. If some of the technologies, such as that for glass and glazed materials, and possibly for early iron working, were already established in court workshops in the Middle Bronze Age, ready to flourish and develop in the greater internationalism of the Late Bronze Age, then the element of continuity would be even more striking.

I have left to the end of this paper any mention of a role in technology transfer at this time for 'travelling tinkers' in particular, transhumant or fully nomadic peoples in general. Such a role for them, indeed their very existence in the Near Eastern Bronze Age, is controversial. Those who posit their existence and significance as mobile artisans have largely done so on the basis of a single famous wall-painting from a tomb, dated about 2000 B.C., at Beni-Hasan in Egypt and projection back in time of a much more recent phenomenum. The makers and menders of arms and tools, the basic artifacts on which warfare and so much material culture depend, have traditionally into modern times led a wandering life in the Near Eastern landscape. The tinker was 'so low in popular esteem that no-one will touch him, and his reward is bare subsistence' (Coon 1952, 210). Albright (1953, 98–9, 109, 200, 205, 210), accepting that the Beni-Hasan painting shows a travelling smith's equipment on the back of a pack-donkey rather than simply a shield and spear, saw in him a representative of the craft tradition believed to be that of the Old Testament Kenites in south Canaan in the Middle and Late Bronze Ages. They have been identified as making a living as metal craftsmen and have been compared to the modern Arab tribe, the *Sleib*, who ply their trade as travelling smiths or tinkers, following more or less regular routes with their asses and tools, supporting themselves by means of craftsmanship, supplemented by divination and music. At present, any evidence from archives that might sustain this as a general hypothesis across the ancient Near East is not available. The significance of a passage cited by Adams, but not yet fully published, in a Middle Assyrian (c. 1400–1200 B.C.) document from Assur is equivocal. It records:

> 'groups of specialised craftsmen travelling with extensive trains of animals, wagons, families, including much baggage and many slaves, and never settling down and carrying on their craft activity in response to the commands of the king or whatever it was' (cf. Moorey 1994, 14).

In reflecting on the history of technology in their recent study of Mediterranean History Horden and Purcell (2000, 288) note of the Classical World that 'the actual diffusion of new techniques in almost completely unknown to the literary (and to a large extent the documentary) traditions, because it happened at a social level below the attention of the writers. If we sometimes hear of failures to disseminate useful ideas – such as the exciting metallurgical discovery in Pontus that was lost because the designer never told anyone about it (Aristotle: *On Wonderful Reports*, 62) – we never hear anything about the wholly unglamorous transmissions of craft expertise.' We may then perhaps be grateful that for an even remoter period, it is possible to indicate, if no more, something of the ways and means by which new techniques may have been diffused. Demonstrating the fact is rather a different matter. This is where increasing application of materials science, to which this conference is primarily directed, is the crucial way forward.

CONCLUDING REMARKS

In closing, I would like to place this presentation of a very particular body of evidence in a wider perspective. When the impact of empires is discussed in the history of technology, notably that of Rome in antiquity or of the British more recently, it is the technical innovations introduced by the overlords that are at the centre of the discussion not, as in this case, the receptivity of the imperial power herself. There may be a special reason why this case is different. Egypt is somewhat isolated geographically even today; it was more so in antiquity. Moreover, it was richly endowed with agricultural and mineral resources. The principal commercial imports before the New Kingdom were tin and lapis lazuli, olive oil and resins, and various timbers. She had no reason to expand territorially and she was very rarely invaded.

In the period immediately preceding the New Kingdom, however, intruders from Palestine, known to the Egyptians as 'rulers of foreign lands', to us from the Greek as the Hyksos, had established their rule over the Delta (Lower Egypt) and into the northern Nile valley. Eventually local rulers at Thebes in Upper (southern) Egypt quelled internal unrest and expelled the foreigners, ushering in the unprecedented prosperity and power of the 18th and 19th Dynasties, which have been the focus of my paper. Thereafter the 'Hyksos Period' was regarded as a national catastrophe, an abomination to be avoided at all costs in the future. Whereas before the Hyksos Egypt had only intervened in Western Asia sporadically when necessary to sustain the products she required from Palestine, after them her rulers not only launched pre-emptive strikes against the foreign peoples who threatened her, but also established imperial controls (cf. Redford 1992, 148). As she was thus drawn further into Western Asia, both diplomatically and militarily, her crafts and industries, as we have seen, benefited considerably from the contact. Had it not been for the shock of the Hyksos, the Egyptians might well have gone on slowly developing traditional indigenous technologies without sudden, extremely significant infusions of foreign knowledge and skill by means which my paper has sought in some measure to elucidate.

BIBLIOGRAPHY

Albright, W. G., 1953, *Archaeology and the Religion of Israel*, Johns Hopkins, Baltimore.

Aldred, C., 1970, 'The foreign gifts offered to Pharaoh', *Journal of Egyptian Archaeology*, 56, 105-16.

Coon, C., 1952, *Caravan: The Story of the Middle East*, Cape, London.

de Vaux, R., 1967, 'Les Hurrites de l'histoire et les Horites de la Bible', *Revue Biblique* 74, 481 ff.

Durand, J-M., 1992, 'Unité et diversité au proche-orient à l'époque amorrite', in Charpin D. and Joannès, F. (eds.), *La circulation des biens, des personnes et des idées dans le proche-orient ancien*, Editions Recherches sur les Civilisations, Paris, 197–128.

Eyre, C. J., 1987, 'Work and organisation of work in the New Kingdom',in Powell, M. A. (ed.), *Labor in the Ancient Near East*, New Haven, Yale University Press, Connecticut, 167–221.

Frankfort, H. 1958, *The Art and Architecture of the Ancient Orient* (2nd revised impression), London: Penguin Books.

Freydank, H., 1976–80, 'Kar-Tukulti-Ninurta: A. Philologisch', *Reallexikon der Assyriologie*, V, de Gruyter, Berlin and New York, 455 ff.

Gelb, I. J., 1973, 'Prisoners of war in early Mesopotamia', *Journal of Near Eastern Studies* 32, 70–98.

Gombrich, E. H., 1993, *Topics of our Time: Twentieth-century Issues in Learning and Art*, Phaidon, London.

Horden, P. and Purcell, N., 2000, *The Corrupting Sea: A Study of Mediterranean History*, Oxford: Blackwells.

Knudtzon, J. A., 1907–1915, *Die El-Amarna-Tafeln*, Vorderasiatische Bibliothek 2, Leipzig.

Kühne, C., 1973, *Die Chronologie der Internationalen Korrespondenz von El-Amarna*, Neukirchner Verlag, Tübingen.

Lilyquist, C. and Brill, R. H., 1993, *Studies in Early Egyptian Glass*, New York: The Metropolitan Museum of Art.

Littauer, M. A. and Crouwel, J. H., 1979, *Wheeled Vehicles and Ridden Animals in the Ancient Near East*, Leiden.

Littauer, M. A. and Crouwel, J. H., 1985, *Chariots and Related Equipment from the Tomb of Tutankhamun*, Oxford: Griffith Institute

Liverani, M., 1979, ''Irrational' elements in the Amarna Trade', in *Three Amarna Essays*, Monographs on the Ancient Near East I(5), Malibu: Undena.

McLeod, W., 1970, *Composite Bows from the Tomb of Tutankhamun*, Oxford: Griffiths.

McLeod, W., 1982, *Self-Bows and other Archery Tackle from the Tomb of Tutankhamun*, Oxford: Griffiths.

Moorey, P. R. S., 1986, 'The emergence of the light, horse-drawn chariot in the Near East, *c.*2000–1500 B.C.', *World Archaeology* 18, 196–215.

Moorey, P. R. S., 1989, 'The Hurrians, the Mittani and technological Innovation' in de Meyer, L. and Haerinck E.(eds.), *Archeologia Iranica and Orientalis: Miscellanea in Honorem Louis Vanden Berghe*, Peters, Gent, 273–86.

Moorey, P. R. S., 1994, *Ancient Mesopotamian Materials and Industries: the Archaeological Evidence*, Oxford: OUP.

Moran, W. L., 1992, *The Amarna Letters*, Baltimore: Johns Hopkins.

Oded, B., 1979, *Mass Deportations and Deportees in the Neo-Assyrian Empire*, Wiesbaden.

Oppenheim, A. L., 1970, 'The cuneiform texts' in Oppenheim, A. L., *et al.*, *Glass And Glassmaking in Ancient Mesopotamia*, Corning: Corning University Press, 1–101.

Oppenheim, A. L., 1973, 'Towards a history of glass in the ancient Near East', *Journal of the American Oriental Society* 93, 259–66.

Oppenheim, A. L., 1978, 'Man and nature in Mesopotamian civilization' *Dictionary of Scientific Biography* 15, Supplement 1, New York, 634–66.

Peltenburg, E. J., 1987, 'Early faience: Recent studies, origins and relations with glass' in Bimson, M. and Freestone, I. C. (eds.) *Early Vitreous Materials*, London: British Museum, 5-29.

Redford, D. B., 1992, *Egypt, Canaan, and Israel in Ancient Times*, New Jersey: Princeton.

Sasson, J. M., 1968, 'Instances of mobility among Mari artisans' *Bulletin of The American Schools of Oriental Research*, 190, 46–54.

Sasson, J. M., 1990, 'Artisans and artists: Documentary perspectives from Mari' in Gunter, A. (ed.), *Investigating Artistic Environments in the Ancient Near*, Washington: Smithsonian Institution, 21–7.

Schaeffer, C. F. A., 1939, 'Une hache d'armes Mittanienne de Ras Shamra', *Ugaritica* I, Geuthner, Paris, 170 ff.

Schulmann, A. R., 1963, 'Egyptian chariotry: a re-examination' *Journal of the American Research Center in Cairo*, 2, 75-98.

Smith, C. S., 1981, *A Search for Structure: Selected Essays on Science, Art and History* Cambridge, Mass: Harvard.

Speiser, E. A., 1950, 'On some articles of armour and their names' *Journal of the American Oriental Society*, 70, 47–9.

Starr, R. F. S., 1937, *Nuzi: Report on the Excavations at Yorgan Tepe near Kirkuk, Iraq 1927–31, I. Text*, Cambridge, Mass: Harvard.

Starr, R. F. S., 1937, *Nuzi: Report on the Excavations at Yorgan Tepe near Kirkuk, Iraq 1927–31, II. Plates and Plans*, Cambridge, Mass: Harvard.

Stech-Wheeler, T., 1981, 'Iron at Taanach and early iron metallurgy in the East Mediterranean', *American Journal of Archaeology*, 85, 245-68.

Stevenson Smith, W., 1965, *Interconnections in the Ancient Near East: A Study of the Relationships between the Arts of Egypt, the Aegean, and Western Asia*, Princeton: Yale.

Sturt, G., 1923, *The Wheelwright's Shop*, Cambridge: CUP.

Van de Mieroop, M., 1997, *The Ancient Mesopotamian City*, Oxford: OUP.

Vogelsang-Eastwood, G. M., 1999, *Tutankhamun's Wardrobe: Garments from the Tomb of Tutankhamum*, Barjesteh van Waalwijk Van Doorn, Rotterdam.

Western, A. C., 1973, 'A Wheel Hub from the Tomb of Amenophis III', *Journal of Egyptian Archaeology*, 59, 91–4.

Wilcke, C., 1969, *Das Lugalbandaepos*, Wiesbaden.

Zaccagnini, C., 1983, 'Patterns of Mobility among ancient Near Eastern Craftsmen', *Journal of Near Eastern Studies*, 42, 245-64.

Chapter 2

Technological Change in the East Mediterranean Bronze Age: Capital, Resources and Marketing

Andrew and Susan Sherratt

Abstract

This paper explores some of the factors of technological change, in terms of socio-political, environmental and economic-structural relationships, (and the dialectics between them). These 'factors of production' only partly coincide. Areas which are in a position to concentrate capital (such as southern Mesopotamia) are often poor in natural resources; areas with particular resources (e.g. the Caucasus) may be too remote from contemporary trade routes; areas advantageously placed for maritime traffic (e.g. the Levant and Cyprus) may be dependent on consumers in other political blocs over which they have no control. Technological change results from the combination and interaction of these elements. Metalworking, for instance, required both the political centralisation capable of supporting specialised workshops, and also control of abundant raw materials. Metallurgical innovation, and the production of secondary materials (e.g. glass) thus took place neither in the 'heartlands of empire', nor in remote metalliferous areas, but in areas such as Cyprus and the Levant where mineral wealth, centrality in maritime trading patterns, and the gradual emergence of political centralisation, slowly came together. (It was this conjunction which was critical in the first development of large-scale ironworking – an advance which had global repercussions.) The paper thus discusses the geography of innovation, comparative advantage, and the factors which favoured it in the Late Bronze Age.

INTRODUCTION

The history of technology is one of the fields of historical inquiry to which archaeology has made a major, and well-recognised, contribution. A landmark in this process was the publication of the seven-volume *History of Technology*, sponsored by Imperial Chemical Industries and published by Oxford University Press between 1954 and 1958 (Singer, Williams and Raper 1954–8): a work which – albeit in one of the slimmer reissues, rather than on the heavy, thick-leaved paper which made the original such a solid monument to scholarship – is still a primary source of information on the subject. Its first two volumes, dealing respectively with the prehistoric and Near Eastern periods, and with classical and medieval times, are both a benchmark of accumulated knowledge and a shop-window of archaeological achievements.[1] It is still a reference book of first resort, whether one is seeking a broad survey of the development of technological competences, or some specific feature such as the forms of early leather vessels (whose significance will appear later in this paper). Consciously or unconsciously, these survey articles and their beautifully redrawn illustrations have informed the work of archaeologists ever since. It seems appropriate for a volume arising from a conference in Oxford to look back to this baseline, and to the work of an earlier generation of writers on this topic; but it is especially useful here to remind us how the history of technology has itself changed, and the changing ways in which technology has been treated as a historical phenomenon. Of the authors of the first two volumes of the Oxford *History*, perhaps only one would have claimed the label of Marxist, and that was Gordon Childe. His contributions were the most self-consciously intellectual, dealing with 'Early forms of society', and 'Rotary motion', approached from the standpoint of that famous quotation from Marx, 'the handmill gives the feudal lord, the steam-mill the capitalist'. But whether or not the other authors subscribed to the wider philosophy that this brand of Marxism implied, there would certainly have been a broad consensus on the role of technology. In the process of reconstruction after World War II, technology was a driving force. The history of mankind was a series of discoveries and inventions, which marked the path of progress – 'from Pithecanthropus to spaceman' (as the blurb to the first [1961] edition of Grahame Clark's *World Prehistory* described it. Economic historians found earlier industrial revolutions in the Middle Ages (Carus-Wilson 1963; Nef 1964), and politicians (like Harold Wilson in the later 1960s) talked about 'the white-hot heat of the technological revolution'. Machinery maketh man.[2] Evolutionary anthropologists such as Leslie A. White (e.g. 1959, Ch. 2) celebrated the increasing control of energy which marked successive stages in cultural evolution, and Childe himself continued to make 'invention' into the chief motor of progress in changing Neolithic farmers into Bronze Age townsfolk (Childe 1936).

It is not necessary nowadays to remark how technology has today been dethroned from this privileged position as the prime mover of cultural change – a role which, in the work of Childe and White, it quite consciously inherited from theology.[3] Green politics, and the increasingly obvious effects of industrial pollution, have caused a sea-change in popular attitudes to technical 'progress'; while archaeologists have found other factors to power the course of social transformation – first in more broadly ecological factors, and latterly in the sphere of ideas and performances.[4] At the same time, though largely in France, a new style of writing about technology has emerged, which might be called (and indeed it has, though not by the French) the 'social construction of technology' (Bijker *et al.* 1987; Pfaffenberg 1988). The work of André Leroi-Gourhan, and ethnographers such as André Haudricourt or François Sigaut, and more recently the school around Bruno Latour and Pierre Lemonnier, have viewed technology as an integral component of social change (Leroi-Gourhan 1943; Sigaut 1975; Haudricourt 1988; Latour and Lemonnier 1994). A recent selection of their work

(Lemonnier 1993) appeared under the title of *Technological Choices*, and the very word 'choice' implies a less deterministic attitude, and a consciousness of alternative paths and traditions, rather than a unilinear sequence. This sense of alternatives is a useful reminder that the paths of technological change are local traditions, and that we should not expect uniformitarian sequences within Childe's (1944) 'archaeological ages as technological stages'.[5]

An example of such cultural alterity is provided by the manufacture of metal vessels, which in the Near Eastern and Mediterranean world from the fourth millennium B.C. onwards were typically made of sheet-metal (Sherratt and Taylor 1989). Alexander the Great's Admiral, Nearchus, who explored the Indus Valley in 326 B.C., commented with surprise that Indian bronze vessels were cast, not forged, so that they broke rather than crumpled when dropped (Strabo *Geography*, xv.1.67; Rajpitak and Seeley 1979). Within the Old World as a whole, moreover, casting technologies for bronze have been more usual than sheet-working: beaten bronze vessels only spread into temperate Europe from the 13th century onwards (at the same time as hollow-cast objects like spearheads were brought south from temperate Europe into the Mediterranean); and bronze vessels in China were from the beginning cast in multiple moulds. Thus one set of civilisations perfected the use of sheet-metal, and in their ceramic skeuomorphs gave us the typical shapes of Grey Minyan ware, Greek vases, and *terra sigillata*; while the other perfected cast metal, and gave us Shang bronzes and the solid, stocky forms of pottery which we associate with the Han dynasty (Vickers 1986; Rawson 1986; Watson 1986). It was in the Tang period that western habits of wine-drinking and Sasanian sheet-worked silverware reached China along the Silk Route, to affect a revolution in pottery design, with the thin, lotus-petal bowls that we now find in Chinese restaurants and think of as quintessentially Chinese. But in producing this new, thin-walled pottery in imitation of sheet-metalwork, the Chinese used the characteristic high-temperature technology they had mastered in bronzecasting, to produce first White wares and then porcelain – which they then exported, first to the Islamic world (Tampoe 1989) and then further west, along with tea. Full circle, to steal a phrase from a recent book on economic history (Jones *et al.* 1993). So these distinctive and culturally specific technological traditions hybridised and interacted, in a manner which is as much a part of the history of consumption as the history of technology, to produce a continuing sequence of novelties; and we shall return to this theme of such dialectical interactions between technological traditions later on in this paper.

The study of ancient technologies in the western Old World has been energetically pursued in recent years, the fruits of such researches being well exemplified in Roger Moorey's *Ancient Mesopotamian Materials and Industries* (1994, reissued 1999) or Paul Nicholson and Ian Shaw's edited compendium on Egypt (1999), which provide detailed contemporary surveys for these two critical areas. Such work has been informed by two sources in particular, of which the first has been the input from material scientists in the analysis of both inorganic and organic remains – the incentive for this volume – and of which the second has been the painstaking textual work of epigraphers and philologists, especially in the Near Eastern field, writing dictionaries and translating texts. These achievements have yet to be fully appreciated by the writers of standard accounts of technological history, who all too often remain wedded to a western narrative which concentrates on the Greeks and Romans (if they manage to consider the ancient world at all: Greene 1994) and on medieval Europe. The sheer scale of industrial production in the ancient East (and here one should include China, though examples from late-third millennium Mesopotamia are equally impressive)[6] puts even the early stages of the 18th century Industrial Revolution into their proper perspective. Yet there has been revisionist writing in the western history of technology, too, and here it is relevant to mention in particular Robert Fox and his associates, who have largely re-written the story told thirty years ago about medieval technology by Professor Lynn White Junior (White 1964; Fox 1996).[7] All these indications point to the timeliness of a radical reconsideration of the role of technology in ancient societies, and the need to 'bring technology back in'. It is especially challenging now that the history of consumption is seen to be as vital as the history of production, and the role of consumer demand and why people want material goods has been raised to such good effect by anthropologists and historians (Douglas and Isherwood 1979; Brewer and Porter 1993).

TECHNOLOGY, ECONOMICS, AND CULTURE

The aim of the following discussion is to situate the study of ancient technologies within a wider interpretative framework dealing with production and consumption in the ancient world, in such a way as to bring together the two sets of assumptions hitherto employed – the notion of a growing complexity in the manipulation of materials, and the idea of particular types of technology as being characteristic of certain kinds of societies. While it relates primarily to the Late Bronze Age and the beginning of the Iron Age in the east Mediterranean, it will be necessary to step outside these limits from time to time, in order to compare and contextualise. Its goal is to try and make explicit some of the rules of thumb that we employ in discussing how and where technological innovations occur, by describing them in a vocabulary derived from economic history of the new, socially conscious kind, which is sensitive to motivations and marketing.

The use of economic terminology requires some justification in this context, since this vocabulary was not invented with the ancient world in mind. This is not the place, however, to review the swings in interpretation concerning the nature of the ancient economy, and the rebound from the 'minimalist' view of trade which characterised the view of Sir Moses Finley (1973), as the intellectual history of this debate has been summarised elsewhere (Sherratt and Sherratt 1998). While the vision of an essentially pre-market model for the ancient world was a useful distancing from the excesses of regarding phenomena such as Phoenician maritime enterprise as essentially just like that of the British or Dutch in the Early Modern period, those using it have sometimes swung to the opposite extreme, and made the Bronze and Iron Ages more like the Stone Age. The approach which is espoused here owes much to a small but remarkable book by the Oxford economist Sir John Hicks (1969), called *A Theory of Economic History*, and it is within this framework that it is useful to employ terms like 'marketing', 'import substitution' and 'added value production' for the ancient Mediterranean, so long as we remain conscious that the scale of overall activity which is described by these terms was very different from that of more recent societies. The principal danger in employing economic terminology stems not from the nature of the models themselves, however, but from the attitudes to material culture which often accompany them, and which project backwards a purely utilitarian view of technology that resembles the attitudes criticised at the beginning of this paper. What is anachronistic is to regard the products of ancient technology and trade simply as decontextualised material objects. When we are talking of lapis lazuli, or even its high-tech substitute, blue glass, we need to re-capture some of the attitudes described for instance by Joan Evans (1922) in her book on *Magical Jewels in the Middle Ages*, or Theophrastus' *De Lapidibus*, On Stones. (Eleanor Robson's paper in this volume does just this.) These give us some hint as to why such materials were transported over thousands of miles: even in the Middle Ages, the lazulite-based pigment was regarded as particularly precious and used for instance to paint the heavenly-blue cloak of the Virgin Mary, so giving its name to the colour 'azure'.[8] This exceptional value also explains why the ability to manufacture a bright blue vitreous material by Late Bronze Age craftsmen was such a breakthrough in import substitution.[9] The medieval alchemists' search for a method of transmuting base-metals into gold was a similar enterprise; and while these experiments may have shown a deplorable ignorance of the properties of Mendeleev's periodic table, they were at the time a sensible piece of attempted import-substitution, with respectable antecedents as far back as the Bronze Age.[10]

The flows of high-value materials such as metals provided both an outline structure for the Bronze Age urban economy and a principal motor of its growth, as sources of supply were sought often at great distances from the urban centres. Inevitably, therefore, the history of this process is to a large extent the history of elites. The point is neatly made for a later period in a contemporary illustration of the *Schatzkammer* of the Habsburg Emperor Maximilian I (reproduced as Figure 2 in Sklenár, 1983) with its accompanying doggerel: *Den grössten Schatz hat er allein/ Von Silber, Gold und Edelstein...*, since the long-distance trade routes reaching out from Mesopotamia, and which were established already in the fourth millennium, were concerned precisely with these three precious substances 'of silver, gold and precious stone' – which in the latter case included lapis lazuli transported all the way from Badakhshan (Bactria). Silver probably came primarily from Anatolia and the Caucasus, gold from Egypt, and these long-distance supply routes acted as the principal axes of economic expansion. By the first millennium B.C. a similar triple axis may be discerned, with the silver of Spain, the gold of the Sudan,[11] and the incense of Arabia (since incense – or for that matter amber – was an organic equivalent of precious stones like lazurite, with the power to travel across whole continents). The Mediterranean world of the 16th century AD described by Braudel – importing the silver of the Americas, the gold of west Africa and the spices of the Orient – had a similar structure but on a larger scale (Braudel 1973). This propensity of certain substances to travel arose from the particular qualities which made them precious, embedded in beliefs about their nature (which are thus culturally specific, despite their increasingly universal appeal as particular patterns of consumption proliferated).[12] A striking example from the Late Bronze Age is the fact that the scarab buried next to the heart of the Pharaoh Tutankhamun turns out to be made of amber, all the way from the Baltic (Hood 1993). So the first point is the simple and obvious one that flows of high-value raw materials were fundamental to the overall structure of the Bronze Age economy, and that the need to acquire them defined the external growth-points of the trading system, as it expanded outwards.

These are materials which possess what Arjun Appadurai (1986) has called *prime value*. Despite Immanuel Wallerstein's dismissal of such substances as mere 'preciosities' (1974), we can recognise (with Jane Schneider: 1977) their critical role in economic growth. But this is only half of the equation, for civilisations are as much concerned with manufacturing, and the creation of *added value* – the qualities added to materials (not necessarily valuable in themselves) by labour. The paradigm example here is textiles and clothing. The wearing of particular forms of elaborate body-covering is not a physiological necessity, and requires the expenditure of much skilled labour-time in making textiles, and specialised livestock-rearing to produce bulky raw materials: yet this is a hallmark of 'civilisation' and its impact on other cultures. (Look at ethnographic photographs – there are some good examples in Mowatt *et al.* (1992) on the uses of native American basketry – around the beginning of the last century: the women are still performing traditional tasks – but now attired in long Victorian dresses!) The motif of the 'discovery of nakedness' in the account of the Fall in the Book of Genesis sums up this transformation as the end of the state of nature and the beginning of civilised norms. It is not surprising that the growth of a textile industry has been a constant accompaniment to the spread of urban life, and that Waetzold's figures (quoted above) for the numbers of people engaged in textile production at Larsa at the end of the third millennium represent a substantial proportion of the inhabitants of the city, mainly women and children. As Robert Adams pointed out in 1974, this growth in manufacturing goes a long way to explaining why the raw materials were able to move so consistently towards the urban centres, as the result of the structural differentiation in production between the suppliers of raw materials and the technologically more advanced centres of manufacturing, where 'added value' was created. Moreover this process provides much of the expansionary dynamic of urban networks, as new centres of production spring up through the process of import-substitution. (Jane Jacobs' *Economy of Cities* [1970] provides an elegant exposition of this model.)

Let us be clear what is happening here. Labels such as 'product innovation' or 'import substitution' describe only the bare economic facts of the process; but these phenomena are predicated on the spread of patterns of meaningful consumption – regimes of value – involving complex sets of ideas and practices concerning appropriate types of buildings, furniture, adornments, dress, food and (especially) drink, which require the mobilisation of increasing volumes of materials.[13] This is not just an economic process but also an aesthetic and even a religious one, since it implies the transmission of a lifestyle and a set of values. 'Added value' has to be culturally created, in a total package of products and practices, which spread in processes typically named after their mother-cultures: 'Hellenisation' or 'Indianisation' in the first millennium B.C., or 'Urukisation' (if you will pardon an ugly but useful expression) in the fourth. All of these involved characteristic consumption-practices, involving the mobilisation of a larger volume of specific types of goods. At one level, such practices were culturally specific: as urban life spread along the northern shores of the

Mediterranean in the first millennium, Greek vases owed much of their appeal, not just to their artistry but to the wine that was drunk from them and the silver vessel-forms they echoed, and the culture they encapsulated. In the same way, the spread of that most boring artefact, the Uruk bevelled-rim bowl, was symptomatic of the adoption of a style of food-preparation and consumption which we see directly represented in the pictographic symbols of the earliest economic texts (themselves developed from clay tokens representing commodities such as wool-bales or prepared liquids in containers), which show the importance of brewing, baking, and the use of liquid commodities like clarified butter and beer.[14] But at another level, these kinds of consumption-practices became international – indeed became symbols of internationalism, like wine-drinking among the Iron-Age Celts. What we are witnessing in these examples is the spread of a high-consumption lifestyle, demanding the mobilisation and processing of an ever-increasing range of goods and commodities. With time, these evolve a local idiom: Egyptian modes of consumption are quite distinct from those of Mesopotamia, and these in their turn from those of the Cretans and the Hittites. But they have a common set of ideas about what constitutes value, and a very clear idea of the commodities which they wish to import. They are, as it were, specific dialects of a common language of consumption – and especially elite consumption, for we are talking about societies in which income-distributions were sharply skewed. The actual consumers of these mobilised goods were a tiny fraction of the total population, but they were the critical players in a common game of high-value consumption. The economic system which provided these goods spreads outwards, as more players joined the game.

Where does technology fit into all this? Since all these processes involved material transformations, they are relatively easy to follow archaeologically. There are certain technological features which characteristically accompany an urban lifestyle, and which were thus notably slow in spreading across Eurasia, because they require an urban system to make sense. (In this respect they form a striking contrast to things like wheeled vehicles, which were very widely adopted and imitated in many different kinds of societies.) Let us list just three of these slow-diffusers, which (in Europe, at least) have almost inevitably accompanied each other: they are cupellation, to produce silver from lead; the use of the sail, to power bulk water-transport;[15] and the use of the wheel in the mass-manufacture of pottery. An isochronic map of the spread of any one of these features across Europe would reliably predict the spread of the other two innovations, from a restricted area of the eastern Mediterranean in the third millennium B.C., to their arrival in Scandinavia in the first millennium A.D. (and this complex of features would also, of course, include the arrival of literacy). Why should these quite different technological phenomena be so closely associated with one another? Silver is important as the common currency of complex systems of exchange; bulk transport is essential for the distribution of commodities; and mass-produced pottery only makes sense within an assured market of a certain size. These features have no direct technological connection with each other, but they go together because of their complementary roles within an economic system; they are all relatively expensive procedures which depend on economies of scale.[16]

Their common context, therefore, is a concentration of capital. To discuss this with any degree of abstraction would keep us all day, so let us consider a down-to-earth example. Literally so, in fact, since it is based on Paul Halstead's recent discussion in the *Bulletin of Sumerian Agriculture* (1995). It concerns the cost of keeping draught-animals in mediterranean environments. Halstead points out that oxen are an expensive item in mediterranean or sub-tropical climates, where grazing is limited; and that they have always been more or less confined to large estates with a production-unit of 5 hectares or more (like the Turkish çiftlik – whose very name comes from çift, 'a pair', in reference to the plough-team). This is supported by the evidence of early texts: barley had to be set aside to feed the oxen during ploughing, and this could be a high proportion of the expected yield – a quantity equal to the amount required for seed-grain.[17] Draught oxen are thus an expensive item to maintain; and moreover their training only began after their third year, so they represented a very considerable investment. (A Sumerian literary text entitled 'a disputation between the plough and the hoe' amusingly contrasts the cheapness and reliability of the latter.) So this kind of high-input farming was in fact predicated upon the existence of institutions capable of making such an investment – probably in the very temple-estates where writing itself began, to record the concentration of commodities indicated in the pictographs (wool, beer, etc.). It is in this kind of context that we should situate technological innovations such as the threshing-sledge and the first wheeled vehicles, even though they subsequently attained a very much wider distribution: the initial capital had to be available to support the R&D ('research and development'). A similar argument could be applied to all the kinds of specialised commodity-production so far mentioned: dairy herds, wool-flocks, and tree-crops for the production of wine or oil; and also for the new forms of transport which arose to move bulky or heavy materials like wool and metals over long distances, such as pack-animals or vessels with sails. They all needed capital to get them going.

THE CONTEXTS OF INNOVATION

Now technological innovation as such is not an inevitable accompaniment of such situations of economic growth. If we go back to spinning and weaving, the main processes of textile-production, we see that very little innovation occurred in the actual methods of production – really down to the Industrial Revolution, and the invention of powered machinery. What was critical in getting a textile industry going was organisation – aggregating the output of vast numbers of individual hand-looms, each operated by one or two workers (as noted above, mostly women and children). That is probably why the fact that these were temple-estates was so important, since the motivation was to offer work to the gods, just as Neolithic farmers conceived of their labours as serving the ancestors, who renewed the fertility of the land. (All management depends on mythology, and that is no less true today).[18] These organisational complexities explain why literacy and accounting was such a consistent feature of such systems – and initially closely associated with religious scribes ('clerics/clerks').

If production methods allowed little scope for improvement, however, there was plenty of scope for development in production techniques. These comprise those tricks of the trade which add quality, and thus desirability and value: methods of treating the yarn, and dyeing or finishing the cloth. Moreover there were some fundamental geographical contrasts, both in the materials used and in the methods used in manufacturing, which bring us to the kinds of 'technological choices' emphasised by Lemonnier and his French colleagues. Material contrasts include, in the first place, the locally prevalent usage of linen or wool – entirely linen in Egypt, initially, and wool in the mountain fringes of the Fertile Crescent; though woolly breeds of sheep spread rapidly along with the urbanisation process itself, and beyond it into prehistoric Europe.[19] But there was also another, cultural, contrast, as Elizabeth Barber has elegantly described: the different weaving traditions based respectively on the ground-loom and on the upright, warp-weighted loom (Barber 1990, Fig. 11.1). The former was the indigenous method in the plains and alluvial valleys of Greater Mesopotamia and Egypt, the latter in Anatolia, the Aegean, and the rest of Europe. The combination of conditions suitable for wool-flocks, and the technology of the upright loom, gave certain areas a critical advantage in producing potentially fine textiles; and Barber has shown how Crete took full advantage of this in providing Egypt, for instance, with an internationally valued commodity that could not be produced locally. Indeed Barber has gone further, in showing how Minoan textile-motifs were transmitted into international circulation, and appear as decorative ceiling-motifs in 12th to 18th Dynasty Egyptian wall painting, echoing the rich textile hangings and awnings which formed the backdrop for royal ceremonies (Barber 1990, Ch. 15).[20]

This unequal distribution of geographical opportunities brings up a second great principle of economics, propounded by David Ricardo: the principle of comparative advantage. Even where two areas are both capable of producing a commodity, it makes sense to specialise if one can do so more cheaply and effectively. Even a small difference can tip the scales if transport conditions allow the easy exchange of goods: and Crete had both the right combination of conditions for textile-production, as well as access to the sea. The fact that water transport was something like 10 times cheaper (because more efficient in bulk) than overland transport, down to the beginning of the railways, is a constant factor in economic development. Although we often tend to think of the Euphrates and the Nile primarily in terms of their provision of water for irrigation, their role as transport arteries was in many ways more significant: indeed it is the conjunction of these two properties which goes a long way to explain the early pre-eminence of these two areas. But while these conditions gave them their initial advantage in capital-formation and the concentration of productive labour for manufacturing, they also went along with what in the Mesopotamian case is a striking disadvantage, namely the absence of many kinds of raw materials – most obviously stones and metals. The development of metalworking raises some critical issues from this perspective.

The fundamental question can be simply stated: where did metallurgical skills develop? At a basic level, the answer is obvious: near to the copper sources, where experience with the pyrotechnology of potting was applied to the casting of metal objects. But with the onset of urbanism, and the genesis of high-input techniques requiring craft skills and a concentration of capital, the question becomes more complicated, for there is a geographical disparity between the two sets of conditions – a tension between the principles of economic theory, and local knowledge and resources. To put the question more concretely: where did the technology of making sheet-metal vessels originate? On the one hand it required the advanced skills of specialist (temple?) workshops in a capital-intensive economy, as with Uruk-period wheelwrighting and accountancy discussed above; on the other, it required intense familiarity with the metalworker's craft and a knowledge of the materials, more likely in a metalliferous area itself and thus outside the early urban heartlands.[21] The archaeological record of such objects is notoriously unrepresentative, both generally because of the re-cycling of metal, and particularly in the fourth millennium B.C., because of the absence of elite graves before the Early Dynastic period. Arguments from skeuomorphism can take us a little bit further, as Grahame Philip and Thilo Rehren have shown in their discussion of the Early Bronze Age Tell Farah silver bowl (1996); and there are some hints in the construction of spouted containers of the Uruk period (for instance the vessel shown in Potts 1997, VI.14) that the design may reflect a two-piece sheetmetal construction (not unlike a Villanovan urn from Early Iron Age Italy).[22] Such metal vessels as do occur – i.e. those of sufficiently low value not to be re-cycled – are commonly of lead, suggesting that this was a cheaper by-product of the production of silver by cupellation (presumably carried out locally from imported galena). This would imply that sheetworking techniques were pioneered in Mesopotamia itself, on the basis of precious metals (perhaps even following experiments using lead), and only later used for the more difficult copper-based alloys. This leads to a rather different interpretation from what we may call the Oxford History of Technology viewpoint ('man's mastery of metals', in a characteristic phrase of the time) and suggests that experimentation with advanced techniques arose from a manipulation of the symbolic properties of certain rare, malleable materials – perhaps to make containers for the equally symbolic and precious organic substances of alcoholic drinks (plentifully exemplified in the ED III grave of Queen Pu-Abi in the Royal Cemetery at Ur in the following millennium). Both may have been imbued with magical properties.[23]

On the other hand, the application of copper-alloy production on a large scale (e.g. for sheet-metal) may well only have taken off when this elite lifestyle reached areas nearer to abundant sources of this metal. It is hard to imagine that a site like late fourth-millennium Arslantepe, for instance – an Urukising polity on the edge of the Mesopotamian world in SE Turkey, but also adjacent to the abundant metal sources a little higher up the Euphrates – did not have a formative role in developing early metallurgical technology, especially in copper-alloys; and indeed there is a famous hoard of swords and spearheads from this site (Frangipane and Palmieri 1983). Vessels of sheet-copper (alloyed with arsenic) became relatively common in the region of the adjacent Kuro-Arax and Maikop cultures, in the metal-rich regions higher up the Euphrates and into Transcaucasia (incidentally, where grape-wine probably had its origins).[24] In this case, one could postulate a conjunction of conditions, where the zone of capital concentration met abundant ore-sources, following the Late Uruk expansion up the Euphrates, and its impact on adjacent cultures. Arslantepe participated fully in the Uruk consumption-pattern, with abundant bevel-rim bowls and liquid-containers; and the scale of its architecture and decorations shows that it was a major regional centre. (Frangipane 1997). The capital-concentration was critical, since such innovations did not take place in metal-rich areas elsewhere; but possession of local resources naturally played a role, once these conditions were present. This situation of a conjuncture between local resources and introduced capital-intensive technologies is a good example of the interplay between these two sets of factors.

There is an interesting analogy here between what happened with another set of lowland innovations, this time in transport: the pack-donkey and wheeled vehicles. One of us has recently argued (A. G. Sherratt in press) that horses were first domesticated in the steppes on either side of the Caucasus, as a result of the arrival of the donkey as a pack animal from the Fertile Crescent (probably along the upper Euphrates on the route past Arslantepe); and a geographically symmetrical argument could be made for domestication of the camel, on the lapis routes through Iran. It is possible to take this argument further, and to propose that the light chariot was invented as a result of continuing interaction in the contact-zone between Near Eastern urbanism and the steppes during the following millennium, in which state-organised Mesopotamian societies experimented with the harnessing of local equids for warfare, but steppe groups found the solution to this problem by developing a practical method for yoking horses and controlling driven animals with bridle-bits. Both elements – Mesopotamian military needs and steppe resources and expertise – were necessary parts of the process of innovation. What we could see here is another example of the interaction between capital-intensive technologies and military needs originating in urban heartlands, and their hybridisation (literally, perhaps, in the case of the donkey and the horse), in areas of distinctive local resources and technological traditions. It seems worthwhile recognising this phenomenon with a specific term, and the phrase 'technological dialectic' serves to convey the two-sidedness of the process – in this case a specific form of dialectic between large-scale economies with abundant capital, and neighbouring areas with particular resources and expertise.

This sort of dialectic, fundamentally related to changing patterns of consumption and material needs, is certain to have involved an ideological component. One expression of this is the way in which innovations which occurred in response to external stimulus came to be seen in some cases as culturally more appropriate in their local contexts than the introduced practices. (The substitution of Anatolian wine for Uruk beer might be seen as just such a rhetorical response on the part of Kuro-Arax populations, and is reflected in their contrasting pottery assemblages.) A more recent parallel for such processes can be seen in the cultural changes which accompanied the spread of Islam from its area of origin in Arabia to the Mediterranean world, which involved both the proscription of all forms of alcoholic drinks, a suspicion of the use of precious-metal drinking-vessels, and a shift from the use of wheeled vehicles to camel-riding (Sherratt 1995, 20–4); all of these practices (and their supporting technologies) had implications for a sense of cultural identity. The spread of 'alien' technologies thus constantly involves not just technological choices but cultural decisions, which may mirror other social and ideological tensions. This is very far from the old view of 'technological progress', but nevertheless this dialectical description allows for the continuing replication of more complex forms of production whilst at the same time recognising that such transfers sometimes involved an element of resistance to the imposition of alien cultural values and patterns, even when this consisted of inventing a local alternative to fill the same role. The paths of technological innovation thus often follow a zigzag pattern, with improvements being added for different motives in different areas; though in the long term the 'good ideas' (like wine, or the chariot) have a tendency to spread between cultural groups – but always with the possibility of proscription in favour of culturally preferable alternatives (as with Islam, which forbade both of them).

In a larger sense, this description of technological dialectic applies also to the broader patterns of interaction across Eurasia with which our discussion began – though here on an intercontinental scale, which ultimately involved urban centres at opposite ends of the Old World. We have suggested that the prominence of sheet-metalworking in the Near East and the Mediterranean arose from the early association of precious metal containers with alcoholic drinks – beer, and then wine – in the first urban societies and their neighbours. Although certain features of advanced metal-forming techniques (in particular the use of the bivalve mould) spread widely beyond the growing Bronze Age urban community of the east Mediterranean, the use of sheet-metal spread only more slowly (Sherratt and Taylor 1989), and new techniques of hollow-casting developed in this outer area in the interval of time between the arrival of bivalve casting and the appearance of sheet-working. A critical element in this process was the development of the metal spearhead (itself a feature generated by the urban style of warfare in the Fertile Crescent), which began in the Uruk period as a tanged form (like the 'poker-butted' types in the Arslantepe hoard) but developed during the third millennium into a form with a forged and wrap-round ('folded') socket, in the area of of eastern Anatolia and the Caucasus which had particular expertise with the sheet-working of copper alloys (Chernykh 1992). While some tanged or slotted types appeared in central Europe in the later third millennium (Gerloff 1993), the folded-socket form became more common on the western steppes, supplied from the newly-established centre of metallurgy in the Urals in the early second millennium (Chernykh 1992). To the north of this, however, in the forest-steppe zone, a surprisingly innovative school of metallurgy developed – from the Urals to the Altai – called the Seima-Turbino complex (Chernykh 1992; Chernykh and Kuz'minykh 1989), which was associated both with the lost-wax casting of figures on the handles of curved knives, and the development of hollow-socketed spearheads and axes formed by using a suspended core in a bivalve mould. Hollow-casting thus began here in imitation of the properties of sheet-metal. These virtuoso pieces of metalwork (in a culture otherwise of 'Sub-neolithic' type, living in semi-subterranean huts) argue for a privileged role for metallurgy, perhaps in a context explicitly associated with shamanic practices, are an outstanding example of the way in which the stimulus of connection with wider networks of contact can lead to technological innovation in surprising places: but the development of this hollow-casting technology seems to have had a decisive influence on the metallurgical traditions both of Europe and China. In Europe, spearheads made by this method became common in Europe in the early second millennium (Reinecke A2);[25] and socketed forms of spearheads and axes also reached China at the same time (the beginning of the Shang dynasty), from whence they spread into south-east Asia (socketed axes appearing for instance at Ban Chiang in Thailand): indeed, it seems likely that the introduction of this technology was a critical factor in the first Chinese hollow castings, which rapidly developed into the multiple-mould technique used to reproduce traditional indigenous vessel-forms, in the form of the famous 'Shang bronzes' for the ritual consumption of millet wine.[26]

The rapid development of metallurgy in the Near East and the east Mediterranean, from the fourth to the second millennium, was due in large part to a fortunate conjunction of circumstances: not just the ecological endowment of cereals and tree-crops which made possible an early development of farming and the subsequent development of organic commodities, but also the plentiful occurrence of a variety of inorganic resources – especially metals – resulting from its location at a major convergent boundary of tectonic plates. Interactions between these factor-endowments – like the association of alcoholic drinks and precious metal vessels[27] – multiplied the effects of these individual advantages, and not only defined much of the peculiar character of Old World civilisation but produced a faster rate of transformation here than anywhere else in the world. This explains why the broad geographical pattern of second-millennium metallurgy had an essentially concentric structure with its core in the Fertile Crescent. Nevertheless this central area had no monopoly of expertise, and indeed benefited from a continuing dialectical relationship with its neighbours, which themselves added crucial technological advances: there were important foci of innovation at critical points within the spreading arc of new techniques – picking up and elaborating certain possibilities on the basis of newly-arrived opportunities.[28] While some of these technological traditions were largely abandoned when metropolitan technologies caught up, others developed into methods which were added to the international repertoire, as with the north-European types of solid-cast slashing-swords which were adopted in the Mediterranean world in the 13th century B.C. (when, in a complementary exchange, the Mediterranean tradition of sheet-bronzeworking was adopted in central Europe for bronze armour and vessels). This dialectic and exchange of techniques continued throughout the Bronze Age, both on a small scale within Europe as well as on a larger scale across Eurasia; and by the later centuries of the second millennium the eastern Mediterranean was both the largest producer and technologically the most advanced user of copper and its alloys. It was also the area in which bronze came to be superseded as the primary metal; and it is to this transition, and its economic context, that we now turn.

CREATING DEMAND: TECHNOLOGY AND SALESMANSHIP

The examples discussed in the previous section were mainly concerned with innovations occuring at the edges of an expanding set of consumption patterns and regimes of value. While the fallout from this process could occasionally affect societies at the other end of the continent, the effects of new production-methods and high-volume consumption are principally evident in the slow extension of the core network of urban communities linked by bulk trade, which by the later second millennium included the Aegean and eastern Mediterranean as well as the older centres of Mesopotamia and Egypt.[29] Underlying this was the process whereby areas on the edge of such trading networks, initially acting as suppliers of raw materials, could transform themselves, through a combined process of organisational change and technological innovation (i.e. 'import substitution'), into independent producers capable of participating in exchanges of manufactured products. This opened the way for a measure of regional specialisation on the basis of local resources and traditions (as in the export of Cretan textiles to Egypt), and to the emergence of certain well-placed areas as hubs of trading activity. This growing scale and internal differentiation gave rise to what may be described as a truly international economy, with interactions – competitive as well as mutually beneficial – between already-established players with their own external contacts and sources of raw materials. It is in this context that concepts such as niche marketing become useful descriptions of the strategic choices made by producers within it.

The process of urban expansion which began in Mesopotamia in the fourth millennium B.C. depended on the continuous stimulation of demand, to induce new players around the perimeter to join the game. The disparity in capital concentration between centre and periphery provided the potential for the extension of urban regimes of value and the outward movement of added-value items. However, as the urban trading area grew larger, and the frontier with the non-urban world moved further out, the opportunities provided by these simple asymmetric forms of exchange gave way to more complex sets of inter-relationships within the central area. One symptom of this was the appearance of imperial power structures, beginning in the later third millennium with the Akkadian attempt to bypass competing centres and tap the outer periphery directly (like Sargon of Agade's expeditions to 'cedar forest' and 'silver mountain': Marfoe 1987), and continuing in more lasting form with the consolidation of Babylonian, Egyptian and Hittite Empires in the second millennium, carving up the central inland territories with their productive centres and overland arteries of supply. An alternative set of strategies lay open to the newly-urbanised polities of the growing maritime world in the east Mediterranean, for these lay outside the reach of imperial domination and could exploit their positions as middlemen on long-distance supply-routes by providing certain types of manufactured items. The production of textiles in Crete shows how local advantages could be used in this way, and the manufacture of perfumed oils in mainland Greece is another well-documented example. Both of these are 'luxury' products based on cash-crops, feeding elite demand for distinctive consumables. However, there were also other opportunities, based on a growing volume of trade and consumption, and especially on new classes of consumers at 'sub-elite' level who nevertheless took their style and tastes from the ruling minority. By the Late Bronze Age, from the sixteenth century B.C. onwards, these formed a significant class of consumers with their own sets of commodities, typically pottery and metalwork – both of which provided appurtenances for the consumption of wine, oil and incense in echoes of the patterns set initially by local rulers a millennium or so earlier. These provided opportunities for 'creative marketing': the active stimulation of demand and the discovery of niche markets. To the extent that such innovation was necessary for continuing prosperity, the economy now had production-led as well as demand-led elements.

This phenomenon can be seen in the production of items specifically for export. These can be unambiguously recognised, not just from their more abundant occurrence in areas outside their centres of production (since this may be due simply to depositional bias in the archaeological record),[30] but from their use of specifically exotic forms, at home in the importing culture. Classic examples are the well-known 6th century B.C. 'Nikosthenic' amphorae, which are Attic pots signed by Nikosthenes but made in specifically Etruscan bucchero shapes and exported to Italy. This is a fairly clear example of export-oriented production between two linked economies, identifying a specific niche and aiming to fill it. In a precisely similar way, the 14th-13th century potters of the Mycenaean Argolid produced pots for sale in Cyprus: chariot-kraters, more plentiful in the east Mediterranean than in Greece itself, and even straightforward copies of Cypriot Base-Ring or milkbowl shapes in Mycenaean fabric. In response, Cypriot potters mimicked Mycenaean in the 13th century, and even sold it on to the Levant. (This involved the adoption of mass-production methods including wheel-made production of some types formerly produced in handmade wares – and other evidence of cost-cutting in manufacturing techniques.) Then, in a gratifying example of the imitator imitated, Levantine potters themselves produced Cypriot-style pots of Mycenaean type – a whole chain of import-substitutions along the entire eastern Mediterranean. 'Pot wars' among rival producers are clearly a long way from elite procurement of precious goods, or even from comparative advantage in textile production, and certainly from earlier and contemporary elite exchanges of high-value items, well known from diplomatic correspondence: 'send me some of your excellent so-and-so' (Liverani 1987) and from episodes depicted in Egyptian tomb-paintings. In contrast with these elite items, the values of pots must have been trivial by comparison with commodities such as metals or textiles – though there is evidence that even these materials were becoming cheaper and more plentiful, and themselves moving downmarket.

The volume of metal in circulation increased dramatically during the Late Bronze Age (as the 354 copper oxhide ingots on the late 14th-century Uluburun wreck dramatically demonstrate); and while the Uluburun ship itself arguably represents a bulk consignment travelling between political elites, there is increasing evidence that metal was beginning to circulate outside these 'official' channels. This is indicated by the growing occurrence of scrap metal, suggesting that the tightly-controlled patterns of distribution which had previously characterised the supply of bulk metals were breaking down, and that bronze was circulating in new ways – probably amongst just those segments of the population which were also customers for the new scale of mass-produced pottery. By the 13th century, bronze was being re-cycled from a multiplicity of sources – even from Urnfield Europe, via Italy, whence new types of solid- and hollow-cast objects were also being introduced: swords, Peschiera daggers, cast spearheads and also fibulae (Sherratt 2000b). The carriers, too, were changing: no longer chiefly the monopolistic, palace-based traders but a new generation of entrepreneurial traffickers, often based in Cyprus or the adjacent coasts of the Levant (Sherratt 1998). The consequence of this new form of trade was that formerly restricted commodities now surged uncontrollably around less centrally organised networks of distribution, carried in small, easily manoeuvrable vessels such as the double-prowed vessels of the Sea Peoples depicted on the Medinet el Habu reliefs, or that reconstructed for the Iria wreck (Vichos 1999, fig. 16), capable of navigating the lower reaches of rivers, and with brailed sails for easy handling into the wind. These new patterns of maritime traffic are reflected in shifts of settlement location (such as the upsurge of newly-prosperous sites which appeared along the Corinthian Gulf and on the Ionian Islands), suggesting that traders now interacted directly with local dealers at critical points along long-distance routes. By joining up bypasses across isthmuses and the linking sea-routes, this new form of distribution circumvented the older, palace-dominated arteries of trade and the segmented patterns of distribution on which they had depended. Palaces, consequently, went into decline and tended to be destroyed and abandoned (S. Sherratt in press). Although this is commonly portrayed as an economic disaster (which of course it was, from the point of view of palatial elites), it is more accurately regarded as a structural shift and organisational change of the kind we would now label 'privatisation', as a consequence of a growing body of consumers that was no longer restricted to a narrow elite. This is very evident on Cyprus, which in many ways was both the cradle and focus of these developments.

Such was the socio-political context: what of technological change? It is in this setting that we might situate the beginnings of iron-production on Cyprus (Sherratt 1994). The development of one technology often produces opportunities for the development of others, sometimes in unexpected ways. New uses may be found for otherwise useless by-products, like the production of cylinder-seals in the later fourth millennium from the drill-cores removed from stone maceheads by the hollow-drilling technique. In similar fashion, Denys Stocks (1993) has suggested that glazed frit was a creative response to the quantities of rock-dust produced by drilling Egyptian stone vases. Such imaginative leaps are fundamental to the kinds of product-innovation described by Jane Jacobs in *The Economy of Cities*, both as import substitution and in more active form as export creation. The invention of glass is a classic example of the way in which expertise in the production of one material helped in the creation of another, since it arose from an accumulated experience in the removal of impurities from primary copper ores, and a longer history of the use of alkaline fluxes in manipulating silicates. It is noteworthy that the new scale of Late Bronze Age trading saw the use of this technology promoted from simply making small ornaments to one of providing containers for precious liquid commodities, and an active search for colourants which themselves carried symbolic meanings (Shortland, this vol.). Cypriot metallurgists had long experience in the smelting of chalcopyrite ores using iron-rich fluxes, which had the potential to produce small quantities of metallic iron as an adventitious by-product – enough, say, to produce objects like small blades; and they were also experienced in forging techniques used in the production of bronze sheet-objects and bladed weapons. It looks as though by the beginning of the 12th century, at places like Enkomi, Cypriot metallurgists may have found some way of regularising the process of producing useful iron as a by-product, using carburisation and quenching to improve their metallic properties, as Pickles and Peltenburg have discussed (1998). From around this time there is a marked increase in the number of iron objects (particularly knives) in Cyprus – many of them apparently carelessly discarded in settlement contexts, indicating that they were not locally treated as items of great value.[31] There is evidence from the 12th century onwards that the potential for producing small iron objects, in the context of the new, devolved trading activities in which Cypriots were engaged, was actively exploited in precisely the kind of niche-marketing described above in relation to pottery. Iron knives and daggers as items of personal equipment were a novel product which could be injected into a wider east-Mediterranean cultural context in which iron was still regarded as rare and magical – 'the precious metal from heaven' – and in which the pharaoh Tutankhamun could be buried with a pair of daggers, one of gold and one of iron, both equally ornamental. This disparity in value, between a local by-product in Cyprus and a desirable status-symbol in the rest of the East Mediterranean, was one of which Cypriot traders took full advantage, by including such items as iron knives in the mix of products which they peddled around the network.

The rest of the story is well known. What began as an exercise in selling a specialised novelty turned into a technological revolution with global consequences. The practice of ironworking spread slowly from its area of R&D to places where (unlike Cyprus) primary iron ores were available; and techniques of exploiting these were developed (probably by the 10th century in the Levant, and by the 9th century in Greece). Iron slowly became 'the democratic metal' available from many more sources than the limited quantities associated with ores of copper. But these future developments should not overshadow our perception of its beginnings: a logical, even inevitable, outcome of the combined processes of economic and technical change, in a context where means, motive and opportunity came together. As with the beginnings of farming or urban life, a conjunction of circumstances resulted in the exploration of a potential already inherent in existing practices and local resources. Small, local innovations initiated a further round of proliferation and dialectical interaction, marked in the archaeological terminology we have inherited from the 19th century as the end of the Bronze Age and the dawn of a new era.

CODA

We have tried in this rapid review of Bronze Age technologies to construct some analogies between a number of well-known episodes, in the hope of distilling the general principles involved. This has involved tracking innovations outwards from their areas of origin, to trace how they interacted with local traditions and opportunities, and also following the consequences of the increasing volume of production and market saturation within the core areas themselves. Both processes involved an intimate interaction between factors of demand – the social meanings of material products for their consumers – and the factors of supply, whether these relate to the occurrence of natural materials or the technologies involved in their transformation. If there is a single moral to be recovered, it is that the story which archaeologists tell is inevitably a composite one. Technical specialisation may require us to explore the unique insights of our separate disciplines; but when it comes to putting the pieces together, the boundaries between technology, economy, society and politics dissolve before our eyes. In a sense, therefore, there is no 'history of technology'; but there is no separate, autonomous history of anything else, either: only the history of the human species, and that much-abused word, culture.

NOTES

1. Contributors included Gordon Childe, Daryll Forde, Edmund Leach, Otto Neugebauer, John Bradford, Harold Coghlan, H. J. Plenderleith, Frederick Zeuner, and such distinguished female scholars as Grace Crowfoot and Peggy Drower, in Volume 1; Martyn Jope, Donald Harden, Herbert Maryon, Richard Goodchild and Rex Wailes in Volume 2.
2. Tools themselves were seen as defining characteristics of humanity – 'Man the toolmaker'. This is no longer the case since toolmaking skills (and specific local traditions) have been recognised amongst chimpanzees.
3. The theological origins of these modes of thought are betrayed by the use of the term 'prime mover' (*primum mobile* – the first mover of the heavens in pre-Newtonian conceptions of cosmology), whose role in cultural evolution was briefly displaced from technology to population growth during the 1960s, before such ways of thinking were called into question during the 1970s. Both Childe and White rebounded from their particular forms of Christian upbringing by espousing the contemporary Marxist emphasis on the power of technology as a force of change: Mary Helms recalls White's emphatic (and unsettling) assertion to Anthropology 101 undergraduates at Ann Arbor in the late 1950s: 'There is no God!' (pers. comm. to AGS, 2000).
4. Though interestingly one eminent Near Eastern archaeologist, R. M. Adams, has in his retirement written a more recent history of western technology, *Paths of Fire* (1996).
5. To be fair to Childe, nor did he: his 1930 book *The Bronze Age* is an elegant exemplification of alternative paths of development in bronzeworking technology, on the basis of the classic ('Montelian') typological sequences; but he downplayed such alterities in favour of a technocentric evolutionary view, from the later 1930s onwards.
6. One striking example is the figures quoted by Waetzold (1972) for Ur III textile production at Larsa: several thousand workers (mainly women and children) were engaged in this task. The importance of China is increasingly being recognised, both in the degree to which the division of labour had been applied in productive processes by the later first millennium B.C. (when, incidentally, coal was already being used in iron-smelting), and in the global significance of the Chinese economy down to the 18th century – a point emphasised in the title of A. G. Frank's most recent book (1998), *ReOrient*.
7. Many of White's theories have been revised in the light of archaeological evidence: see for instance the essays by Bert Hall and Richard Holt in Fox (1996).
8. The use of powdered lapis lazuli as paint (including cosmetic use), incidentally, is one possible reason for its relative rarity in the archaeological record outside deliberate deposits in the areas where it was consumed: like precious metals, it could be recycled to the point of invisibility. On processing sites in Iran and Afghanistan, however, its debris is relatively abundant.
9. That the two were seen as equivalent (though differing in absolute value) is shown by Egyptian inscriptions in which the term for 'lapis lazuli' is used for blue glass, while the term 'royal lapis lazuli' is used for the real thing. (A good example occurs on the relief of Tuthmosis III at Karnak, showing the booty from his Palestinian conquests, which was discussed by several contributors to the Conference.)
10. So too, of course, was tomb-robbing – an activity positively encouraged by medieval potentates in search of precious metal.
11. Or Bactria, by this time.
12. China is again a relevant comparandum, since its valuation of different materials was not the same as those of its western contemporaries, and moreover changed through time. Jade was especially valued from late Neolithic times onward, and bronze from the beginning of

the Bronze Age; but gold and silver only appeared in any quantity from the later first millennium onward. (It is a pleasing symmetry that one of the principal early sources of Chinese jade was at Hetian (Khotan) in southern Sinkiang, only 1000 km from the sources of lapis lazuli – even though separated by the mass of the Pamirs; for three millennia the former went eastwards, the latter westwards, before the two routes were linked into the continuous artery of the Silk Road.) The high valuation of gold and silver in the Near East and Europe also contrasts with indigenous African systems of value, where the powerful reddish colour of copper made it more valuable than the paler metals; initially the trans-Saharan metal trade (copper for gold) brought pure advantage to both parties in the transaction-chain.

13. Of course, these practices were largely confined to elites; and it is this restricted body of consumers which is the most evident contrast with the mass markets, or even bodies of bourgeois consumers, about whom economic theories have been written. But the limited number of elite consumers was balanced by the (to us) extraordinary concentration of income, so that relatively small numbers of people wielded extraordinary economic power – which of course also coincided closely with political, and also military power.
14. Wine made from grapes was a contribution of the northern neighbours of the urban Uruk world, probably the Transcaucasians of the Kuro-Arax culture – though its promotion to a standard element of ritual consumption was arguably a response to the spread of Uruk practices. This pattern of development through metropolitan stimulus and local substitution is a pattern which underlies several features of technological innovation discussed below. Cultivated vines spread both eastward and westward, as wine-drinking was adopted by the urban elites of the mediterranean vegetation-zone, stretching from the Aegean to Fars in the later third millennium, while the prestige of non-grape-based alcoholic drinks spread even further. It was intimately associated with sheet-metal vessels and their pottery imitations.
15. This is itself has implications for the use of bulk containers, such as pottery types like Canaanite and collar-rim jars (or their first-millennium equivalents, amphorae and dolia), used for long-distance maritime transport in economies capable of mobilising goods on this scale.
16. Cupellation of silver from lead typically precedes the others by a short length, since it is advantageous for core economies to spread this technology to the supply-zones on their periphery, in order to increase the supply of silver which they can obtain from them at (what are to them) advantageous rates, since their valuation of silver – which is a principal medium of exchange – is higher than that prevailing in non-urban economies.
17. Archive of Lugalanda of Lagash, 2400 B.C. (Nissen *et al.* 1993, 63 and 68).
18. The local temple-seminary of management here in Oxford is dedicated to its donor, the arms-dealer Wafic Said.
19. Just as beer and wine (or clarified butter and olive-oil) had different areas of consumption, in the alluvial river-valleys and mediterranean hill-country respectively
20. Barber's later volume on textiles in the ancient world (1994) is called *Womens' Work*, and is an appropriate reminder of the relative contributions of the sexes to the development of techniques and their sustained application to material production. The point came home to one of us in visiting the Viking Ship Museum in Oslo, admiring the boatbuilding skills of Viking shipwrights. But what the label reminded visitors admiring this typically male craft was that as much work went into the sails (which, being of wool, survive even less well than the wood) as into the hull of the vessel itself; and that this essential contribution was of course a female one. Just as the bones of male-hunted animals survive better in archaeological contexts than female-gathered seeds, so the archaeological record tends consistently to be stacked in favour of the male contribution to technology.

21. One could construct similar arguments for high-tech jewellery techniques, such as those involving filigree and granulation, which required specialist workshops.
22. There may be an analogy here with the development of wheelmade pottery, since the earliest forms made on the wheel, in the Uruk period, are liquid-containers for the new commodities such as beer and oil. These mass-produced containers would have been used for storage and transport, while more elegant and precious versions of the same shapes were used for serving these substances in consumption rituals.
23. Note that for the ancient Andean civilisations, gold was the 'sweat of the sun', silver the 'tears of the moon'. This emphasises the important point (mentioned above in connection with the different valuation of gold *versus* copper in Africa north and south of the Sahara), that *colour* is often a primary property in the valuation of metalwork. Thus arsenical alloying may initially (or even primarily) have been more important for its 'silvering' effect (by surface enrichment through inverse segregation) than for its mechanical properties, such as work-hardening: see Sherratt (2000a, 72–3) for Early Cycladic examples.
24. Arslantepe itself (level VIA) has produced some of the first evidence for morphologically domesticated grapes (five 'wild' and five 'domesticated' pips: Belisario *et al.* 1994); on the basis of the occurrence of tartaric acid, wine is thought to have been produced earlier, around 6500 B.C. at Hajji Firuz on Lake Urmia (McGovern *et al.* 1996). The Arslantepe VIA assemblage has many Uruk forms, and probably therefore a 'lowland' cuisine and consumption habits (beer?); the following level VIB has a Kuro-Arax assemblage, with many of the cups and jugs of general types which spread through Anatolia and the Aegean in association with the domestic vine and (presumably) wine-production. It looks as if wine, rather than beer, had become the elite beverage of choice for the regions north and west of Mesopotamia.
25. The types of slashing-swords developed in central Europe were solid-cast, rather than involving extensive forging, as in the Mediterranean. This technology of cast metalwork continued to be elaborated in northern Europe, to produce such spectacular items as cast belt-boxes and lurs in the Nordic Late Bronze Age. One additional element in this line of developent may have been the ready availability of lots of beeswax for casting in Boreal-forest environments (since wax was a traditional northern export in ancient and medieval times), and the Scandinavian tradition of adding spiral ornament to cast surfaces used this method. Curiously, socketed axes only spread into Europe in Urnfield times, replacing the indigenous palstave and its derivatives.
26. Robert Bagley (1987) has suggested that there existed an earlier Chinese tradition of sheet-metalworking, on the basis of pottery skeuomorphs of jugs based on a sheet-construction (with imitation rivets) belonging to the Qijia culture in Gansu in the later third millennium. The form of argument is sound, but neglects an alternative possibility for the material being copied. It seems much more likely that these were in fact *leather* vessels (in shapes not dissimilar to those illustrated in John Waterer's article on 'Leather' in the Oxford *History of Technology* Vol. 2, 147–87), more appropriate to a culture of China's western desert-margin at this time.
27. This interaction did not take place in China, which continued to elaborate the casting tradition, in part perhaps because of the relative rarity of precious metals in use in Shang times, and the consequent absence of cupellation. Silver and gold only became common in China in the later first millennium B.C., and bronze itself was the main currency metal; the high-value commodity which might be said to correspond to precious metals in the West would be jade. (Metals are not essential to the emergence of urbanism – New World civilisations used obsidian and jade as their privileged materials – though metals have the advantage of liquidity, both literal and metaphoric; but textiles were everywhere important

as media of added value.) The Indian technique of making cast metal vessels (which so surprised Nearchus) probably derives from this east-Asian tradition.

28. Such a description would also apply to other novel forms of technology and consumption-practices generated by the rapidly-urbanising societies of the Near East, involving animals and plants as well as inorganic materials: animal-harnessing techniques were applied successively to cattle, horses, and reindeer, as this technology spread northward, and ultimately (in the Arctic) to sleigh-dogs; the preparation of intoxicating beverages came to involve not just beer and wine but kymiss and kava.
29. The trading community around the Persian Gulf, linking Mesopotamia to the Indus valley, had undergone a severe collapse in the earlier second millennium, so that the Mediterranean had become the principal axis of expansion.
30. This is a common phenomenon – for instance the occurrence of classical Greek precious metal vessels in Scythian and Thracian tombs, although they do not survive in Greece itself.
31. They are listed in the Appendix to Sherratt (1994).

REFERENCES

Adams, R. Mc. 1974, Anthropological perspectives on ancient trade, *Current Anthropology*, 15(3), 239–58.

Adams, R. Mc., 1996, *Paths of fire: an anthropologist's inquiry into Western technology*: Princeton University Press, Princeton, NJ.

Appadurai, A., 1986, *The Social life of things: commodities in cultural perspective*, Cambridge, Cambridge University Press.

Bagley, R. W., 1987, *Shang ritual bronzes in the Arthur M. Sackler collections*, Arthur M. Sackler Foundation, Washington, D.C.

Barber, E. J. W., 1990, *Prehistoric textiles: the development of cloth in the Neolithic and Bronze Ages with special reference to the Aegean*: Princeton University Press, Princeton, NJ.

Barber, E. J. W., 1994, *Women's work: the first 20,000 years: women, cloth, and society in early times*, Norton, New York and London.

Belisario, M. V., Follieri, M. and Sadori, L, 1994, Nuovi dati archaeobotanici sulla coltivazione di *Vitis vinifera* L. ad Arslantepe (Malatya, Turchia), in Milano, L. (ed), *Drinking in ancient societies: history and culture of drinks in the ancient Near East: papers of a symposium held in Rome, May 17-19*, Sargon, Padova, 77–90

Bijker, W. B., Hughes, T. P. and Pinch, T. (eds), 1987, *The social construction of technological systems: new directions in the sociology and history of technology*, MIT Press, Cambridge MA.

Braudel, F., 1973, *The Mediterranean and the Mediterranean world in the age of Philip II*, Collins, London.

Brewer, J. and Porter, R. (eds), 1993, *Consumption and the world of goods*, Routledge, London.

Carus-Wilson, E. M., 1963, *Essays in economic history*, Edward Arnold, London.

Chernykh, E. N., 1992, *Ancient metallurgy in the USSR: the early metal age*, Cambridge University Press, Cambridge.

Chernykh, E. N. and Kuz'minykh, S. V., 1989, *Drevniaia metallurgiia Severnoi Evrazii: seiminsko-turbinskii fenomen*, Nauka, Moskva.

Childe, V. G., 1936, *Man makes himself*, Watts & Co, London.

Childe, V. G., 1944, Archaeological ages as technological stages (Huxley Memorial Lecture), *Journal of the Royal Anthropological Institute*, 74, 7–24.

Childe, V. G., 1930, *The Bronze Age*, Cambridge University Press, Cambridge.

Clark, G., 1961, *World prehistory, an outline*, 1st edn.,Cambridge University press, Cambridge.

Douglas, M. and Isherwood, C., 1979, *The world of goods: towards an anthropology of consumption*, Allen Lane, London.

Evans, J., 1922, *Magical jewels of the middle ages and the renaissance, particularly in England*, The Clarendon Press, Oxford.

Finley, M. I., 1973, *The ancient economy*, Chatto & Windus, London.

Fox, R. (ed), 1996, *Technological change: methods and themes in the history of technology*, Harwood Academic, Amsterdam.

Frangipane, M., 1997, A 4th-millennium temple/palace complex at Arslantpe-Malatya. North-South relations and the formation of early state societies in the northern regions of Greater Mesopotamia, *Paléorient*, 23(1), 45–73.

Frangipane, M and Palmieri, A. (eds), 1983 [published 1988], Perspectives on protourbanisation in eastern Anatolia: Arslantepe (Malatya). An interim report on the 1975–1983 campaigns, *Origini*, 12(2), 287–668.

Frank, A. G., 1998, *ReOrient: global economy in the Asian Age*, University of California Press, Berkeley and London.

Gerloff, S., 1993, Zu Fragen mittelmeerländischer Kontakte und absoluter Chronologie der Frühbronzezeit in Mittel- und Westeuropa, *Prähistorische Zeitschrift*, 68, 58–102.

Greene, K., 1994, Technology and innovation in context: the Roman background to medieval and later developments, *Journal of Roman Archaeology*, 7, 22-33.

Halstead, P., 1995, Plough and power: the economic and social significance of cultivation with the ox-drawn ard in the Mediterranean, *Bulletin on Sumerian Agriculture*, 8, 11–22.

Haudricourt, A-G., 1988, *La technologie, science humaine: recherches d'histoire et d'ethnologie des techniques*, Éditions de la Maison des sciences de l'homme, Paris.

Hicks, J. R., 1969, *A theory of economic history*, Clarendon Press, Oxford.

Hood, S., 1993, Amber in Egypt, in Beck, C. W. and Bouzek, J. (eds), *Amber in Archaeology: Proceedings of the Second International Conference on Amber in Archaeology, Liblice 1990*, Institute of Archaeology of the Czech Academy of Sciences, Prague, 230–35.

Jacobs, J., 1970, *The economy of cities*, Cape, London.

Jones, E. L., Frost, L. and White, C., 1993, *Coming full circle: an economic history of the Pacific Rim*, Westview, Boulder, Col. and Oxford.

Latour, B. and Lemonier, P., 1994, *De la préhistoire aux missiles balistiques: l'intelligence sociale des techniques*, La Découverte, Paris.

Lemonnier, P., 1992, *Elements for an anthropology of technology*, Museum of Anthropology, University of Michigan, Ann Arbor.

Lemonnier, P., 1993, *Technological choices: transformation in material cultures since the Neolithic*: Routledge, London.

Leroi-Gourhan, A., 1943, *L'homme et la matière*, Albin Michel, Paris

Leroi-Gourhan, A., 1945, *Milieu et techniques*, Albin Michel, Paris.

Leroi-Gourhan, A., 1971, *Evolution et techniques*, Albin Michel, Paris.

Liverani, M., 1987, The collapse of the Near Eastern regional system at the end of the Bronze Age: the case of Syria, pp. 66–73 in Rowlands *et al.* 1987.

Marfoe, L. 1987, Cedar Forest to Silver Mountain: social change and the development of long-distance trade in early Near Eastern societies, pp. 25–35 in Rowlands, M, Larsen, M. and Kristiansen, K. (eds), *Centre and Periphery in the Ancient World*, Cambridge University Press, Cambridge.

McGovern, P. E, Glusker, D. L., Exner, L. J. and Voigt, M., 1996, Neolithic resinated wine, *Nature*, 281, 480–1.

Moorey, P. R. S., 1994, *Ancient Mesopotamian materials and industries: the archaeological evidence*, Clarendon Press, Oxford.

Mowatt, L, Morphy, H. and Dransart, P., *Basketmakers: meaning and form in native American baskets*, Pitt Rivers Museum Monograph No. 5, Oxford.

Nef, J. U., 1964, *The conquest of the material world*, University of Chicago Press, Chicago and London.

Nicholson, P. T. and Shaw, I. (eds), 1999, *Ancient Egyptian materials and technology*, Cambridge University Press, Cambridge.

Nissen, H. J., Damerow, P. and Englund, R. 1993, *Archaic bookkeeping: early writing and techniques of the economic administration in the ancient Near East*, University of Chicago Press, Chicago and London.

Pfaffenberg, B., 1988, Fetishised objects and humanised nature: towards an anthropology of technology, *Man*, NS 23, 236–52.

Philip, G. and Rehren, T., 1996, Fourth millennium BC silver from Tell esh-Shuna, Jordan: archaeometallurgical investigation and some thoughts on ceramic skeuomorphs, *Oxford Journal of Archaeology*, 15(2).

Pickles, S. and Peltenburg, E., 1998, *Metallurgy, society and the bronze/iron transition in the east Mediterranean and the Near East*, Report of the Department of Antiquities, Cyprus 1998, 67–100.

Potts, D.T., 1997, *Mesopotamian civilization: the material foundations*, Athlone Press, London.

Rajpitak, W. and Seeley, N., 1979, The bronze bowls from Ban Don Ta Phet: an enigma of prehistoric metallurgy, *World Archaeology*, 11(1), 26–31.

Rawson, J. 1986, Tombs or hoards: the survival of Chinese silver of the Tang and Song periods, seventh to thirteenth centuries AD, in Vickers, M. J., *Pots & pans: a colloquium on precious metals and ceramics in the Muslim, Chinese and Graeco-Roman worlds*: Oxford University Press for the Board of the Faculty of Oriental Studies, University of Oxford, Oxford, 31–56.

Schneider, J., 1977, Was there a pre-capitalist world system? *Peasant Studies*, 6(1), 20–9.

Rowlands, M, Larsen, M and Kristiansen, K. (eds), *Centre and Periphery in the Ancient World*, Cambridge University Press, Cambridge.

Sherratt A., 1995, Alcohol and its alternatives: symbol and substance in preindustrial cultures, in Goodman, J., Lovejoy, P.E., and Sherratt, A. (eds), *Consuming Habits: Drugs in History and Anthropology*, Routledge, London, 11–46.

Sherratt, A., in press, The horse and the wheel: the dialectics of change in the circumpontic region and adjacent areas, 4500–1500 BC, To appear in McDonald Institute Conference Proceedings.

Sherratt, A. and Sherratt, S., 1998, Small worlds: interaction and identity in the ancient Mediterranean, in Cline, E. H. and Harris-Cline, D. (eds.), *The Aegean and the Orient in the second millennium. Proceedings of the 50th Anniversary Symposium, Cincinnati, 18–20 April 1997*, 329–43. Aegaeum 18, Liège and Austin.

Sherratt, A. and Taylor, T., 1989, [repr. in Sherratt, A., 1997, *Economy and Society in Prehistoric Europe: changing perspectives*, Edinburgh: Edinburgh University Press, Ch 17] Metal vessels in Bronze Age Europe and the context of Vulchetrun, in Best, J. and de Vries, N. (eds), *Thracians and Mycenaeans*, Brill, Leiden, 106–34.

Sherratt, S., 1994., Commerce, iron and ideology: metallurgical innovation in 12th-11th century Cyprus, in Karageorghis, V. (ed.), *Proceedings of the International Symposium 'Cyprus in the 11th century B.C.', organized by the Archaeological Research Unit of the University of Cyprus and the Anastasios G. Leventis Foundation, Nicosia, 30–31 October 1993*, Nicosia, 59–106.

Sherratt, S., 1998, 'Sea Peoples' and the economic structure of the late second millennium in the eastern Mediterranean, in Gitin, S., Mazar, A. and Stern E. (eds.), *Mediterranean peoples in transition, thirteenth to early tenth centuries BCE*, Jerusalem 292–313.

Sherratt, S., 1999, Eppur si muove: pots, markets and values in the second millennium

Mediterranean, in Crielaard, J. P., Stissi, V. and van Wijngaarden G. J. (eds), *The Complex past of pottery: production, circulation and consumption of Mycenaean and Greek pottery (16th to early 5th centuries B.C.), Proceedings of the ARCHON International Conference, Amsterdam 1996,* Gieben, Amsterdam, 163–211.

Sherratt, S., 2000a., *Catalogue of Cycladic antiquities in the Ashmolean Museum. The captive spirit,* Oxford.

Sherratt, S., 2000b, Circulation of metals and the end of the Bronze Age in the eastern Mediterranean, in Pare, C. F. E. (ed.), *Metals make the world go round. The supply and circulation of metals in Bronze Age Europe. Proceedings of a conference held at the University of Birmingham in June 1997,* Oxford, 82–98.

Sherratt, S., in press, Potemkin palaces and route-based economies, in Killen, J. and Voutsaki, S. (eds.), *The Mycenaean palatial economy,* Cambridge.

Sigaut, F., 1975, *L'agriculture et le feu: rôle et place du feu dans les techniques de préparation du champ de l'ancienne agriculture européenne,* Mouton, Paris.

Singer, C. J., Williams, T. I. and Raper, R., 1954-1984, *A history of technology,* Clarendon Press, Oxford.

Sklenár, K., 1983, *Archaeology in Central Europe: the first 500 years,* Leicester University Press, Leicester.

Stocks, D. A., 1993, Making stone vessels in ancient Mesopotamia and Egypt, *Antiquity,* 67, 596–603.

Tampoe, M., 1989, *Maritime trade between China and the West: an archaeological study of the ceramics from Siraf from the 8th to 15th centuries A.D,* BAR international series, 555, Oxford.

Vichos, Y., 1999, The Point Iria wreck: the nautical dimension, in Phelps, W., Lolos, Y. and Vichos, Y. (eds), *The Point Iria Wreck: interconnections in the Mediterranean ca. 1200 B.C. (Proceedings of the International Conference, Island of Spetses, 1998),* Hellenic Institute of Marine Archaeology, Athens, 77–98.

Vickers, M. J. (ed.), 1986, *Pots & pans: a colloquium on precious metals and ceramics in the Muslim, Chinese and Graeco-Roman worlds, Oxford 1985,* Oxford University Press for the Board of the Faculty of Oriental Studies, University of Oxford, Oxford.

Waetzoldt, H., 1972, *Untersuchungen zur neusumerischen Textilindustrie,* Roma.

Wallerstein, I., 1974, *The Modern World-System: capitalist agriculture and the origins of the world-economy in the 16th century,* Academic Press, New York.

Watson, W., 1986, Precious metal – its influence on Tang earthenware, in Vickers, M. J., 1986, *Pots & pans: a colloquium on precious metals and ceramics in the Muslim, Chinese and Graeco-Roman worlds, Oxford 1985,* Oxford University Press for the Board of the Faculty of Oriental Studies, University of Oxford, Oxford, 161–74.

White, L. A., 1959, *The Evolution of Culture: the development of civilisation to the fall of Rome,* McGraw-Hill, New York.

White, L. T., 1964, *Medieval technology and social change,* Oxford University Press, Oxford.

Chapter 3

Technology in Society: Three Textual Case Studies from Late Bronze Age Mesopotamia

Eleanor Robson

Abstract

There is a bewildering variety of textual evidence for Late Bronze Age technological practices, practitioners, and terminology. This paper provides three examples of how technical information and societal context can be gleaned from the cuneiform record.

INTRODUCTION

The second half of the second millennium saw a complete restructuring of the political landscape, as ephemeral city-states were increasingly dominated by stable territorial polities – epitomised for us by the so-called 'Amarna age', in which a small group of 'brother' kings controlled vast swathes of the Middle East, both directly and through the rule of their vassals. This paradigm shift in territorial governance was fed by and fed into the technological revolution of the Late Bronze Age (LBA): improvements in transport, communications, and militaria, as well as in the availability and quality of a huge variety of manufactured articles.

The Amarna correspondence (Moran 1992) inevitably dominates our interpretations of the 500 year period, even though we know it spans a fraction of that time. Not only does it provide us with unrivalled information about contemporary international affairs (e.g. Cohen and Westbrook 1999) but the dowries and tribute lists in particular yield fascinating and detailed data on the manufacture and distribution of luxury goods throughout the region. Although the Amarna correspondence is, of course, greatly skewed towards the high élite and reveals little if anything of more demotic technologies, the evidence it provides is typical in at least one respect: it is written in cuneiform on clay tablets. The vast distribution of cuneiform writing systems across the Middle East, from Anatolia to Egypt to western Iran, during the LBA is unparalleled at any other period of history (Walker 1987, 34). As historians of ancient technology we thus have an unrivalled databank of written evidence for the later second millennium: although cuneiform documentation is sparser in its Mesopotamian heartland than during almost every other period of its use (Van De Mieroop 1999, 12), at no other time was it in such widespread use by neighbouring civilisations. Thus the adaptability, ease of use, and long-term durability of this typically LBA technology has provided us with a potential gold mine of insights into other technologies of the time.

Many of the pertitent archives and corpora have now been accessibly translated into modern European languages (see Bibliographical Appendix). It is no longer necessary to rely so heavily on the Amarna Letters, attractive and richly detailed though they are, now that other large bodies of textual evidence are becoming available. The three studies presented here, for example, are all based on material available in English translation and none of them calls upon the Amarna archives for support. Indeed, it could be argued that the Amarna material has skewed our interpretation of LBA technology towards élite, palatially-based international trade in luxury goods. The sensitive use of temple records, familial archives, and ancient literary and scholarly works from a wider geographical area may lead us to a more rounded and nuanced understanding of the impact and spread of technologies in the ancient Near East.

Traditionally textual studies and the technical examination of material culture have interfaced simply as a means of lexicographic identification: a question of locating the ancient names for materials and manufactured objects (e.g. Dalley 2000). That task is still necessary and interesting today – as can be seen from the large number of untranslatable or unreadable words in the extracts I give below. It is also a complex and delicate procedure which has to recognise that ancient semantic and functional word categories rarely coincide with modern ones, particularly when one is dealing with technical and classificatory terms which today are based on organisational and scientific principles developed in the last 150 years. For instance, the Akkadian word *abnu* is usually translated 'stone'. But as well as the range of meanings we associate with the word 'stone' – its natural state, as both ores and gems, as well as large fruit seeds, gallstones, and hailstones – it can also signify metal, glass, and metrological weights made from any material. Interestingly it is also bi-gendered: a rare and always significant phenomenon in Akkadian.

> '**abnu A** s[ubstantive] masc[uline] and fem[inine]; 1. stone (in natural form and location), 2. stone (prepared for specific use), 3. precious colored (shaped and polished) stone, 4. stone weight, weighing stone, 5. pebble, counter (for accounting), 6. hailstones, 7. stone (as med[ical] term referring to bladder stones), 8. stone or seed of a plant, 9. glass, [...] from O[ld] Akk[adian period, *ca.* 2350 B.C.] on; rarely fem[inine], see [meanings] 3, 4, 6 and 7, [...] wr[itten] syll[abically] and NA_4 [...].' (Oppenheim *et al* 1956, A I 54)

We shall return to *abnu* later, but first I shall present two other case studies. There has been much interest recently in 'embedded technologies' and 'technological choices' (e.g. Sillar and Tite 2000). While ancient textual sources rarely tell us directly and unambiguously about technical terms processes they often speak eloquently of the social context of technologies and the lives of technological practitioners. My first example explores the mis-match between royal rhetoric and the actuality of religious building in mid-thirteenth century Ashur, illustrating the pitfalls of using textual evidence to reconstruct the architectual history of built structures without accounting for their cultural meaning. Then I examine the socio-economic status of temple brewers in Ur some 50 years later, with some surprising results. Finally I discuss why LBA innovations in glass-making entered the later Mesopotamian scholarly tradition while other, arguably more important, new technologies did not.

WHAT'S IN A NAME? SHALMANESER I'S RECONSTRUCTION OF THE ASHUR TEMPLE

As the Hurrian kingdom of Mittani disintegrated in the mid-fourteenth century B.C., the governors of the city of Ashur began to establish themselves as independent rulers on an equal footing with the other major powers of the Near East (Harper *at al.* 1995, 23; Moran 1992, no. 15). From the time of Ashur-uballit I onwards, as their territory and wealth grew steadily, the Assyrian kings devoted increasing attention to the renovation and beautification of their ancient city. The temple and ziqqurrat of Ashur's eponymous deity was extensively restored under Ashur-uballit's great-great-grandson, Shalmaneser I (1273–1244 B.C.). Exemplary German excavations of Ashur in the early 20th century included an investigation of the cultic precinct (Andrae 1977), which not only revealed the detailed architectural history of the remains (Haller and Andrae 1955) but also recovered the monumental inscriptions of the kings who commissioned the work (Grayson 1987, 109–327).

Five different accounts of Shalmaneser's building work were found in and around the temple complex of Ashur (Table 3.1). There were two primary versions: a long description, preceded by an account of Shalmaneser's military campaigns, written in at least 20 copies on stone tablets (Inscription 1), and a shorter redaction of the building work alone on at least 60 clay cones written in a different year (Inscription 2). The other three exist in single exemplars on stone and clay tablets, with Inscriptions 3 and 5 dating to a third year and Inscription 4 (more of which below) coming from a fourth. Sadly, the order in which the Middle Assyrian year-name officials held office is not yet known and therefore their chronological sequence can be deduced only from internal evidence (and which determines the sequence in which they are presented here).

Inscriptions 1 and 2, which describe the history of the temple from its foundation until Shalmaneser's time, greatly facilitated the excavators' task of dating and attributing rebuilding phases (Haller and Andrae 1955, 38–39). But the temple's supposed founder Ushpia has left no traces in the archaeological or historical record except as the penultimate entry in the famous enumeration of "17 kings who lived in tents" at the start of the Assyrian King List (Kuhrt 1995, 85). The earliest datable excavation levels belong to the reign of Shamshî-Adad I (1813–1781 B.C.), below which "earlier buildings came to light but were so meager they could not be more precisely dated. The excavators designated them 'prehistoric' structures with no cult function" (Harper *et al* 1995, 37; cf. Haller and Andrae 1955, 6–8). How, then, to explain this discrepancy between the archaeological and the textual evidence: have the architectural remains been misinterpreted or do the inscriptions exaggerate the age of the building?

Towards the end of Inscription 1, Shalmaneser describes how he anointed and reburied the monumental inscriptions of his ancestors who had themselves worked on the temple. These too have come down to us. Erishum (1939–1900 B.C. – Kuhrt 1995, 82) had bricks and stone elements of the Ashur temple inscribed with accounts of his

Table 3.1: Shalmaneser I's monumental inscriptions recording his rebuilding of the Ashur temple.

Inscription 1	Inscription 2	Inscription 3	Inscription 5
When E-hursang-kurkura, the temple of Ashur my lord	**When E-hursang-kurkura, the** ancient **temple**	**When E-hursang-kurkura, the temple** of Ashur my lord	[**When**] **E-hursang-kurkura, the temple** of Ashur [my lord]
- which Ushpia, vice-regent of Ashur, my (fore)father, had previously built and which became dilapidated and which Erishum, my (fore)father, vice-regent of Ashur, rebuilt. 159 years after the reign of Erishum had passed and that temple became dilapidated and Shamshî-Adad, vice-regent of Ashur rebuilt (it). 580 years passed and that temple which Shamshî-Adad had built became old and decrepit.	- which Ushpia, my (fore)father, vice-regent of Ashur, had previously built, which became dilapidated and which Erishum, my (fore)father, vice-regent of Ashur, rebuilt. 159 years passed and that temple became dilapidated again and Shamshî-Adad, also my (fore)father, vice-regent of Ashur rebuilt (it). 580 years passed and	- which since earliest times the kings my (fore)fathers had built but which had become old and dilapidated -	- which since earliest times [the kings] my (fore)fathers had built-
A fire broke out in it. The temple, its sanctuary, chapels, shrines, daises, cult platforms, seats: all the property of the temple of Ashur my lord burnt in the fire:	the temple and its sanctuary were ruined by the fire-god Girra:	that temple was ruined by the fire-god Girra:	that temple was ruined by the fire-god Girra:
At that time, I cleared away that temple. I had its earth cleared away, and **I reached the bottom of its foundation pit.**	**I reached the bottom of its foundation pit.**	At that time, I cleared away the temple of Ashur my lord to its enclosure wall. I had its earth cleared away, and **I reached the bottom of its foundation pit**.	I cleared away [the temple of Ashur my lord] to its enclosure wall. I had its earth cleared away, and **I reached the bottom of its foundation pit.**
I laid its foundation with strong stone **like the bedrock of a mountain.**	**I laid down its foundation like the bedrock of a mountain.**	**I laid its foundation** with strong stone **like the bedrock of a mountain.**	**I laid its foundation** with strong stone **like the bedrock of a mountain.**
I skilfully built the pure temple, the elevated shrine, the exalted dais, the awesome inner sanctum, which greatly surpasses its ancient predecessors.	I built in addition 2 towers that had not been built before. I greatly extended and enlarged the forecourt of the god Nunamnir and the open space of the forecourt of Ashur my lord.	I extended and enlarged the temple of Ashur my lord. The towers of the Kalkal Gate and the towers for entering the fore-court of Nunamnir from the Step Gate: in addition I built these two towers which had not been built before. I greatly enlarged the forecourt of Nunamnir.	Inside it I built the house of the divine judges, my lords. I restored it. I made E-hursang-kurkura, the temple of Ashur my lord, much larger than before.
It is worthy of praise; it is dedicated befittingly to his great divinity; it is greatly suited to his lordship. I strove to build (it) for Ashur my lord.			
At its foundation I set down stones, silver, gold, iron, copper, tin, aromatic plants upon aromatic plants. I mixed oil, sweet oil, cedar resin, honey, and ghee into its wall plaster.			

Inscription 1 (ctd.)	Inscription 2 (ctd.)	Inscription 3 (ctd.)	Inscription 5 (ctd.)
I built from its foundations to its parapet.		I built the temple of Ashur my lord to its enclosure wall, from its foundations to its parapet.	I built from its foundations to its parapet.
	I laid down seats and cult platforms in their shrines and estab-lished all the gods of E-kur inside it.	I established the gods who dwell in E-kur, my lords, inside it and	I established all the gods, those of the E-kur, inside it.
I deposited my monumental inscriptions. I held a celebratory ceremony.	**I deposited my monumental inscr-iptions** and foundation documents.	**I deposited my monumental inscriptions.**	**I deposited my monumental inscriptions.**
When Ashur the lord enters that temple, and joyfully occupies his exalted dais, may he see the festive activities of that temple and rejoice. May he receive my petitions, may he hear my appeals. With his mighty voice may he greatly speak a fate of well-being for my vice-regency and the vice-regency of my seed, and abundance during my reign for ever after.			
I anointed with oil the monumental inscriptions of earlier kings. I made a libation. I restored them to their places. In future days may a later prince, when that temple becomes old and dilapidated, hear of my heroic acts and tell of my praise-worthy skill. As I restored the monumental inscriptions of earlier kings to their places, may he anoint with oil my monumental inscriptions, make a libation, and restore them to their places. Ashur and Ninlil will hear his prayers.		May a later prince restore my inscribed name. Ashur will hear his prayers.	May a later prince, when that temple becomes dilapidated and he rebuilds (it): as I did not alter the monumental inscriptions of Ashur-uballit my (fore)father but restored them to their places, may he restore my monumental inscriptions. Ashur and the divine judges will hear his prayers.
He who alters my inscriptions and discards my written name: may Ashur, the exalted god who dwells in E-hursang-kurkura, all the Igigi of heaven and the Anunaki of the underworld frown angrily at him and furiously curse an evil curse. May they eradicate his name and his seed from the land. May a king who is his enemy remove his throne, and rule his land before his very eyes.	He who alters my inscriptions and my name: may Ashur my lord overturn his kingship and eradicate his name and his seed from the land.	He who alters my inscriptions and my name: may Ashur my lord overturn his kingship and eradicate his name and his seed from the land.	
Extract from a 168 line inscription on double-columned stone tablets. 22 exemplars, 17 with known findspots, 5 with dates (Grayson 1987: 185 A.0.77.1 112-167).	*Extract from a 26 line inscription on clay cones. 67 exemplars, 27 with known findspots, 21 with dates* (Grayson 1987: 189 A.0.77.2 5-24).	*Extract from a 48 line inscription on a dated stone tablet with known findspot* (Grayson 1987: 190-1 A.0.77.3 5-38).	*Extract from a 37 line inscription on a dated stone tablet with known findspot* (Grayson 1987: 193-4 A.0.77.5 6-35).

building works there (Grayson 1987, 19–37, A.0.33.1–14); he makes no mention of Ushpia or any other earlier builders.

> 'Erishum, vice-regent of Ashur, [son of Ilu]-shumma, [vice-regent of] Ashur, [......] he dedicated for Ashur his lord. I mixed ghee and honey into the mortar of every wall and then laid one layer of bricks. The name of the temple is Wild Bull. [......] If in the future the temple should become [dilapidated and] old and a prince of [my] status wishes to rebuild the temple: he must not [disturb] the clay cone but [restore] the clay cone to [its place].' (Grayson 1987, 32–33 A.0.33.10)

Similarly Shamshî-Adad I's stone tablets and stamped bricks refer only to Erishum as a previous builder of the temple (Grayson 1987, 47–63, A.0.39.1, 9, 11).

> 'The temple of Enlil which Erishum, son of Ilu-shumma, had built had become dilapidated so I abandoned it. I built the temple of Enlil my lord, the fearful dais, the large inner sanctum, the seat of Enlil my lord, which were methodically made by the skilled work of the building trade within my city Ashur. I roofed the temple with cedar. In the rooms I erected cedar doors with silver and gold stars. Under the walls of the temple (I put) silver, gold, lapis lazuli, and carnelian. I mixed cedar resin, best oil, honey, and ghee into the mortar. I methodically made the temple of Enlil my lord, and called it E-am-kurkura, "Temple Wild Bull of the Lands": the temple of Enlil my lord, within my city Ashur.' (Grayson 1987, 49 A0.39.1 ll. 18–58)

Coming back to Shalmaneser's historical summary, we see that it is only from the time of Erishum onwards that the years elapsed between construction events are recorded (based perhaps on eponym lists). Ushpia, for Shalmaneser, seems to have existed in some primordial time before the quantifiable past.

All four inscriptions continue with an account of the fire which destroyed the temple and the resultant total clear-out of the ruins, reaching "the bottom of its foundation pit". Stone foundations "like the bedrock of a mountain" were laid in its place. However, while there is clear archaeological evidence of that fire, and of the new stone foundations, neither the fire nor Shalmaneser's workmen made the clean sweep that the inscriptions imply (cf. Haller and Andrae 1955, 17 Abb. 2; 50 Abb. 14). Why, then, were claims made of total renovation? Is this a further example of empty royal rhetoric, like the reference back to Ushpia?

Shalmaneser called the Ashur temple by the Sumerian name **e-hursang-kurkura** "Temple Hill of the Lands" and earlier in Inscription 1 gave it the epithets "residence of the gods, mountain of the lands" (Grayson 1987, 182 A.0.77.1 lines 6–7). Erishum, however, had called his Ashur temple by the Akkadian name *rîmum* "Wild Bull" (Grayson 1987, 32 A.0.33.10 line 12); Shamshî-Adad, while syncretising Ashur with the pre-eminent Sumerian deity Enlil, had designated the temple in both Sumerian and Akkadian **e-am-kurkura** *rîm mâtâtim* "(Temple) Wild Bull of the Lands" (Grayson 1987, 49 A.0.39.1 lines 52–53). So Shalmaneser not only rebuilt the Ashur temple but gave it a fundamentally new identity: no longer a wild animal, it was now a holy hill rising up to link earth to heaven. Although Shalmaneser abandoned the explicit identification of Ashur and Enlil, paradoxically he adopted the metaphor of temple-as-mountain – traditionally associated with Enlil's temple E-kur (Temple Mountain) in Nippur, as exemplified by the Sumerian-language Hymn to the E-kur from the eighteenth century B.C.

> 'The great house is as great as a mountain. The house of Enlil is as great as a mountain. The house of Ninlil is as great as a mountain. The bedchamber is as great as a mountain. The house which knows no daylight is as great as a mountain. The house at the Lofty Gate is as great as a mountain. The house at the Gate of Well-being is as great as a mountain. The courtyard of Enlil is as great as a mountain. The Hursang-galama (Hill of the Country) is as great as a mountain. The holy Renowned Gate is as great as a mountain. The Gate From Which Grain Is Never Diverted is as great as a mountain. The Ubshu-unkena is as great as a mountain. The Ga-gish-shua is as great as a mountain.' (Black *et al.* 1998–, 4.80.4 lines 1–12)

Although Shamshî-Adad had equated Ashur and Enlil , he chose not to align the Ashur temple with the E-kur but to maintain Erishum's temple name "Wild Bull". Shalmaneser, however, even refers to the E-kur in three of his four inscriptions.[1]

The holy mound (**dul kug**) or hill (**hursang**) was considered to have been the home of the gods in the mythological time before the natural world was created. In the Sumerian-language Debate between Sheep and Grain, again from the Old Babylonian period, the holy hill is the birthplace of the gods, of sheep, and of grain.

> 'When, upon the hill of heaven and earth, An spawned the Anuna gods, since he neither spawned nor created Grain with them, and since in the Land he neither fashioned the yarn of Uttu (the goddess of weaving) nor pegged out the loom for Uttu — with no Sheep appearing, there were no numerous lambs, and with no goats, there were no numerous kids, the sheep did not give birth to her twin lambs, and the goat did not give birth to her triplet kids; the Anuna, the great gods, did not even know the names Ezina-Kusu (Grain) or Sheep. ...
>
> At that time, at the place of the gods' formation, in their own home, on the Holy Mound, they created Sheep and Grain. Having gathered them in the divine banqueting chamber, the Anuna gods of the Holy Mound partook of the bounty of Sheep and Grain but were not sated; the Anuna gods of the Holy Mound partook of the sweet milk of their holy sheepfold but were not sated. For their own well-being in the holy sheepfold, they gave them to mankind as sustenance.' (Black *et al.* 1998–, 5.3.2 lines 1–11; 26–36)

Does the resonant new name E-hursang-kurkura help to explain the apparent inaccuracies in Shalmaneser's inscriptions? I would argue that it does. Naming, according to discourse analysis, is a performative act (e.g. Austin 1999, 64), and in ancient Mesopotamia it was *the* performative act *par excellence*: to name something was tantamount to bringing it into being (Bottéro 1992, 97). If the Ashur temple was to be the Hill of the Lands it needed to have had a primeval existence in the world of the gods; yet as a building it also had to inhabit real time and space. This contradiction could be reconciled by attributing its foundation to a mythologically ancient ancestor, namely Ushpia, who was somehow both of natural time and out of it. Similarly, as the Hill of the Lands, the temple should be as solid as a rock. The word (rather clumsily) translated here as "bottom of foundation pit" is Akkadian *dannatu*, derived from the verb *danânu* "to be strong". Its other primary meaning is "fortress, fortified place" but also has nuances of famine, hard times, and extreme cold (Oppenheim *et al.* 1956–, D 87–91) – all of which have mountainous associations. The temple-as-hill must rest on a stone foundation "like the bedrock of a mountain" and that is what Shalmaneser provided, in both word and deed.

The four inscriptions diverge once they start to detail the actual works carried out. The only phrases which are repeated verbatim in all four passages translated in Table 3.1 concern the new name of the temple, its *dannatu* and its mountain bedrock, and the installation of monumental inscriptions. This is not mere coincidence: the naming, the foundation of the mountain, and the written record of the temple's new identity comprise the conceptual core of the temple's renewal. Shalmaneser's building inscriptions, then, are not just an occasionally inaccurate key to interpreting the structures in which they were placed but expressions of the temple's cultural identity above and beyond the bricks and mortar from which it was made. In this view, we should expect the material and textual perspectives to be somewhat divergent.

THE BREWERS AND BURGLARS OF UR

The fourth of Shalmaneser's five inscriptions about his rebuilding of the Ashur temple, attested on one clay tablet found *in situ*, describes the construction of an elaborate monumental doorway and the renovation and enlargement of a beer-brewing facility.

> 'When I built high towers in E-hursang-kurkura, the temple of Ashur, my lord, I erected a bronze entrance in the holy gate of Kalkal, (and) large steps. I installed lintels, divine discs, divine emblems, and bronze doors.
>
> At that time, the old breweries which the kings, my (fore)fathers had previously built: to expand the wall and to enlarge the open area, I destroyed the breweries. I cleared away the whole lot. I enlarged the open area by 16+ cubits; I thickened its internal wall by 10 bricks, its external wall by 5 bricks. I built the *erinakku*. I restored the beer-vats and pipes. I made them larger than before. I built and perfected (the breweries) from their foundations to their parapets, and I deposited my monumental inscriptions.' (Grayson 1987, 192 A.0.77.4 lines 19–39)

The brewery had apparently been an integral part of the complex since its foundation, according to Erishum's building inscriptions and stamped bricks.

> 'From the Sheep's Gate to the People's Gate, I set aside land for Ashur my lord. I built all of the temple outbuildings. I constructed twin beer vats [...], 2 geese weighing 1 talent next to them. I plated two moon-disks with bronze and I set 2 *argabu* (birds?) weighing 1 talent on top of them.' (Grayson 1987, 20 A.0.33.1 lines 9–16; cf. A.0.33.11, A.0.33.13)

In one of the storerooms along the south-west side of the temple's outer courtyard, constructed under Shalmaneser's building works, the excavators found ten large pottery jars containing around 650 administrative tablets with dates spanning the twelfth century B.C. (Pedersén 1998, 84). Three of the vessels bear inscribed labels, two of which describe their contents.

> 'Document-holder of the accounts of the brewers of the Ashur temple, under the supervision of Ezbu-lêshir, the offerings-overseer of the Ashur temple, servant of Tiglath-Pileser (I).
>
> Of the victuallers and oil-pressers of the Ashur temple, under the supervision of Ezbu-lêshir, the offerings-overseer of the Ashur temple, servant of Tiglath-Pileser (I).' (Postgate 1986, 170)

Ezbu-lêshir, it appears, was responsible for receiving regular provincial offerings (*ginû*) of grain, fruit, honey, and oilseeds and redistributing them to the temple personnel. As well as the brewers (*sirashû*), victuallers (*kakardinnû*), and oil-pressers (*sâhitû*), there were threshers (*dâ'ishû*), grinders (*sâmidû*), and bakers (?) (*alahinnû*) working under him, all presumably processing the offerings into food-stuffs (Postgate 1980, 69; 1986, 171).

The majority of Ezbu-lêshir's tablets are yet to be published, so it is impossible at the moment to describe the food processors' relationship with the temple in any detail. However, we can glean a good deal of information about a different group of food technologists: the temple brewers who served the moon god Sîn in the southern city of Ur. The Dayyânâtu family occupied a brick-built house close to the E-kishnungal temple precinct (Woolley 1965, pl. 63A), leaving there an archive of about seventy tablets spanning the period *ca.* 1240–1175 B.C. (Gurney 1983, 1). Three family members held the title 'brewer of Sîn'. They had dealings with at least eleven other men with the same professional designation and a further eight who handled beer products in other capacities (Gurney 1983, 2).

Because 'private citizens kept only specific types of records, namely those that documented that something was owed to them, or that proved their ownership' (Van De Mieroop 1999, 18–19), we have very little evidence of the Dayyânâtus' actual brewing activities – though enough clues survive to show that they did indeed brew and weren't simply holders of prebends or entitlements to regular temple income. But the archive does give a good idea of their social status and wealth, and their lives outside the brewery. First, although we know almost nothing about the dwelling they occupied as only traces of its foundations were found (Woolley 1965, 78–79) and no documentation of ownership survives, the fact that it was less than 100m from the temple complex is surely significant. A memo recording property stolen from the house gives a flavour of how it was furnished.

> 'Breach in Shamash-êtir's house, which Martuk son of Mu-[......] and Iliya son of [......] burgled:
>
> 2 tables, 2 *ishhenabe* garments, 2 *muhtillû* garments, 3 spotted (?) ox hides, 2 ...... hides, 5 sheep-skins, and 1 millstone, in [Martuk's (?)] hand. He had stolen [them and they were found] in his very hands. [His wife (?) stole] 9 minas of wool. He (Shamash-êtir) gave her to Shamash-nâsir's custody but did not prosecute her.
>
> Property of Shamash-êtir which Martuk and Iliya burgled.' (Gurney 1983, 115 no. 40)

Table 3.2: Shamash-êtir's dated property records

Year	Transaction	Silver equivalent	Tablet
1245-33	Buys calf for 5 shekels of silver	5 shekels	Gurney 1983: 58-59 no. 14
1237	Denies accusation of stealing cow		Gurney 1983: 60-62 no. 15
1230	Buys 1 ox for 10 gur of grain	12 shekels	Gurney 1983: 98-99 no. 31
1230	Buys slave girl for 4 gur grain and 5 garments	32 shekels	Gurney 1983: 84-86 no. 25
1229	Robbed of deliveries: 2 1/2 minas wool and 32 precious stones	16 shekels	Gurney 1983: 72-74 no. 20
1226	Robbed of 3 oxen	36 shekels?	Gurney 1983: 28-31 no. 3
1224	Buys unsatisfactory slave boy for 9 shekels of gold	36 shekels	Gurney 1983: 22-28 no. 2
1223	Buys cow and calf for 25 gur 8 ban of grain and 1 garment	10 shekels	Gurney 1983: 104-106 no. 34
1223	Accused of stealing cow		Gurney 1983: 70-71 no. 19
1222	Buys slave boy for 1 donkey, 2 shekels of red gold, 1 gur of grain	30 shekels	Gurney 1983: 74-76 no. 21
1222	Buys same slave boy for 1 donkey, 4 shekels of red gold, 2 gur of grain	44 shekels	Gurney 1983: 77-79 no. 22

Table 3.3: Shamash-êtir's undated property records

Contents	Silver equivalent	Tablet
Sends 2 gur of malt, 1 gur of beer-bread, plus other items now missing, to brewing colleague	...	Gurney 1983: 113-114 no. 39
Pays damages (?) for loaned ox which died in his care	...	Gurney 1983: 125-127 no. 45
Receives marriage gift of grain, wool, 1 sheep, and other items now missing	24 shekels	Gurney 1983: 143-145 no. 53
Receives elderly donkey in return for gift of goods now missing	10 shekels?	Gurney 1983: 151-153 no. 58
Gives grain, aromatic plants, wool, cress, oil, 1 ox, 1 cow, and other items now missing	32 shekels	Gurney 1983: 153-154 no. 59
Gives vessels and other objects now missing	...	Gurney 1983: 154-156 no. 60
Gives gifts of grain, wool, oil, and garments	...	Gurney 1983: 156-158 no. 61
Receives at least 1 gur grain	2+ shekels?	Gurney 1983: 164-165 no. 65

We can trace the movable property transactions and legal cases of one family member, Shamash-êtir over at least twenty years, from roughly 1237 to at least 1215 B.C. (Table 3.2, Table 3.3). He buys four cattle over that period, but is robbed of three and accused of stealing two others. Of the four servants he buys, he returns one as unsatisfactory, and negotiates two different prices for another (for reasons which remain unexplained). His total recorded outgoings amount to nearly 4 minas of silver, though usually paid in a mixture of other commodities. Cattle appear to have been highly vulnerable to theft: a younger family member, Abu-tâbu, is attested in seven tablets from the archive; three record his purchase and sale of livestock, while the remaining four document court cases in which he and accomplices are accused of stealing at least ten sheep, cows, and oxen (Gurney 1983, nos. 16, 29, 30, 43, 44, 57, 73). One of the older tablets records the brewer Ilu-iddina's successful release of his wife Inbûsha from debt-slavery as a weaver (Gurney 1983, 17–22 no. 1). Further documents relate to a long-running dispute between the brewers and a temple official named Sîn-leqe-unninnî who was responsible for issuing them with barley for brewing (Gurney 1983, no. 68). In 1233 B.C. he appears to have deprived them as a group of the regular offerings (*ginû* again) which had been their right for 150 years.

> '[Regarding the X gur] of barley – the royal regular offering for Shamash-bêl-kitti, the brewer of Sîn: Sîn-leqe-unninnî took it and the regular offerings ceased.
>
> Regarding the annual 62 gur of barley – the backlog of the brewers of Sîn which king Kurigalzu established and which they were receiving until year 11 of Shagarakti-shuriash: in year 12 of Shagarakti-shuriash Sîn-leqe-unninnî removed it.
>
> [... *long gap* ...]
>
> Regarding the 14 gur 2 ban of barley – [the ...... of somebody], the brewer of Sîn: Sîn-leqe-unninnî took it and the regular offerings ceased. 1 shekel of red gold and 8 sheep were received from him.' (Gurney 1983, 160–161 no. 63)

The brewers must have taken Sîn-leqe-unninnî to law, for an intermediate agreement was reached in his favour a few years later in 1230 B.C. (Gurney 1983, 158–159 no. 62) and a final settlement made at least ten years after the dispute first broke out.

> 'Damages which the brewers of Sîn had inflicted on Sîn-leqe-unninnî. They presented themselves to Adad-shuma-usur the king, and he ordered their settlement. He sent Tâb-silli-Marduk son of Arad-Ea with them and they measured out for Sîn-leqe-unninnî 10 gur of barley by the 5 sila measure from the regular offerings of Shamash-êtir. Ditto (i.e. Sîn-leqe-unninnî) received 2 shekels of red gold from Shamash-êtir. Sîn-leqe-unninnî received 1 shekel of red gold from Irim-Shamash. Sîn-leqe-unninnî received 1 shekel of red gold from Sîn-shuma-êrish, X sheep from Iddin-Nergal, and 1 shekel of gold. Ditto (Sîn-leqe-unninnî) received 2 ...... from Sîn-êtiranni. He received 1 shekel of gold from Bunni-Shamash. Sîn-leqe-unninnî will receive this regularly from the brewers.' (Gurney 1983, 116–118 no. 41)

We get, then, an unexpectedly complex picture of the fortunes of the Dayyânâtu family, brewers of Sîn. On the one hand, they occupy and perhaps own a brick-built house in the centre of Ur, with servants and livestock (which implies that they also owned, or at least worked, some agricultural land). On the other hand, they are prone to burglary and cattle theft, and may be guilty of it themselves. At times their family members may be sold into debt-servitude. But their family identity is not inseparable from their profession: only a few of them hold the title "brewer of Sîn", and many of their colleagues apparently have other kinship ties. The brewers do have some degree of corporate identity though, but most strikingly they have strained and litigious relations with the temple which provides their livelihood.

THE NATURE OF GLASS

Various types of glassy substances were manufactured in the ancient Near East and Egypt from the early third millennium B.C.. True glass was made by heating quartz or sand with potash and limestone at very high temperatures until they liquefied. Other additives were used to colour the glass deep blue, turquoise, red, yellow, green, black, or white. Techniques for manufacturing small containers developed in the mid-second millennium. Core-formed vessels were made by winding heated rods of glass around a clay-based core, a process rather like coiling a pot. It was an extremely laborious procedure, as the glass had to be heated and reheated at frequent intervals, and then the finished product reheated and smoothed down. At this point extra bands of colour could be added, and fluted patterns made by dragging a pointed object over the hot surface (Moorey 1999, 204, fig. 13; e.g. Harper *et al.* 1995, 107 nos. 69, 70). Mosaic glass vessels, on the other hand, were made from rods of different coloured glass which were cut into thin sections, and placed in a mould, which was then kiln-fired (Moorey 1999, 205, fig. 14; e.g. Collon 1995, 113 no. 92; Harper *et al.* 1995, 109 no. 73). Glass-making reached a peak of activity in the third quarter of the second millennium B.C.: core-formed and mosaic glass remains from this period are found all over the Near East (Moorey 1999, 192–198; Oppenheim *et al.* 1970, 135–153).

About a dozen or so known tablets give instructions on the manufacture of glass. The earliest of are contemporaneous with the mosaic glass of the late second millennium: one from Hattusha, one from Babylon, and one from an unknown location (Oppenheim *et al.* 1970, 22–23).

> 'For a mina of *zukû* glass: 10 shekels of lead, 15 shekels of copper, half (a shekel) of *anzahhu* glass, 1/2 (a shekel) of antimony: these are the ingredients of Assyrian red glass.
>
> For a mina of *zukû* glass: a sixth (of a mina) of lead, 14 (shekels) of copper, 10 shekels of antimony, a shekel of *anzahhu* glass: these are the ingredients of Akkadian red glass.
>
> You remove it twice from the kiln (?). On your third removal you throw leaves (?) on it and take it out. You pour (it in) again and it will boil. You look at its crust: do not be anxious if (it is like) alabaster. You take up Akkadian and Assyrian (glass) equally and you fuse them together. After you have fused them together, you grind together and pour into the hot (glass) a shekel and a half of *zukû* glass, 7 1/2 grains of *anzahhu* glass, 7 1/2 grains of copper, 7 1/2 grains of lead to a mina of cast (glass). You remove it once. You

> take it out and cool it. .... should be clear and the your fire should be high. You pour it out and you sprinkle "cumin seed" stone onto it and your cooking vessel should be .... With the tip of your poker (?) you test (?) the glass and you immerse it and boil it. You cool its crust. You look at it: do not be anxious if the crust (is like) alabaster – inside its casing it is red (glass).
>
> If that red (glass) inside its casing is (like) copper dust, there is an excess (?) of copper. For each mina you mix 10 shekels of *zukû* glass, 15 grains of copper, 15 grains of lead, 15 grains of *anzahhu* glass – do not introduce any antimony – and look at it. In its mixing ......, and you take it out.' (Oppenheim *et al.* 1970, 63)

Most of the 'recipes', however, come from the seventh century library of king Ashurbanipal in Nineveh (668–627 B.C.). Oppenheim (1970, 28) judges them to be *n*th generation copies of tablets, at least half a millennium younger than their Middle Assyrian (LBA) originals.

> 'If you want to produce lapis lazuli-coloured glass, you grind separately 10 minas of *immanakku*-stone, 15 minas of glasswort ashes, and 1 1/2 minas of 'white plant'. You mix these together. You put them into a cold kiln which has four openings and arrange (the mixture) between the openings. You burn a good and smokeless fire until the 'stone' becomes fritted. You take it out and cool it off. You grind it again. You collect (the powder) on a clean slab. You put it into a cold 'doorway' kiln. You burn a good and smokeless fire until it (the 'glass') becomes golden. You pour it onto a baked brick. Its name is *zukû*-glass – 'clear'.
> You place 10 minas of ... copper on a clean slab. You put it into a hot 'doorway' kiln and close the door of the kiln. You burn a strong and smokeless fire until the copper compound glows. You crush and grind 10 minas of *zukû*-glass. You open the door of the kiln and throw (the ground glass) upon the copper compound and close the door of the kiln again. Once the *zukû*-glass dissolves onto the copper compound you stir it once, twice, three times with a poker until you see a 'frost' (of molten glass) on the tip of the poker. If the 'stone' looks like ripe grapes you boil (it). You pour it onto a baked brick. Its name is *tersîtu* – 'preparation'.
>
> You collect 10 minas of preparation, 10 minas of *bûsu*-glass, filtered glasswort ashes, not returned (?), 1/2 mina of mother-of-pearl (?), 1 2/3 minas of washed *anzahhu*-glass on a new slab. You put it into a kiln with 4 openings and place it on a support. The base of the slab must not touch the kiln. You burn a good and smokeless fire [......]. Once your mixture is melted you cool (it). You take (it) out and grind (it). [You collect (it) on a clean slab. You put (it) into a cold 'doorway' kiln. You burn a good and smokeless fire. Not until the 'stone' glows do you close the kiln door. Once it is glowing you close the kiln door and stir it once in front of you until it becomes golden. After it has become golden you will see 'frost' (forming at the tip of the poker). If the 'stone' is solidifying you pour it onto a new slab and lapis lazuli-coloured glass will come out of the cool kiln.' (Oppenheim *et al.* 1970, 32–4 Tablet A 13–42)

But glass-making technology had long-since moved on, with the introduction of casting and cold-cutting in about 900 B.C. (Moorey 1999, 205–206, fig. 15). Around the same time, it became fashionable to add antimony to the glass mixture, making it an almost transparent shade of yellow or green in imitation of rock crystal (Moorey 1999, 207; e.g. Collon 1995, 174 nos. 139a, d). Why, then, were Neo-Assyrian royal scribes continuing to copy detailed instructions for an out-of-date technology – scribes who were otherwise concerned solely with high literature, religion, and the esoteric scholarship of healing and divination (cf. Brown 2000, 33–52) and who apparently made no effort to record other technical processes?

The answer, I would argue, lies in the cultural associations of glass. We have seen that early glasses were manufactured in deep, rich colours. They aimed to imitate precious stones such as carnelian, agate, and lapis lazuli, which were not native to Mesopotamia. Indeed, the commonest designation for glass in Akkadian was *aban kûri* 'stone of the kiln', or simply *abnu* 'stone', as we saw earlier. Now precious stones were not only expensive and beautiful; they were also considered to have special properties. These properties were recorded in a scholarly work called *abnu shikinshu*, 'the nature of stone' (Horowitz 1992) and other similarly esoteric compositions (Reiner 1995, 119–132). Sadly, very little survives of *abnu shikinshu* but enough remains to show that stones were imbued with what we could call beneficial magic. Lapis lazuli – the stone most imitated in glassware – begins the list. Some of the more legible parts show that the properties of stones are deeply embedded in their names.

> 'The stone whose nature is like fish roe: [its name is] *igizaggû*.
> The stone whose nature is like the eye of a fish: [its name is] fish-eye stone.
> The stone whose nature is like ...... gold: [its name is ......] fish-eye stone.
> The stone whose nature is like the eye of a pig: [its name is] pig-eye stone.
> The stone whose nature is like the eye of a snake: [its name is] snake-eye stone.
> The stone whose nature is like snake-skin: [its name is] snake stone.'
> (Gurney and Finkelstein 1957, pl. 137 no. 108 lines 27–32)

Abnu shikinshu was more mainstream than it might seem: the Neo-Assyrian kings Sennacherib and Esarhaddon (704–681; 680–669 B.C.) regularly quoted it in inscriptions concerning the construction of their palaces (Reiner 1995: 119–121), while many first millennium rituals for averting evil or curing sickness prescribed the use of necklaces composed of particular beneficial stones (Reiner 1995, 124–129).

> 'Carnelian, lapis lazuli, yellow obsidian, *mekku* glass, *egizangû*, agate, *papparminû*, *lamassu*, antimony, jasper, magnetite, breccia, *abashmu*, twelve (!) stones if a man has an ill-wisher. You string the stones and phylacteries on a [linen (?) thread]; you set in place a holy-water vessel; you purify the stones and the phylacteries; you place the stones before the Goat-star (i.e. Lyra); you set up a censer with aromatics; you libate beer; (you recite) the incantation:
>
> O bright one, let your angry heart be appeased,
> Let your innermost relent, O Gula, exalted Lady.
> You are the one who created mankind, who bestows lots, food portions, and
> food offerings.
> Be present at my lawsuit, let me obtain justice through your verdict,
> Because of the sorceries, spittle and spatter, evil machinations

Of my adversary, let his evil doings turn back against him and affect his head and body,
And I, your weary servant, will sing your praises.'
(from Late Babylonian Uruk; after Reiner 1995: 129)

In the Sumerian-language myth *The exploits of Ninurta*, in circulation from the early second millennium onwards, the warrior god Ninurta defeats the stones in cosmic battle. He then assigns fates for each of them, punning so that their names embody their very destinies.

> 'My King stood before the sangkal stone, he addressed the gulgul and sangar stones. Ninurta son of Enlil fixed their destiny:
>
> "Sangkal stone, since you flew up against me ......; gulgul stone, since you sparked lightning against me ......; sangar stone, since you shook your head at me, since you ground your teeth at me, the Lord! The sangkal stone will smash you, sangar stone, young brave, and the gulgul stone will destroy (**gul**) you. You will be discarded as contemptible and valueless (**sang nukala**). Be a prey to the famine (**shangar**) of the Land; you shall be fed by the charity of your city. You shall be accounted a common person, a warrior among slave-girls. They shall say to you 'Be off with you, hurry!', it shall be your name. And now, by the destiny fixed by Ninurta, henceforth you shall be called a bad lot in the Land. So be it.' (Black *et al.* 1998–, 1.6.2 lines 448–463)

So we begin to see why glass-making methods might have been recorded by the scholarly experts of Mesopotamia, who even compiled them into a series of tablets called "Door of the Kiln" (Oppenheim *et al*, 1970, 25), just as they compiled the series of omens, medical prescriptions, and other specialist collections: the manufacture of artificial precious stones touched upon the realms of their expertise in magic, medicine, and religion. Thus, even when it was no longer fashionable to manufacture coloured glass, for the scholars it remained important to maintain a codified tradition of that knowledge, couched in a textual structure and technical language familiar to them from other genres of their works. This becomes increasingly apparent when we look at the text which precedes the seventh century glass-making instructions quoted above. In similar tones to the technical paragraphs which follow it – and to the hundreds of medical prescriptions and ritual instructions which have come down to us – it describes how to prepare the glass-making kiln. An ominously favourable day of the year must be chosen, and the finished kiln must be furnished with ritual figurines. Offerings must be made on the day of inauguration, and even the type of fuel for the kiln is controlled.

> 'When you set up the foundation of a kiln for glass, you search in a favourable month for a propitious day so that you can set up the kiln's foundation. As soon as you have completed the kiln and built it, you install divine *kûbu*-images there. No outsider or stranger should enter; unclean persons should not pass in front of them. You regularly sprinkle strewn offerings before them. On the day when you place the 'metal' inside the kiln you make a sheep sacrifice before the divine *kûbu*-images. You place juniper incense on the censer; you make a fire in the bottom of the kiln and place the 'stone' in the kiln. The persons whom you allow to come near the kiln must be ritually clean so that you can

> let them come down to the kiln. The wood you burn in the bottom of the kiln should be thick, peeled poplar wood: logs without knots, bound together with leather straps and cut in the month of Abu (i.e., July-August). This wood should go in the hearth of the kiln.' (Oppenheim *et al.* 1970, 32–4 Tablet A 13–42)

The boundaries between science and religion, medicine and magic were always blurred in the ancient Near East: the spiritual was inseparable from the rational. For the scholars of Mesopotamia coloured glass was not simply an attractive synthetic material but had powerful medico-magical properties. In other words, the nature of coloured glass was that it resembled precious stone both physically and conceptually.

CONCLUSIONS

I have presented three case-studies which use very different methods to test the cultural beds on which ancient technologies rest. With close reading I showed that official descriptions of monumental structures do not just summarise their physical appearance and practical function but also reinforce their symbolic meaning. Then, using a more traditional socio-economic approach I explored the archival remains of a group of technological practioners to reveal some unexpected aspects of their lifestyle and working relations. Finally, I brought together a wide range of ancient literature and scholarly writings to explore the the cultural values associated with coloured glass in later Mesopotamia.

Although I focussed on Assyria and Babylonia, these approaches are also applicable to their western neighbours of the LBA, for large corpora of similarly wide-ranging materials are now becoming increasingly available. In no circumstances, however, will textual evidence reveal all we wish to know about ancient technologies: they were not written for us to read.

> 'The cuneiform texts are not books of history that only need to be deciphered, translated, put in order, and then consulted for the answers to whatever questions seem important to a modern reader. They are artifacts of ancient life and *like tools or pots or other artifacts* they have attributes of style and use. They are evidence of the purposes for which they were meant, of the processes and techniques by which they were formed, of the circumstances in which they were used, deposited, preserved, and discovered, and of the status, organisation, or expectations of the people who produced them.' (M. Stolper 1992, 255; emphasis added)

If we allow for this, and expect the unexpected, then the historical record – when handled sensitively – can add an extra depth to the study of ancient technologies, helping to situate them, however approximately, in their original social contexts.

BIBLIOGRAPHICAL APPENDIX

An ever increasing number of textual sources are being published, especially from the fourteenth century onwards, in editions accessible to the non-cuneiformist. Pedersén (1998, 13–125) has recently made a site-by-site survey of the context and contents of all

LBA tablets found *in situ* in 'libraries' and 'archives'. It is by no means an exhaustive treatment, however: the vast numbers of tablets with no secure archaeological provenance are excluded, as are inscribed artefacts such as monumental inscriptions which fall outside his definition of 'archives and libraries', whether or not they have recorded findspots. Further, many of the tablets he discusses are not yet published. More generally, an accessible treatment of Kassite Babylonia is still conspicuously lacking but historical syntheses of, and cuneiform corpora from, many important LBA centres have appeared in the last few years: Hattusha and the Hittites (Beckman 1996; Bryce 1998), Emar (Arnaud 1985–87; Chavalas 1996), Ugarit (Watson and Wyatt 1999), Ashur (Grayson 1987; Harper *et al.* 1995, 81–124), Nuzi and the Hurrians (Wilhelm 1986; Owen and Wilhelm 1999), Susa and Elam (Harper *et al.* 1992; Potts 1999, 188–258), as well as Amarna itself (Moran 1992; Izre'el 1997).

The *Concise Dictionary of Akkadian* (Black *et al.* 1999) is the first complete and affordable English-language dictionary of one of the major languages of the ancient Near East. The Italian periodical *Orientalia's* annual Keilschriftbibliographie (cuneiform bibliography) aims to lists all recent archaeological, textual, and historical studies and reviews of the Mesopotamian area, indexed by subject; the Austrian journal *Archiv für Orientforschungen* provides a similar bibliography, with substantial indices of realia, Akkadian and Sumerian words, and cuneiform texts cited. Finally, no survey of the current literature would be complete without mention of *ABZU* (Jones 1994–), the canonical collection of websites and other internet resources on ancient Near Eastern topics.

TECHNICAL NOTE

All translations from Akkadian are my own, except the quote from Reiner (1995) on pages 52-52; all translations from Sumerian are taken from Black *et al.* (1998–). Sumerian words are in **bold** type; Akkadian words in *italics*. The following weight units are used throughout the paper: 180 grains = 1 shekel; 60 shekels = 1 mina; 60 minas = 1 talent. 1 mina is approximately 0.5 kg. 1 cubit is approximately 0.5 m. Middle Babylonian capacity measures are a little more complicated. 1 sila is approximately 1 litre, but although 1 gur = 300 sila was the usual standard, sometimes one finds, for instance, 1 gur = 150 sila ("by the 5 sila measure") or 1 gur = 360 sila ("by the 12 sila measure") (cf. Powell 1987–90, 498). All dates follow Walker 1995 unless marked otherwise.

ACKNOWLEDGEMENTS

This paper owes a great deal to Sarah Paynter, who got me started on this sort of study by asking me difficult, interesting and often unanswerable questions about the cultural meaning of glass and glazes in Mesopotamia over the last few years. Discussions with other workshop participants led me to express my intentions more explicitly. Niek Veldhuis cast his ever critical and constructive eye over the final draft. Andew Shortland, kind but firm, has been an exemplary conference organiser and editor. All errors and infelicities that remain are my own work, however. This project was begun under the auspices of the British Academy and finished at All Souls; to both institutions I am enormously grateful.

NOTES

1 Van Driel (1969, 10-11), however, rejects the early syncretism between Ashur and Anlil, concluding that the temple Wild Bull of the Lands must have been a separate shrine to Enlil within Ashur's temple. But that theory does not explain why Shalmaneser refers to Ashur's shrine as E-kur, the name of Enlil's temple at Nippur.

REFERENCES

Andrae, W., 1977, *Das wiedererstandene Assur*, Verlag C. H. Beck, Munich.

Arnaud, D., 1985–87, *Recherches au pays d'Ashtata: Emar*, 4 vols., Éditions Recherche sur les Civilisations, Paris.

Austin, J. L., 1999, How to do things with words, in *The discourse reader* (eds. A. Jaworski and N. Coupland), Routledge, London, 63–65.

Beckman, G. M. 1996, *Hittite diplomatic texts*, Scholars Press, Atlanta GA.

Black, J. A., Cunningham, G. G., Robson, E., and Zólyomi, G. G., 1998–, *The electronic text corpus of Sumerian literature*, http://www-etcsl.orient.ox.ac.uk/, Oriental Institute, Oxford.

Black, J. A., George, A. R., and Postgate, J. N., 1999, *A concise dictionary of Akkadian*, Harrassowitz, Wiesbaden.

Bottéro, J., 1992, *Mesopotamia: writing, reasoning, and the gods* (trans. Z. Bahrani and M. Van De Mieroop), University of Chicago Press, Chicago.

Brown, D. R., 2000, *Mesopotamian planetary astronomy-astrology*, Styx, Groningen.

Bryce, T., 1998, *The kingdom of the Hittites*, Clarendon Press, Oxford.

Chavalas, M., 1996, *Emar: the history, religion, and culture of a Syrian town in the Late Bronze Age*, CDL Press, Bethesda, MD.

Cohen, R., and Westbrook, R., 1999. Amarna diplomacy: the beginnings of international relations, Johns Hopkins University Press, Baltimore, MD.

Collon, D., 1995, *Ancient Near Eastern art*, British Museum Press, London.

Dalley, S. M., 2000, Hebrew *tahash*, Akkadian *duhshu*, faience and beadwork, *Journal of Semitic Studies*, 45, 1–19.

Grayson, A. K., 1987, *The royal inscriptions of Mesopotamia. Assyrian periods, volume I: Assyrian rulers of the third and second millennia B.C. (to 1115 B.C.)*, University of Toronto Press, Toronto.

Gurney, O. R., 1983, *The Middle Babylonian legal and economic texts from Ur*, British School of Archaeology in Iraq, Oxford.

Gurney, O. R., and Finkelstein, J. J., 1957, *The Sultantepe tablets*, I, British Instititute of Archaeology at Ankara, London.

Haller, A., and Andrae, W., 1955, *Die Heiligtümer des Gottes Assur und der Sin-Shamash-Tempel in Assur*, Verlag Gebr. Mann, Berlin.

Harper, P. O., Aruz, J., and Tallon, F., 1992, *The royal city of Susa: ancient Near Eastern treasures in the Louvre*, Metropolitan Museum of Art, New York.

Harper, P. O., Klengel-Brandt, E., Aruz, J., and Benzel, K., 1995, *Assyrian origins: discoveries at Ashur on the Tigris: antiquities in the Vorderasiatisches Museum, Berlin*, Metropolitan Museum of Art, New York.

Horowitz, W., 1992, Two *abnu shikinshu* fragments and related matters, *Zeitschrift für Assyriologie*, 82, 112–122.

Izre'el, S., 1997, *The Amarna scholarly tablets*, Styx, Groningen.

Jones, C. E. 1994–, *ABZU: guide to resources for the study of the ancient Near East available on the internet*, http://www.oi.uchicago.edu/OI/DEPT/RA/ABZU/, Oriental Institute, Chicago.

Kuhrt, A., 1995, *The ancient Near East, c. 3000–330 B.C.*, Routledge, London.

Moorey, P. R. S., 1999, *Ancient Mesopotamian materials and industries: the archaeological evidence*, 2nd ed., Eisenbrauns, Winona Lake.
Moran, W. L., 1992, *The Amarna letters*, Johns Hopkins University Press, Baltimore, MD.
Oppenheim, A. L., Reiner, E., *et al.*, 1956–, *The Assyrian Dictionary of the Oriental Institute of the University of Chicago*, Oriental Institute, Chicago, IL.
Oppenheim, A. L., Brill, R. H., Barag, D., and von Saldern, A., 1970, *Glass and glassmaking in ancient Mesopotamia: an edition of the cuneiform texts which contain instructions for glassmakers with a catalogue of surviving objects*, Corning Museum of Glass Press, Corning, NY.
Owen, D. I., and Wilhelm, G., 1999, *Nuzi at seventy-five*, CDL Press, Bethesda, MD.
Pedersén, O., 1998, *Archives and libraries in the ancient Near East, 1500–300 B.C.*, CDL Press, Bethesda, MD.
Postgate, J. N., 1980, Review of Freydank, H., 1976, *Mittelassyrische Rechtsurkunden und Verwaltungstexte*, Akademie Verlag, Berlin, *Bibliotheca Orientalis*, 37, 67–70.
Postgate, J. N., 1986, Administrative archives from the city of Assur in the Middle Assyrian period, in *Cuneiform archives and libraries* (ed. K. R. Veenhof), Nederlands Historisch-Archaeologisch Instituut te Istanbul, Leiden, 168–83.
Potts, D. T., 1999, *The archaeology of Elam: formation and transformation of an ancient Iranian state*, Cambridge University Press, Cambridge.
Powell, M. A., 1987–90, Masse und Gewichte [in English], in *Reallexikon der Assyriologie und vorderasiatischen Archäologie*, 7 (ed. D. O. Edzard), Walter de Gruyter, Berlin and New York, 457–517.
Reiner, E., 1995, *Astral magic in Babylonia*, American Philosophical Society, Philadelphia.
Sillar, B., and Tite, M. S., 2000, The challenge of 'technological choices' for materials science approaches in archaeology, *Archaeometry*, 42, 2–20.
Stolper, M., 1992, Cuneiform texts from Susa, in Harper *et al.* 1992, 255–276.
Van De Mieroop, M., 1999, *Cuneiform texts and the writing of history*, Routledge, London.
Van Driel, G., 1969, *The cult of Assur*, Van Gorcum, Assen.
Walker, C. B. F., 1987, *Cuneiform*, British Museum Press, London.
Walker, C. B. F., 1995, Mesopotamian chronology, in Collon 1995, 230–238.
Watson, W. G. E, and Wyatt, N., 1999, *Handbook of Ugaritic studies*, Brill, Leiden.
Wilhelm, G., 1989, *The Hurrians*, Aris and Phillips, Warminster.
Woolley, C. L., 1965, *Ur excavations, VIII: the Kassite period and the period of the Assyrian kings*, The Trustees of the British Museum and the Trustees of the University Museum, London and Philadelphia.

Chapter 4

Egyptians, Hyksos and Military Technology: Causes, Effects or Catalysts?

Ian Shaw

Abstract

Weaponry is an area of ancient Egyptian technology where there are undoubtedly key links with political and social change, particularly in the Second Intermediate Period and New Kingdom (*c.*1800–1100 B.C.). Among the most crucial innovations in the late Second Intermediate Period was the introduction of the horse-drawn chariot (Egyp. *wrrt* or *mrḳbt*). Apart from its value as a piece of military technology, the chariot was of paramount social and political significance. Many other innovations in Egyptian warfare seem to coincide with the immediate aftermath of the 'Hyksos period', such as body armour, a smaller type of shield, the more elastic composite bow, a new form of dagger which gradually developed into a weapon resembling a short sword, and the *ḫpš*, a scimitar-like weapon with a curved blade modelled on an Asiatic form. This paper discusses the complex web of relationships between technological innovation and various aspects of political and social change in Late Bronze Age Egypt and Syria-Palestine, and examines the possibility that the Hyksos represented an *obstacle* to Egyptian military innovation rather than a source of inspiration.

INTRODUCTION

The arms races of the 20th century led the American humorist Will Rogers to observe 'You can't say civilization don't advance – in every war they kill you in a new way'. Although the pace of technological change in military hardware was generally considerably slower in the distant past, there nevertheless seem to have been crucial phases when certain cultures went through very rapid processes of 'catching up' with their rivals. The Egyptian Second Intermediate Period and early New Kingdom seem to have constituted just such a phase. Among the most crucial Egyptian innovations ascribed to the late Second Intermediate Period were the introduction of the horse-drawn chariot (Egyp. *wrrt* or *mrḳbt*) and the composite bow. The chariot in particular was of paramount social and political significance in the early New Kingdom, quite apart from its value as a piece of military technology (see Figs. 4.1 and 4.2).

One of the Amarna Letters (EA22) includes a list of chariot components forming part of the dowry of a daughter of the Mitannian king Tushratta sent to marry Amenhotep III (Moran 1992, 51–61). Similarly, Papyrus Anastasi IV mentions 'beautiful chariots of *bri* wood' and then goes on to list eleven chariot parts, each of which is followed by discussion of its material or place of origin (P. BM10249; Caminos 1954, 200–201). Finally, the so-called *Poem concerning the royal chariot*, preserved on an ostracon in Edinburgh and dating to the Ramessid period, includes a list of the various elements of the chariot and its associated weapons, and each of these individual pieces is linked with some particular aspect of the king's character, usually employing some kind of play on words (Dawson and Peet 1933).

The chariot was not only used in battle by the *marryanu*, an élite corps of the Egyptian army in the New Kingdom, it was also regarded as an essential part of the royal regalia. Depictions of the king charging enemies in his chariot became a common feature of the exterior walls of temples as symbols of 'the containment of unrule', roughly comparable with the more ancient theme of the king smiting foreigners with a mace

In the context of the development of Late Bronze Age Egyptian and Near Eastern military technology, there are good grounds to believe that we are dealing with a very complex web of social and technological factors. A similarly elaborate scenario was encountered by Colin Renfrew in his reanalysis of the archaeological remains at Varna (a cemetery near the Black Sea coast of Bulgaria, dating to the late 5th millennium B.C.). In his discussion of the Varna material, Renfrew (1986, 146) argues that 'the decisive innovation in the development of a new commodity is generally social rather than technical. Often the technology is already there'.

The tendency in the past has been to discuss the origins of various innovations in Egyptian warfare in the late Second Intermediate Period and early New Kingdom in terms of contact between different cultural groups resulting in the transmission of military ideas and hardware. Studies have therefore concentrated on trying to find out when weapons or ideas first appear in the archaeological record, and, if they do not appear to have evolved locally, where their likely source was located.

In this paper, however, it is argued that the process of innovation – or the process of adoption of new technology – is invariably much more complicated, involving not merely the acquisition of technological 'packages' or inventions but also, in each case, the emergence of a sympathetic set of social and economic conditions (see Table 4.1). In addition, it is argued that military equipment and military strategy need to be seen as symbiotic and mutually influential – sometimes particular strategies are sparked off by the acquisition of new types of weapons, and sometimes the adoption of new methods of warfare result in the development or adoption of weapons that can enhance and facilitate such methods.

THE EMERGENCE OF EGYPTIAN CHARIOTRY

The first Egyptian textual reference to chariotry is on the second Kamose stele, which mentions the chariots used by the Hyksos. The earliest examples of spoked wheels in Egypt are found on a small model carriage in the Theban tomb of Ahhotpe, mother of

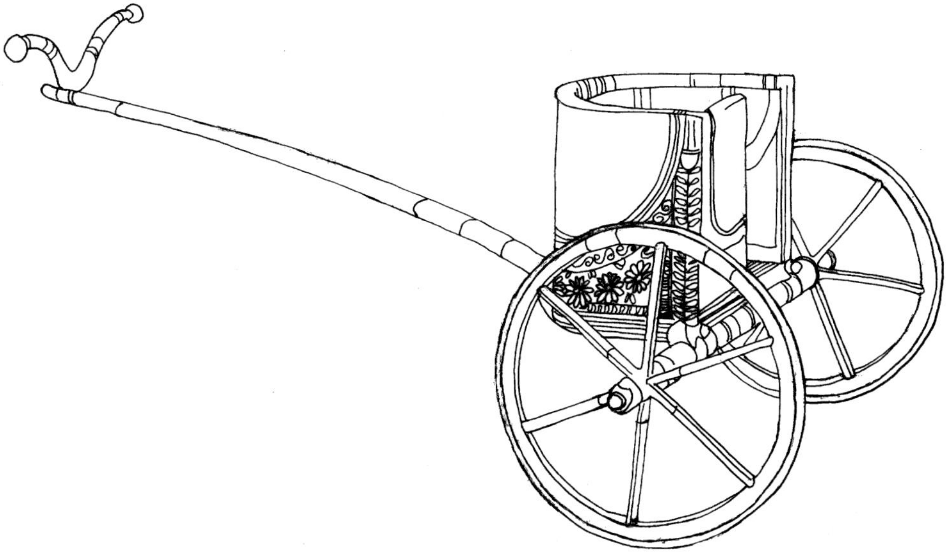

Figure 4.1: Drawing of a typical Egyptian chariot (drawing by Ann Jones)

Figure 4.2: Two Egyptian soldiers in a chariot (drawing by Ann Jones, after Littauer and Crouwel 1979)

King Ahmose. There are only a few previous depictions of the use of any kind of wheel at all, including siege-ladders running on solid wheels (one in the 5th-Dynasty tomb of Kaemheset at Saqqara and the other in the 11th-Dynasty tomb of Intef at Thebes: TT386) and a sledge with four wheels pulled by oxen (in the 13th-Dynasty tomb of Sobeknakht at Elkab). This dearth of any real sign of development towards a sophisticated horse-drawn wheeled vehicle makes it almost certain that the chariot was not invented in Egypt but introduced from the outside world, almost certainly from western Asia.

As long ago as 1951, Torgny Säve-Söderbergh argued that the chariot was not introduced into Egypt by the Hyksos, but that both Egyptians and Hyksos began to use it at about the same time at the end of the Second Intermediate Period. The importance of the social and economic contexts of technological change with regard to the Egyptians' adoption of chariotry can be compared with a study of the introduction of the heavy horse-drawn plough into medieval northern Europe (White 1962). This study demonstrated that the time taken for this process of technological change lay not so much in the innovation itself but in the occurrence of the necessary changes in the social agricultural practices of the societies concerned, so that they were able to use the plough and to benefit significantly from its use. This case indicates that delays in the introduction of new technology are often to be found in social and economic processes of change rather than the ability to invent or replicate new technology. Donald Spratt (1989, 255–6, Fig.12.5) has provided good evidence to suggest that earlier societies took longer to adopt innovations than later ones, presenting a graph which shows 'delay times' of various innovations from the bow and arrow in the 4th millennium B.C. to polyethylene in the 20th century AD. Another good archaeological example of this phenomenon is the delay in the appearance of techniques for casting bronze rapiers in Middle Bronze Age Britain (Rowlands 1976), although it should be mentioned, in fairness, that Stuart Piggott (1969) argues the exact opposite point of view: that no such socio-economic delay occurred in the introduction of wheeled vehicles and the plough into the cultures of Neolithic Europe.

Preconditions for the emergence of Egyptian chariotry

In order for a new piece of technology to be adopted, several basic factors need to be in place: (1) access to necessary resources or materials, (2) knowledge of methods of manufacture, (3) availability of suitably skilled craftsmen both to make and use the artefact, (4) a social need or political requirement for the technology in question, (5) a suitable social or economic context within which the technology can be deployed. If we study these in the case of the chariot and other military items supposedly introduced by the Hyksos we can get some sense of the full complexity of the situation.

In the case of Egyptian chariots, these basic factors correspond to (1) raw materials such as wood and leather, as well as live horses, (2) several different aspects of wood, leather, bone and metal-working technology, (3) access not only to craftsmen specialised in the manufacture of the various components of chariots, harnesses and bridles, but also to craftsmen with the necessary skills and knowledge to convert all these parts into the finished product, (4) military plans and strategies that would significantly benefit from the inclusion of chariot corps, and (5) plans for empire-building campaigns in Syria-Palestine and Nubia, as well as an increasingly militarised ruling group whose life-style and aspirations matched those of the Indo-European *maryannu* – an élite group for whom the chariot was not merely a piece of equipment but an important status symbol.

As Rosemarie Drenkhahn (1976, 130) has already noted, the groups of workers producing chariots must have comprised one of the most complex of all ancient workshops, both because of the diversity of materials involved and the wide range of technological skills required. Several New Kingdom tombs show groups of craftsmen, such as carpenters, joiners and leather-workers working on the different parts of a chariot. Drenkhahn points out that all of the six tombs containing scenes of chariot production between the reigns of Hatshepsut and Thutmose IV show the activities taking place in the workshops of the Temple of Amun at Karnak, and four of the six occupants of the tombs had been employed by the Temple of Amun. It is not clear why there should be this connection with one specific temple estate, although Drenkhahn suggests that temples – as regular recipients of foreign booty and prisoners of war – might have had more ready access to exotic timber and Asiatic skilled craftsmen. From the late 18th Dynasty onwards, similar chariot-making scenes appear in private tombs at Saqqara, and Papyrus Anastasi I (British Museum, EA10247; Gardiner 1911, 1–34, 2–81) provides an intriguing insight into the maintenance of chariotry with a description of an Egyptian charioteer's visit to a repair shop in the Levantine coastal city of Joppa.

To deal with the materials first: the principal materials used for a high quality chariot were wood, leather, rawhide, textile, bone, ivory, copper alloy, gold, gypsum plaster, faience, glass, stone and glue. The two most important materials were wood and rawhide. For the basic chassis of the vehicle, as well as its axle, wheels, pole and yoke (and sometimes, as in the case of Tutankhamun, the blinkers), various different types of wood were required – particularly ash (*Fraxinus excelsior*), imported from outside Egypt, which was used for the axle, felloes and frame. Two other imported woods were also used: field maple (*Acer campestre*) was used for the floor of the Florence chariot, and silver birch bark (*Betula pendula*) was used to cover an axle of a Tutankhamun chariot and also the axles of the Florence example. The purpose of the birch bark was almost certainly to waterproof the glue and rawhide holding the chariot together; it seems to have been used in the same way to waterproof composite bows. Willow (*Salix subserrata*, which *is* found in Egypt) was used for the Florence pole, and elm (*Ulmus minor*, an import) was used for its yoke, handrail and wheel spindles. Elm was also used in the chariot of Amenhotep III. Plum-wood (not found in Egypt) was sometimes used for the spokes. Locally available acacia was used for the front of the chassis.

The craftsmen would have needed to use the steam-bending technique to actually manufacture the chariot chassis and wheels, using strips of unseasoned timber. Egyptian carpenters were clearly already familiar with this technology at least as early as the 12th Dynasty, since there are scenes in the tombs of Amenemhat and Baket II at Beni Hasan (BH2 and BH15, see Newberry 1893, pls VII and XI) showing steam bending being used for the manufacture of bows. In both scenes, the craftsmen are shown firstly holding wood over a pot of hot water to allow the steam to penetrate and soften the cellular structure, and secondly actually bending the stick into a hoop shape, with the ends buried in the ground to keep the shape. There are also 5th-Dynasty scenes in the tombs of Ptahshepses (Abusir) and Ty (Saqqara) showing the bending of dampened unseasoned poles of timber.

Leather was used not only for the many thongs and pieces of harness but also to sheath any surfaces likely to suffer stress or abrasion. In addition, rawhide was shrunk over the composite wheels to hold them together and provide a form of tyre, and leather traces were used to lash the central pole to the yoke. Since we know that there were few changes in leatherworking technology between the Middle Kingdom and New Kingdom (see Van Driel-Murray 2000), it follows that Egyptian leatherworkers would already have been capable of creating such components as tyres, traces and blinkers well before the arrival of the Hyksos.

Textiles were also used in chariots, primarily for 'kickboards', floor matting and housing. Two of the chariots from the tomb of Tutankhamun had a length of painted cloth – a kind of durable matting which would have been placed between the floors of chariots 120 and 122 and the side walls of their chassis. Gillian Vogelsang-Eastwood (2000, 276, 292) notes that one of these mats is made from a special kind of cloth involving a dense layer of long loops; she suggests that the 'springiness of the looping' would probably have helped to cushion the riders from the shaking of the vehicle.

Ivory, bone, stone, copper alloy and faience were used for the many small components holding the chariot and the harnesses together, such as copper alloy nails, stone saddle knobs, and copper alloy bits and linchpins. An important find among the pieces discovered in site Q1 at Qantir was a copper alloy 'nave cap', which prevented the linchpin from grating down on the wooden nave of the chariot. The nave cap of the

Table 4.1: Minimum preconditions and requirements for the Egyptian adoption of chariotry

Technology	Imports
Steam-bending Lathe turning Weft-loop weaving Increased use of leather Mortise and tenon joints	Woods Foreign craftsmen Asiatic/Nubian horses
Social Change	**Military change**
Marryanu class Appearance of 'general-kings' Adoption of horse-riding	Emergence of 'professional' army Diversification of military hierarchy Use of composite bows, arm-guards and plate armour

Florence chariot was made from wood. Faience and glass were used for inlay on chariot bodies, as well as for yoke ends, yoke saddles and bridle bosses. Ivory was used mainly for charioteers' whip-stocks.

The net result of all this is that we can be fairly sure that the Egyptians already had access to nearly all of the necessary materials and technology for chariot production by the Middle Kingdom (see Table 4.1). They did, however, lack several important elements: first, they may not yet have had ready access to such imported woods as ash, elm and birch; secondly, they had most of the necessary skills and craftsmen to produce the components but probably did not acquire the methods and personnel to transform these into a finished chariot until the early New Kingdom; thirdly, there is very little indication of access to horses before the New Kingdom.

Horses

Where did the Egyptians' horses (*Equus caballus*) come from? The earliest physical remains of a horse so far found in an Egyptian context are the bones discovered at Buhen fortress and dated to around 1680–1640 B.C. This horse measured 125 centimetres high at the shoulders, but the yoke measurements of surviving chariots suggest that the average Egyptian horse was probably slightly larger than the Buhen example (about 135 cm at the shoulders), although still smaller than the average modern Arabian horse (c.145 cm). Another horse, with a shoulder height of about 142 cm, was buried beside one of the tombs of Senenmut (TT71), the well-known chief of works in the reign of Hatshepsut. Whereas the Buhen horse appears simply to have been encased in the mud-brick walling built over the spot where it died, the Senenmut horse was actually wrapped in linen bandages and placed in a 2.5–metre-long wooden coffin. A third horse was found in the 18th-Dynasty necropolis at Soleb in Nubia, and its height was between about 1.34 and 1.38 at the shoulder. Three horses were also buried in a re-used tomb at Saqqara, but these can only be dated very vaguely to some point between the 20th Dynasty and the Ptolemaic period.

Both the Buhen and Senenmut horses appear to be Arabian in type, suggesting that the earliest examples were probably imported or plundered from Western Asia. According to the Karnak Annals, Thutmose III captured more than 2000 horses after the battle of Megiddo, and the 18th-Dynasty Theban tomb of Amunedjeh shows horses being brought by Syrians as tribute.

The first textual reference to horses in Egypt is the Kamose stele, which mentions those used by the Hyksos. The fact that the term used for a span of horses '*ḥtr*' is an Asiatic loan-word suggests that both the animals and the techniques for training and utilising them derived from western Asia. It should be noted, however, that there are also strong connections between the Nubians and horses. Although the clearest evidence for Nubia as a widely recognised international source of horses and trainers begins with the 25th Dynasty, there are also earlier indications of Nubian horsemanship, therefore Nubia should not be ruled out as a possible source of horses even as early as the Second Intermediate Period, which is probably when the Buhen horse should be dated. It may be significant in this regard that two of our very small group of surviving early Egyptian horses were actually found in Nubia. This is particularly interesting when we bear in mind that the Thebans were evidently as much cut off from the Kerma culture in Nubia as they were from Syria-Palestine and the Mediterranean during the Second Intermediate Period.

Given the evidence for the high value ascribed to Nubian horses and horsemen – and particularly their eventual use by the Assyrians for their chariotry and cavalry in the 7th century B.C. (see Dalley 1985), where they are specifically referred to as Kushite – the Egyptian adoption of chariotry by the end of the Second Intermediate Period might have been as much a question of Egyptians' access to suitable horses as the process of technological innovation implied in the use of chariots themselves – in other words the recognition of the need for technological change is ineffectual without access to the necessary materials and resources.[1]

THE TRANSITION FROM SIMPLE TO COMPOSITE BOWS: MILITARY TECHNOLOGY AS AN INTEGRATED SYSTEM

Many other innovations in Egyptian warfare seem to coincide with the immediate aftermath of the 'Hyksos period'. It is often difficult to determine the extent to which each of these changes and innovations are linked in any causal way. Probably the most significant and obvious military development apart from chariots was the adoption of the composite bow. The technique of glueing strips of horn and sinew to a wooden self bow produced the more elastic 'composite bow'(see Fig. 4.3), which came in two types – recurved and triangular – and had a considerably greater range than the self bow. From the point of view of the overall process of military change, the adoption of chariots and composite bows must have been very closely linked, given that there is general agreement that chariots served above all as highly mobile bases from which archers could pick off the enemy from a distance.

Throughout the pharaonic period acacia wood was often used for bows and arrows, but the emergence of the composite bow led to a demand for different woods. The only known specimens of ash from ancient Egypt are (1) the wood of a composite bow from the tomb of Tutankhamun (see Nicholson and Shaw 2000, 341) as *F. excelsior*, rather than *F. ornus*, and (2) that used for the axle, the felloes and part of the frame of the floor of the Florence chariot. As mentioned above in the case of chariots, neither ash nor birch occur naturally in Egypt and both of these imports were used primarily for construction and embellishment of chariots and composite bows. Although presumably most horn used in Egypt was derived from native – or at any rate African – species, circumstantial evidence suggests that the horns of the Cretan wild goat (*agrimi*) might have been imported for making Egyptian composite bows (Warren 1995, 7 with references).

The introduction of the composite bow coincided with – and perhaps directly influenced – changes in the form of arm guards and quivers. New types of arm guards were introduced in the early New Kingdom, almost certainly because of the increasing use of the composite bow. The funerary depictions of the brightly coloured arm guards show them tied at both wrist and elbow. Unlike earlier guards, they cover much of the lower arm, but no actual examples seem to have survived, perhaps because they were made of something that would not have preserved well, such as padded textile.

Tapered, round-bottomed styles of quivers replaced the Middle Kingdom tubular

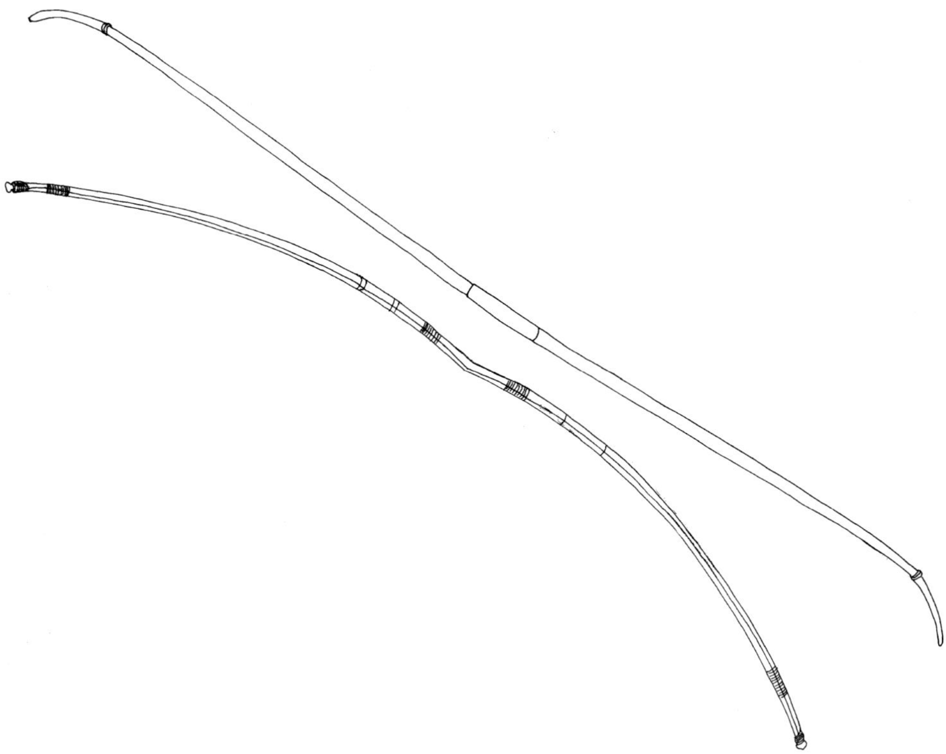

Figure 4.3: The two major types of Egyptian bow: a self bow (top) and a composite bow (bottom, drawings by Ann Jones)

types. A red quiver in the Ägyptisches Museum, Berlin, now almost invisible under gelatinous decay, is decorated with open-work appliqués and couchwork with panels of superimposed coloured strips in red, white, green and black, with typical New-Kingdom-style green, pinked edgings. A quiver from the 18th-Dynasty tomb of Maiherpri in the Valley of the Kings (KV36, reign of Thutmose III) is unusually well-preserved, retaining its fine, crisply executed designs in raised relief (perhaps block-stamped), revealing a quality of workmanship which has otherwise rarely survived (Daressy 1902, pl.X, nos 24071–2). The new style of quiver was almost certainly the result of a change in the status of bowman and the need for both archers and their equipment to be incorporated into the design of the chariot. The arrow and javelin quivers were attached to the side-panel of the chariot in such a way and at such a precise angle that the charioteer could easily remove an arrow or javelin from the quivers when needed.

Body armour (small bronze plates riveted to linen or leather jerkins), was introduced by the early New Kingdom, and a smaller type of shield, with a tapered lower half, began to be used. Both of these must presumably have been connected with the need to protect soldiers against the composite bow.

Here we have a very clear set of cases where, in each instance, the form of one piece of technology was affected by the need to incorporate it into another. This phenomenon is discussed by Donald Mackenzie and Judy Wajcman in their introductory essay to *The social shaping of technology* (1985, 11), when they mention the tendency for technology to be shaped not so much by science as by other technology: 'existing technology is more than just a precondition of new technology, but is an active shaping force in its development'. In other words, each individual weapon is part of an organic and dynamic system, so that individual pieces of military technology constantly interact with one another and adapt to the system, operating as components in a complex and slowly changing overall military strategy.

CHANGING SYSTEMS OF MILITARY REWARD

Turning away from military hardware to software, it is clear that innovations in Egyptian weaponry in the early New Kingdom were accompanied by important changes in the way that the army was organised, and indeed the role that it played in Egyptian society.

David Lorton (1974) has suggested, on the basis of comparisons between the funerary inscriptions of the 6th-Dynasty official Weni at Abydos and the early 18th-Dynasty soldier Ahmose son of Ibana at Elkab, that there may have been no system of reward for individual soldiers in the Old Kingdom, and that complex systems of recompense might not have been introduced until after the Hyksos period, possibly as a direct result of the adoption of Hyksos customs of warfare. Thus, he points out that the term *sḳr ʿnḫ* was used indiscriminately in the Old and Middle Kingdoms to refer both to enemies captured on the battlefield and to those taken while plundering defeated settlements, whereas from the late 18th Dynasty onwards it was used specifically to refer to enemy soldiers captured during the battle itself, while the term *ḥꜣḳt* was used to describe people or goods captured only in the course of post-battle plundering, taken to the king and then redistributed as rewards to individual soldiers, and finally the term *ḥꜣḳ* was used in the 18th Dynasty to designate both objects and people taken in the course of the actual battle and the subsequent plundering.

As with the weaponry, we should beware of suggesting that all changes that emerged in the early 18th Dynasty can be ascribed to contact with the Hyksos. It is equally likely – perhaps more likely – that the adoption of more complex systems of military recompense resulted from Egyptian involvement in Syria-Palestine and the consequent absorption of the 'language' of Asiatic warfare, just as the Egyptians clearly adopted Akkadian as the language of diplomacy by the middle of the 18th Dynasty (see Cohen and Westerbrook 2000).

CONCLUSIONS

Many innovations in Egyptian military equipment and strategies resulted directly from population movements, exchange of ideas between different ethnic and cultural groups, and processes of social change brought on both by large-scale environmental and political influences. Such innovations in military hardware went on to have

Figure 4.4: Detail of the reliefs decorating the exterior of the Temple of Ramesses II at Abydos, showing a Hittite chariot (photograph by the author).

significant effects on the social and political contacts between Egypt and the Levant, and between Egypt and the rest of north-eastern Africa. The chronology of all this, however, suggests that the turning point for the Egyptians was not the arrival of the Hyksos but their departure.

It seems as if the Hyksos may have represented a significant barrier between the Egyptians and access to more sophisticated weaponry such as chariots, composite bows and body-armour. It was only after the removal of the Hyksos that the Egyptians were able to gain access both to some of the necessary imported materials and probably also, initially at least, foreign craftsmen, so that they could begin to manufacture these new weapons. Many of the materials (such as acacia-wood, copper alloy, bone, stone, faience and rawhide) and much of the technological ability (such as steam-bending, lathe-turning and drilling) had already been available for some time within Egypt itself, but the crucial changes were the social and political ones, whereby Egyptians gained ready access to the resources of western Asia. Once they were able to import materials, craftsmen, horses and experienced charioteers, they could begin to create their own vehicles and to incorporate chariotry and horse-riding 'scout' divisions into their own military strategies.

POSTSCRIPT: DIVERGENT SYSTEMS OF CHARIOTRY

In conclusion it is worth noting that an intriguing aspect of Egyptian chariotry – and one that is particularly evident in the Qadesh battle reliefs – is the number of ways in which Egyptian chariots appear to have differed from their Hittite and Syrian counterparts. Whereas the Egyptian chariots had a two-man crew and are shown with quivers attached for the archer, the Hittite and Syrian chariots are shown with three occupants, comprising a shield-bearer in front of the driver and a spearman behind him, and apparently no quivers for arrows or javelins (see Fig. 4.4).

If the Qadesh reliefs are a genuine reflection of differences between Egyptian and Asiatic chariot crews and their weapons, then we may have a situation comparable with that of the American and Soviet development of ballistic missiles after the war. Both sides started off with a fundamentally similar technological prototype – in the case of the Egyptians and Hittites, the basic chariot of the 17th and 16th centuries B.C., and in the case of the Americans and Soviets, the V2 rocket developed by the Germans during the Second World War. Mackenzie and Wajcman (1985, 11-12) describe the situation as follows: 'American and Soviet missile designers...developed significantly different missiles, despite their shared use of the V-2 as departure point'.

In the case of both modern and ancient divergences in the use of technology, whether we are talking about chariots or missiles, there are clear indications that even cultures or ethnic groups sharing a common paradigm will find that their individual technological trajectories can vary considerably as a result of social, political and strategic factors.

NOTES

1 Dalley (1985, 47) argues that 'the late 8th century was a time when the Assyrians were increasingly aware of the importance of equestrian technology. Suddenly during that period cavalry in particular developed into a newly powerful weapon of war. Innovation in the form of breeds of horses, methods of harnessing and of importing foreign experts, in particular from Nubia and Samaria for chariotry, from Urartu for cavalry, contributed to that development. It was a time when Nubia was ruled by kings to whom horses were one of life's chief delights...and free trade in horses flourished. Nubia benefited enormously from an expanding market, and reached a peak of power and prosperity.' Dalley also makes the point that Samaria gained great prestige at this date from its role as entrepot supplying Nubian horses to the Assyrians for their chariotry (while the people of Urartu were supplying smaller horses along with their trainers, for use in the Assyrian cavalry).

REFERENCES

Caminos, R. A., 1954, *Late Egyptian Miscellanies*, Oxford University Press, London.

Cohen, R. and Westerbrook, R., 2000, *Amarna diplomacy: the beginnings of international relations*, Johns Hopkins University Press, London and Baltimore.

Dalley, S., 1985, Foreign chariotry and cavalry in the armies of Tiglath-Pileser III and Sargon II, *Iraq*, 47, 31–48.

Daressy, M. G., 1902, *Fouilles de la Vallée des Rois (1898–1899)*, Service des Antiquités de l'Egypte, Cairo.

Dawson, W. R. and Peet, T. E., 1933. The so-called Poem on the King's Chariot, *Journal of Egyptian Archaeology*, 19, 167–74.

Drenkhahn, R., 1976, *Die Handwerker und ihre Tätigkeiten im alter Ägypten.*, Harrassowitz, Wiesbaden.

Gardiner, A. H., 1911, *Egyptian hieratic texts, transcribed, translated and annotated*, August Pries, Leipzig.

Littauer, M. A. and Crouwel, J. H., 1979, *Wheeled vehicles and ridden animals in the Ancient Near East*, E.J. Brill, Leiden and Cologne.

Littauer, M. A. and Crouwel, J. H., 1985, *Chariots and related equipment from the tomb of Tut'ankhamun*, Griffith Institute, Oxford.

Lorton, D., 1974, Terminology related to the laws of warfare in Dynasty XVIII, *Journal of the American Research Center in Egypt*, 11, 53–68.

Mackenzie D. and Judy Wajcman, J. (eds), 1985, *The social shaping of technology*, Open University Press, Milton Keynes.

McLeod, W. E., 1970, *Composite bows from the Tomb of Tut'ankhamun*, Griffith Institute, Oxford.

McLeod, W. E., 1982, *Self bows and other archery tackle from the tomb of Tut'ankhamun*, Griffith Institute, Oxford.

Moorey, P. R. S., 1986, The emergence of the light, horse-drawn chariot in the Near East c.2000–1500 B.C., *World Archaeology* 18/2, 196–215.

Moran, W. L., 1992, *The Amarna Letters*, Johns Hopkins University Press, London and Baltimore.

Newberry, P. E., 1893, *Beni Hassan* II, Egypt Exploration Society, London.

Nicholson, P. T. and Shaw, I. (eds), 2000, *Ancient Egyptian Materials and Technology*, Cambridge University Press, Cambridge.

Petrie, W. M. F., 1917, *Tools and weapons*, Egypt Research Account, London.

Piggott, S., 1969, The first wagons and carts: twenty-five years later, *Bulletin of the Institute of Archaeology* 16, 3–18.

Renfrew, C. A., 1986, Varna and the emergence of wealth in prehistoric Europe, in A. Appadurai (ed.) *The social life of things: commodities in cultural perspective*, Cambridge University Press,Cambridge, 141–168.

Rowlands, M. J., 1976, *The organization of Middle Bronze Age metalworking*, British Archaeological Reports, Oxford.

Säve-Söderbergh, T., 1951, The Hyksos rule in Egypt, *Journal of Egyptian Archaeology* 37, 59–60.

Schulman, A. R., 1980, Chariots, chariotry and the Hyksos, *Journal of the Society for the Study of Egyptian Antiquities* 10, 105–153.

Spratt, D. A., 1989, Innovation theory made plain, in van der Leeuw, S. E. and Torrence, R. (eds), *What's new? A closer look at the process of innovation*, Unwin Hyman, London, 245–257.

Van Driel-Murray, C., 2000, Leatherwork and skin products, in Nicholson, P. T. and Shaw, I. (eds) *Ancient Egyptian Materials and Technology*, Cambridge University Press, Cambridge, 299–319.

Vogelsang-Eastwood, G., 2000, Textiles, in Nicholson, P. T. and Shaw, I. (eds), *Ancient Egyptian Materials and Technology*, Cambridge University Press, Cambridge, 268–298.

Warren, P. 1995, Minoan Crete and Pharaonic Egypt, in Davies, W. V. and Schofield, L. (eds), *Egypt, the Aegean and the Levant: Interconnections in the Second Millennium B.C.*, British Museum Press, London, 1–18.

Western, A. C. and McLeod, W., 1995, Woods used in Egyptian bows and arrows, *Journal of Egyptian Archaeology*, 81, 77–94.

White, L., 1962, *Medieval archaeology and social change*, Oxford University Press, Oxford.

Chapter 5

Stone Vessel Production: New Beginnings and New Visions in New Palace Crete

Jacke Phillips

Abstract

The end of the Old Palace and beginning of the New Palace period on Crete saw, amongst other key events in Minoan history, fundamental changes in the popularity of certain ceramic vases, a statement also true for stone vessels. Amongst other developments was the revival of the importation and, presumably, direct use of Egyptian stone vessels. The imports also began to be employed as models (to varying degrees) both for some new Minoan vessel forms and integrated with existing Minoan forms to produce new hybrid shapes, as well as being physically converted into popular Minoan types. Although some practices are found to an extremely limited degree at the beginning of Minoan stone vessel manufacture, they had begun to die out by the early Old Palace period. The New Palace revival of foreign stone vessel importation and inspiration fits within a larger picture of social and technological development on the island.

INTRODUCTION

As has long been recognised, Late Minoan I (LM I) is the period during which *tour de force* shapes and technical brilliance in stone vessel manufacture are at their height on Crete; one thinks especially of the vessels in the LM IB cult 'treasury' of Kato Zakro (N. Platon 1972; 133–49). This brilliance also is a measure of the social conditions in which they could be produced. The immediate origins of this virtuosity lie in Middle Minoan III (MM III), in a slow but possibly recognisable process, through LM IA, that is the focus of this paper.

On Crete, the end of the Old and beginning of the New Palace period is associated with the destruction of the earlier palaces especially at Knossos and Phaestos, the event that effectively separates MM IIB and MM III. This event also is recognised at Malia but apparently is not found at Kato Zakro. MM II is little recognised beyond the immediate palace sites, elsewhere MM IB seems mostly to directly precede MM III both ceramically and stratigraphically (Cadogan 1983). As this paper concerns the influence of Egyptian stone vessels, I concentrate on material actually recovered in

MM III–LM IA contexts, in order to isolate MM Ill–LM IA from the succeeding LM IB period. Few distinctly MM HI and even LM IA contexts can be recognised, and few of these contained stone vessels, but isolating them from both earlier and later contexts is instructive. It is clear that certain distinctions can be made between the end of the Old Palace and the early New Palace periods in other media. Virtually all relevant vessels are recovered at a single palatial site, Knossos. In Egyptian terms, MM III–LM IA ranges essentially over the majority of the Second Intermediate Period through very early 18th Dynasty.

HISTORICAL AND SOCIAL CONTEXT

At Knossos a major earthquake completely destroyed the palace at the end of MM IIB or possibly the very beginning of MM IIIA, and seems also similarly to have affected Phaestos and Malia at approximately the same time. This represents the division between the Old and New Palace periods, in Minoan chronological terms. A second, larger palace above the first was then constructed, a feat that entailed considerable planning and labour expenditure, and even more social and organisational hierarchy, to achieve. Whether one believes this was achieved in stages or as a single concerted effort is immaterial here; suffice to say that it must have been done mostly within a generation or two.

The MM IIB palace destruction was followed by a series of smaller earthquakes (Driessen and MacDonald 1997; 17; Hood 2000a; *passim).* Each of these in turn meant further reconstruction or consolidation of newly damaged areas at different times throughout MM III and LM IA, whilst the second palace continued under construction. One in MM IIIA caused considerable destruction, including much burning, especially on the eastern side. Another major destruction at the very end of MM IIIB/beginning of LM IA was caused perhaps by another earthquake, after which much new ashlar construction is evident. Yet another major earthquake caused even more damage late in LM IA.

At Phaestos, the old palace also was destroyed at the end of MM IIB, and again another built above it in MM III. At Malia, the existence of an earlier palace remains problematic, but the élite areas that may represent separate palatial 'blocks' also were destroyed at the end of MM IIB, and the 'second' palace was constructed in MM III. Kato Zakro appears to be an entirely different situation, as the accumulated evidence indicates the major earthquakes (or at least destructions) date to during or at the end of MM IIA, and that the 'first palace' building was not destroyed until the end of LM IA, with no evidence for an intervening destruction (L. Platon 1999).

It is clear that the immediate post-destruction period in central Crete, at the beginning of the New Palace period, is one of extensive social, material and technological change (Walberg 1992;7–8). We can look back and glibly say "the old palaces were destroyed by an earthquake that affected virtually the entire island at the end of MM IIB, and then new palaces were constructed in MM III," but this conveys nothing of the actual situation from the perspective of those who lived through the quake and its aftermath. The immediate and lingering effects to the palatial sites and their inhabitants must have been quite profound, both materially and psychologically. These effects would have lasted for several generations at minimum, over the succeeding period we call MM III which is, by all estimates, well less than a century in length. Consider a very peripheral modern comparison, the psychological and material effects of WW II or, if you prefer, WW I on our grandparents and parents, and the still-lingering effects upon ourselves. Or the Japanese after Hiroshima and Nagasaki. This was emphasised more directly to me by television coverage of the 17 August 1999 earthquake centred just outside Istanbul, and speculation of how much worse – and for how much longer – the subsequent conditions could have been for the survivors without the assistance of extensive international foreign aid, and the modern technological means to immediately implement it.

This is a period of reconstruction, reassessment, restructuring. Development in a vast array of material goods, including pottery, seals, writing, iconography, architecture, all indicate substantial reassessment of old and introduction of new attitudes, undoubtedly the result of circumstances and the need for quick solutions. As in the aftermath of any earthquake in any populated region, the survivors were confronted with immediate problems. Existing priorities suddenly were altered: the need for housing, food, water, disposal of the dead, prevention of disease, salvaging usable goods, reproducing unusable and lost goods and, depending on the time of year the disaster occurred, initially to prepare for, and survive, winter. In less material terms, to maintain social stability and restore order, with attendant shifts up and down in status of certain individuals and social groups. To the stable and sophisticated Old Palace urban society, this event must have galvanised innumerable changes in priority and responsibility, most of which we cannot recognise in the archaeological record but which must have lingered on into the succeeding generations.

But none of these later earthquakes (or, rather, 'destructions' – are they all caused by earthquakes?) entirely destroyed the new palace and levelled its surrounding buildings, as had happened at the end of MM IIB. The new palace survived and continued to be enlarged. Nonetheless, having survived the MM IIB destruction, the later less destructive destructions each must have worried the inhabitants considerably, and much thought must have gone into the improvement of architectural features to minimise even farther the devastation of future events. Shifts in population can be recognised: Malia declines in scale and never recovers, the second Phaestos palace is smaller than the first. At Knossos, despite much excavation, the picture remains unclear but both the palace and its surrounding occupation area seem to have expanded, perhaps due less to a larger population base than to a need for the surviving population to spread out farther to rebuild.

Priorities also shifted in the production and use of necessary goods. Administrative control tightens, records apparently become more detailed. At Knossos, the limited hieroglyphic/pictographic script dies out and the more complex Linear A is adopted from Phaestos, where it already was in use by MM II. Seals now emphasise the face design, the specific identity of its authority, rather than the seal shape. Plain seal forms come into almost exclusive use, lentoids in particular, whilst the multiplicity of elaborate shapes disappears (Yule 1980; Pls. 35–36). In contrast, face designs now include detailed figures and background landscapes, many having overtly (and to our eyes, now more understandable) religious themes (Yule 1980; Pl. 36). Religion itself is given more emphasis, specific cult spaces are set aside in town and palace buildings, and the first structures are constructed at many formerly open air cult sites. The tholos tomb is replaced by the technically less complex but still impressive chamber tomb for the élite. However, older communal tombs continue to be used, probably accompanied by fewer grave goods as little MM III–LM I material is recovered in them; this in itself is indicative of the new attitude. Few New Palace tombs are known.

Ceramics are characterised by larger, simpler forms and decoration, both less carefully produced. Elaborate shapes decline, fabrics are coarser, vessel walls thicker and less well-fired (Walberg 1976, 1983, 1992; Betancourt 1985). The famous 'Classical Kamares' eggshell ware, the epitome of Minoan ceramic technology, disappears, and polychromy declines rapidly. The vessel base becomes emphasised, sturdier, often raised, the vessel less liable to being overturned. Some at least of the MM III 'fine wares' may be immediately pre-earthquake vessels that had survived to continue in use, rather than 'new' post-quake products. This may explain in part the overlap into MM IIIA from MM II of Walberg's (1976; 125 and 1983; 2) period 3 or 'Classical Kamares' forms and decoration. Certain forms are no longer found, others appear at this time. There is a surge in large, and then even larger, storage vessels. Pithoi become massive in scale, as storage floor space declines. All are classic signs that priority had shifted from the aesthetic to the practical in MM III. By the beginning of LM IA, an emphasis on the aesthetic already had begun to return, to cumulate again by LM IB. The only exception to all this MM III practicality is the rhyton, significantly the standard ritual vessel type, which is produced in a new variety of shapes, some quite elaborate.

This change in emphasis is paralleled in stone vessel production (Warren 1969, *passim*). A large number of stone vessel forms die out in the Old Palace period, all quite small in scale. Multiple new forms are introduced in MM III, like their ceramic counterparts being larger in overall scale and with base emphasised. These generally are practical forms, designed for everyday and long-term use. Few are embellished with any decoration or technical virtuosity, at least initially. Many clearly imitate clay and metal forms of the same date, a contrast to MM II when clay forms had imitated metal and, in some cases, stone. Whilst we cannot be very precise, as stone vessels can survive intact and continue in use much longer than clay, it is the MM III period when all this begins to occur. Again, only the overtly cultic forms such as the rhyton are provided with much elaboration.

THE EGYPTIAN VESSELS

Egyptian influence on Crete during the Old Palace period generally is characterised by a very limited adoption and adaptation of Egyptian iconography. Material recognisably embraced by the Minoans is almost exclusively cultic, the standing hippopotamus deity and the squatting ape figures being the most prominent adaptations. These are iconographic images, not physical objects. Scarab seals also were imported and adopted, and then the presentation adapted to a purely Minoan style, also possibly for cultic reasons. All these first appear in the Pre-Palatial period, and the already

Table 5.1: Egyptian and 'egyptianising' Minoan stone vessels

Type	Site	Context	No.	Context date	Comments
MMA IIIA or earlier					
Egyptian spheroid jar	Knossos	NW palace fill	2.a.	EM III or MM III	body fragment
Minoan spheroid bowl	Knossos, Gypsades tomb	in lower layer?	E.1	MM IIB-IIIA	fragment, 'half preserved'
MM IIIB					
Egyptian alabastron	Knossos, HH area D5	no context info	1.a.	MM IIIB	body fragment
Minoan conical rhyton	Knossos	no context info	D.1	MM IIIB	fragment, Egyptian travertine
Minoan conical/pear rhyton	Knossos, HH area G7B	no context info	D.2	MM IIIB	fragment, Egyptian travertine
MM IIIB-LM IA transitional					
Egyptian high shouldered jar	Katsamba	'votive deposit'	3.a.	MMIIIB-LMIA	complete
LM IA					
Egyptian alabastron	Knossos, RRS(W)	no context info	1.d.	LM IA (early?)	body fragment
Minoan spheroid bowl	Knossos, Gypsades tomb	upper layer/ossuary	E.2	LM IA	fragment
Egyptian spheroid jar	Knossos, RRN	no context info	2.b.	LM IA	fragment, sawn and drilled
Minoan baggy alabastron/ewer	Mycenae, GC A	grave V	1.e.	LM IB	converted
Minoan duck bowl	Mycenae, GC B	grave O	F.1	LM IB	Egyptian image
Minoan chalice	Thera	no context info	D.4	ntl LM IA late	Egyptian travertine
MM IIIB-LM IA					
Egyptian closed vessel	Knossos, KSM	no context info	4.a.	MM IIIB-LM IA	rim fragment
Egyptian lid	Knossos, KSM	no context info	5.a.	MM IIIB-LM IA	rim fragment
Minoan spheroid bowl	Poros	settlement	E.4	MM IIIB-LM IA	rim/body fragment
Minoan tall jug	Mycenae, GC A	grave IV	D.3	MM IIIB-LM IB	Egyptian travertine
MMIII-LMI					
Egyptian baggy alabastron	Knossos, Isopata tomb?	no context info	1.b.	MMIII-LMI	complete
Egyptian 'drop-shape' alabastron	Knossos, Isopata tomb?	no context info	1.c.	MMIII-LMI	bottom fragment
Minoan spheroid bowl	Knossos, Mavro Spelio tomb	no context info	E.3	MMIII-LMI	complete
workers' waste	Knossos, 1957-61 excavations	no context info	6.a.	MMIII-LMI	Egyptian travertine

recognisable Minoan manifestations are developed during the Old Palace period. Imported stone vessels, as imported scarabs, also are found in Pre-Palatial contexts, but few of these imports seem to have continued either to be adapted by the Minoans or even to be imported by MM II. Their influence was either on the wane or had died out completely by its end; possibly indicative is that no Middle Kingdom forms are recovered in Crete at this time.

All this changes just before or just after the MM IIB earthquake, and continues (so far as we can tell) at least until the end of LM IB but possibly even beyond. Evidence for external contact now appears in (relative) abundance as imports again are utilised on the island, and it is under these circumstances that we find a resurgence of contexts with imported Egyptian vessels, and of Minoan stone vessels having some relationship to the imports.

Egyptian forms recovered in MM III–LM IA contexts are limited in type, quantity and geography, but generally are the same as those found in greater quantity in the more numerous LM IB contexts. It seems, then, that their importation to Crete was by deliberate choice rather than chance. Only certain forms were needed, or wanted, by the Minoans. Why these particular forms and how did they affect Minoan stone production? These questions may or may not remain insoluble even at some future point, but they are important to consider. Unsurprisingly, the few Minoan vessels with Egyptian associations that we can isolate in MM III and LM IA contexts are limited almost entirely to Knossos and some immediately associated sites. None are recovered in the equivalent contexts at Phaestos or Malia, or elsewhere on the island. Therefore, it seems, we are not discussing importation to Crete during this period, but rather importation only to Knossos, and material needed or wanted by the Knossians or at least accessible only to Knossos and Knossians. Whilst it is entirely possible that other vessels from LM IB or later contexts elsewhere already were in use in MM III–LM IA, this issue is not considered here. Nor are the numerous vessel fragments without context from Evans's excavations, some of which conceivably arrived at this time.

Imported stone vessels are limited to the following (all recovered at Knossos unless otherwise stated):

1) *Alabastra. of banded travertine:* A handless round-bottomed vessel with small aperture, in a variety of profiles ranging from near-globular ('flask-shaped') to quite sloping ('baggy').

1.a. KSM (1957–61). Body fragment, in an unpublished MM IIIB occupation context west of 'Hogarth's Houses' (Warren 1969; 112–13 Type 43:I; see also Driessen and MacDonald 1997; 165 and MacGillivray 1998; 53). Warren notes this is the earliest context of an Egyptian alabastron on Crete.

1.b. HM L1583. Restorably complete 'baggy' vessel, part of a deposit dated to MM III–LM 1 on Isopata hill, together with other objects. The deposit is considered to be from an unlocated tomb (Evans 1914; 3–4, Fig. 3).

1.c. HM L1584. 'Flask-shaped' body fragments, from the same context (Evans 1914; 3–4, Figs. 2, 4; note both these figures depict the same fragments).

1.d. KSM 71/203. 'Baggy' body fragment, in an undescribed LM IA or earlier context, possibly an early? LM IA building phase in an unpublished occupation area on the south side of the Royal Road (Warren 1989; 3 Figs. 1, 4, and Cline 1994; 166 #269; see also Warren 1972; 628).

1.e. Mycenae: NAM 829. Complete 'baggy' vessel in Shaft Grave V, grave dated to LH IB (= latter half of LM IA) and vessel therefore manufactured and exported not later than late LM IA (Karo 1930–33; 147 #829, PI. CXXXVII and Warren 1969; 104 Type 42.B; see also Dietz 1991; 248–9).

2) *Spheroid jars:*[1] A low squat profile, flat collar rim and restricted mouth, often with two horizontal lug handles (sometimes pierced) on the shoulder and a generally restricted interior capacity. The type dates no later than the Old Kingdom, and therefore clearly is an antique by MM III.

2.a. Body fragment, in an unpublished fill context associated with the construction of either an MM III (New Palace) or EM III wall at the north front of the palace, recovered in 1987 excavations (Catling 1988; 69). Probably to be considered earlier than MM III, in either event. Generally similar fill excavated by Evans nearby at the north-west corner of the palace yielded a complete but battered example and a body fragment of the same general type (Evans 1921–35; I, 65 Fig. 32; II.1, 30 Fig. 12), as well as fragments of three deep open bowls (Evans 1921–35; I, 30; II.1;30–31, Fig. 28) and a 'moustache cup' (Evans 1921–35, II.1; 57–8, Fig. 27.b). These may also be related to the ambiguously dated 1987 context, but are not included here. Evans dated this fill much earlier, but no other material is published and it too remains ambiguous (Hood 2000b; 21).

2.b. KSM (1957–61) RR/61/22 (LA 78). Two non-joining body fragments, in an unpublished LM IA context on the north side of the Royal Road, exhibiting both saw and drill marks (Warren 1969; 109 Type 43.A9 and Cline 1994; 191 #505; see also Driessen and MacDonald 1997; 156–7).

3) *High-shouldered jar:* The 'high-shouldered' jar form, handless and again squat in appearance, but with a distinctive high shoulder and restricted mouth. Again, this is an Old Kingdom form, and therefore an antique by MM III. This jar form has many of the same basic characteristics as the spheroid jar, and probably was imported for the same reasons.

3.a. Katsamba HM unnumbered (Katsamba 1957). Complete jar of 'white marble,' recovered in a rock-cut chamber as part of a 'votive deposit' dated to MM IIIB–LM IA transitional (Warren 1969; 110 Type 43.D.2, P597, D321; see also Driessen and MacDonald 1997; 133). This is one of only two relevant vessels not found at Knossos, but the site is nearby, lying in fact between Knossos and its coast. Warren considers this an import. Its profile is Egyptian, but Egyptian use of marble is rare for vessels; perhaps it is recrystalised limestone (Aston 1994; 39–40, 55–6, compare Pl. 9.a with Warren 1969, P597). Nonetheless, the 3rd – 4th Dynasty date of this vessel is contemporary with Egyptian use of both stones for vessels.

4) *Closed necked vessel (jug?) of banded travertine:*

4.a. KSM SEX/81/1381. Rim fragment with beginning of shoulder, flaring neck and exterior thickened rim grooved on exterior edge, in an unpublished MM IIIB–LM IA context of the Stratigraphical Museum excavations, a mostly occupation site (Warren 1989; 5 #20, Fig. 3.20, Pl. 6.top right and Cline 1994; 204 #630; see also Warren 1981). Possibly a Syro-Palestinian vessel, as the grooved rim edge is atypical of Egyptian vessels but occasionally is found in the Levant.

5) *Lid? of banded travertine:*

5.a. KSM SEX/80/1804. Small rim fragment, flat with beginning of inset underside, in another unpublished MM IIIB–LM IA context of the Stratigraphical Museum excavations (Warren 1989; 5 #21 and Cline 1994; 210 #681; see also Warren 1981). Warren (1969) lists no Minoan lids of this material, and (1989) considers this piece an import. Such lids are found throughout the Dynastic period from the Old Kingdom period on.

6) *Raw banded travertine stone:*

6.a. Undescribed workers' manufacturing waste, recovered in the 1957–61 excavations, no details published but implied as being in MM III–LM I context(s) (Warren 1969, 125–126). Presumably some at least were recovered in MM III–LM IA levels.

THEIR USE AND INFLUENCE ON CRETE

These imported stone vessels exhibit a variety of actual, probable and potential uses, which have some bearing on the conditions under which they were imported to Knossos, or for which certain related Minoan vessels were produced there. These uses can be traced further, and in greater quantity, in LM IB and later contexts.

A. As a probable transport container

At least two and probably all three distinctive forms of Egyptian alabastron ('flask,' 'drop-shaped' and 'baggy') are recovered in contexts of this period. All are handless, have a rounded bottom, restricted mouth, and no associated lid. Some possibly are more or less contemporary with their Minoan context, whilst others may be antiques; their Egyptian date range generally covers the latter part of the 12th Dynasty through the entire Second Intermediate Period, although the 'flask' form appears not to survive much beyond the early SIP.

These are the only vessels appearing in any quantity that may be generally contemporary with their Minoan context. Why would they have been imported, and what could they have provided? Probably their contents, in most cases. The travertine alabastron was the ultimate Bronze Age container for oils, unguents and perfumes before the later invention of glass.

It is possible, of course, that all Egyptian imports were transported primarily for their contents rather than themselves. However, I find it difficult to believe that 'antique' vessels would have been generally employed for this purpose when exported abroad. The heavy spheroid and high-shouldered jars, whilst possibly employed as container vessels, instead may have been transported as ballast weight to be discarded on arrival at Knossos, together with any other imported goods that have not survived and to make room for any goods to be transported from Crete.

The alabastra inspire no Minoan artisan to produce any similar vessel before LM IB, when the 'tall alabastron' was introduced and quickly became popular throughout Crete (Betancourt 1985; 151, 170). The decoration on several LM IB clay alabastra clearly indicates their derivation from, and direct imitation of, the imported alabastron of banded travertine (Hawes *et al* 1908; 51 #15, Pl. VII:15, from Goumia; Halbherr, Stefani and Banti 1977; 95 #7, Fig. 61, from Aghia Triadha; Levi 1967–1968; 127, Fig. 78.b, from Phaestos). Alabastra of all three forms continue to be imported in some quantity after LM IA, but the Minoans continued to prefer only the early forms rather that the later (18th Dynasty) type, even through into LM III.

B. As religious and funerary offerings

However, the alabastra and other vessels would have been put to use once they had arrived. Two alabastra are associated with a funerary deposit, another is in an unpublished but apparently occupation context, and the high-shouldered bowl at Katsamba with a 'votive deposit'. Whilst the last may or may not have been imported, deposition in these contexts suggests that their very exoticism may have imbued the vessels with some sacred character. They were at least considered worthy of association with a deity or deities and with the afterlife of the (presumably élite) dead, however these were defined in the Minoan mind at this time. Whether the vessel or its contents is the actual offering is immaterial; if the latter, the vessel must have been chosen for the purpose.

Both aspects continue in LM IB and later, where exotic stone vessels and other objects are recovered in both tombs and ritual contexts at Knossos and elsewhere. The clearest LM IB examples are the 'Treasury' vessels at Kato Zakro and, at LM II–IIIA Knossos, the 'Central Treasury' and Isopata 'Royal Tomb' collections employing vessels whose sacred associations had continued after LM IB (Evans 1921–1935; II.2, 820–26 and IV.2, 771–76).

Derivative Minoan spheroid bowls, discussed in more detail in section E (below), have been recovered in MM III–LM IA tomb contexts at Gypsades and Mavro Spelio. Locally produced derivative vessels were employed for the same purpose(s), possibly in the absence of an imported 'original'. None, however, are found in votive or other religious contexts.

C. For innovative conversion and embellishment to Minoan forms

The treatment of another alabastron provides much insight into the Minoan artisan's mind and ability to work with his material. One 'baggy' alabastron was physically converted by a Minoan artisan (presumably at Knossos) into a Minoan ewer, and then it was exported to Mycenae where it was recovered in Shaft Grave V of Grave Circle A. It is the earliest clear example of this practice, which continued into LM IB and perhaps later. Shaft Grave V itself dates to LH IB, contemporary with late LM IA on Crete. The alabastron therefore was imported to Crete from Egypt, converted and subsequently exported to Mycenae no later than LM IA. Whilst unique in its conversion, it is but one of several demonstrably Minoan stone vessels exported to Mycenae and recovered in Grave Circle A in this period. Presumably these were diplomatic gifts, directly between élite personages, from Knossos to Mycenae; see also section D (below).

It takes a certain artistic 'eye' to look at a finished vessel, and 'see' in it a completely different form. It also shows a certain presence of mind and practicality: why hollow out a completely new vessel, when all that needs to be done is cut out the base and plug the rim with the bit removed. Then turn upside down, add handles, spout and rim embellishment, and the new vessel is finished. Later, especially in LM IB, this direct conversion of various different types of imported vessels reaches extraordinary heights of imagination and invention (Warren 1997). This practice in the East Mediterranean Bronze Age apparently is unique to the artisans of Crete; it is not found in Cyprus or the Levant (Sparks 1998 and pers. comm.) or Egypt. It may be a by-product or parallel development of the Minoan practice of making multi-component vessels, which apparently also begins in this period. Both can be viewed as *tour de force* examples of Minoan stone vessel manufacture.

D. As raw material for direct production of Minoan vessels

Fragments of a spheroid jar were recovered in an LM IA (mature) context in the Royal Road excavations. It had been deliberately broken up for reuse of the stone for another purpose; drill and saw marks are clearly visible. It perhaps was imported, either as a whole vessel or already in fragments, to be systematically reduced to raw stone scrap for secondary use. The excavation itself was of limited scale and remains unpublished so context details are lacking, but the intentional destruction of the vessel into fragments is clear. A possible purpose is unknown but the vessel is anorthosite porphyry, a particularly hard stone that might have been intended for further use because of this property. Possibly an unexcavated stone vase-making workshop is to be located immediately nearby, and the jar fragments may have originated from that area. It is perhaps hard to understand why the vessel would have been so treated, except possibly it had been accidentally broken and the fragments were still considered useful.

In addition to imported vessels being reduced to scrap, a number of Minoan vessels are known in Egyptian banded traverfine, indicating that this raw material was being imported at this time for use by Minoan artisans since the stone is mined only in Egypt. This may also suggest that vessel fragments may have been imported as raw stone, but the question must remain open. If the banded travertine waste fragments are from the same excavation area and level as the sawn and drilled spheroid jar fragments, the probability of a stone vasemaker's workshop nearby is strengthened. Undoubted Minoan vessels of Egyptian travertine are recovered on Crete only at Knossos before LM IB, but also beyond the island at Mycenae and on the island of Thera. A total of nearly forty such vessels, as well as the manufacturing waste pieces, were noted by Warren in 1969 (125–6), and others have been recovered since; "up to ten" more are reported from the Unexplored Mansion excavations alone (Evely 1984; 236). The vast majority are in LM IB and later levels, but those recovered in MM III–LM IA contexts include:

D. 1. Conical rhyton fragment, recovered in an unspecified and unpublished MM IIIB context in the 1957–61 excavations at Knossos (Warren 1969; 84 Type 34.A, 186).

D.2. KSM (1957–61) G 7 B, conical or pear-shaped rhyton fragment, recovered in an unpublished MM IIIB context near 'Hogarth's Houses' (Warren 1969; 88 Type 34.A or B). This may be the same fragment as the last; neither is illustrated.

D.3. Mycenae: NAM 592; tall jug with vertical handle, recovered in Shaft Grave IV, the grave dated to MH IIIB–LH IB (= MM IIIB–later LM IA), and the vessel therefore manufactured and exported to Mycenae not later than late LM IA (Warren 1969; 48 Type 22.D; see also Karo 1930–33; PI. CXL and Dietz 1991; 248).

D.4. Thera: NAM 3964, tall plain footed chalice, without context but which must have been manufactured and exported to Thera not later than late LM IA when the island was destroyed (Warren 1969; 37; see also Zervos 1956; PI. 489). Similar chalices, made of limestone, were recovered in Shaft Graves IV and V at Mycenae (Sakellarakis 1976; 177, PI. II.3, stone misstated as 'alabaster'), corroborating the early date of the Thera chalice.

It seems that banded travertine was the only stone that the Minoans imported from Egypt for vessel manufacture. Other exotic raw stones had been imported to Crete at least by MM II, and possibly as early as MM IB (at the beginning of the Old Palace period). These include carnelian-sand, agate-onyx-sardonyx, amethyst, jasper and lapis lazuli, all but the last likely to possibly originating from Egypt (Yule 1980; 193–99 *passim).* They had arrived in small quantities or at least in small pieces and were used only for seals, but now for the first time very large chunks of stone also were being imported from Egypt. This seems to be a new addition to the repertoire of 'large chunks' imported by the Minoans for stone vessel manufacture, for Egypt was not the only source of raw exotic stone. The Minoans already had been importing other stones from elsewhere in large pieces, for manufacture into otherwise entirely Minoan vessels. Other imported stones in use for vessel manufacture before LM IB include:

a) Spotted obsidian from Gyali island in the Dodecanesos, imported by the Old Palace period. A carinated bowl following an Egyptian profile is without context, but was in use at Knossos in MM II (Evans 1921–1935; I, 86–87 Fig. 55.e and Warren 1969; 75 Type 30.C). Waste pieces were found in MM IIA and III contexts at Knossos, and large blocks in context at MM II Knossos and MM I Malia (Warren 1969; 136, P627);

b) Antico rosso from Laconia. An MM IA circular table was recovered at Knossos (Warren 1969,126), and the LM IA hybridised vessel from Gypsades discussed as E.2 (below).

c) Obsidian, non-Melian and possibly from southern Anatolia. No clear context is known before LM I, when a pear-shaped rhyton was found at Tylisos (Warren 1969; 135–136), but an ewer dated to MM III was recovered at Knossos (Warren 1969; 43, 136);

d) Lapis lacedaemonius from elsewhere in Laconia. Possibly not in use before LM IB when workers' waste was recovered at Knossos (Warren 1969; 133); and

e) Rock crystal, *perhaps* imported from abroad. Warren dates a bowl fragment from the

Domestic Quarter area to MM III–LM I, but the majority of material recovered there apparently is LM IA in date (Evans 1921-35, II.1; 409–410, Fig. 272.a–b; Warren 1969; 79 Type 32.C and see Driessen and MacDonald 1997; 145–46), as is the duck-shaped vessel discussed as F.1, below.

No Minoan vessels of banded travertine are known before MM IIIB, suggesting that importation of the stone is a new phenomenon at this time or, possibly, somewhat earlier although not before MM III. It appears to be an new addition, introduced together with the Anatolian? obsidian and possibly the Lapis lacedaemonius, to the spotted obsidian and antico rosso already being imported for vessel manufacture. Clearly, these particular stones were chosen for some specific reason(s), upon which we may only speculate but which surely at least partly relate to their pleasing visual appearance. Nonetheless, their origins underline the extensive distances travelled for their acquisition, either by Minoans or other importers. Whilst importation of the other stones is understandable as no comparable Cretan stones are known, several indigenous stones visually similar to travertine already were known and used on Crete, including calcite and limestone. The specific appealing factor of travertine must have been its distinctive banding. Other visually appealing stones found and exploited at this time in Egypt are not imported to Crete for vessel manufacture. Again, this strongly suggests specific and deliberate Minoan choice, rather than a limited selection offered them by the Egyptians or any hypothetical middlemen. Apart from the spotted obsidian at Old Palace period Malia, finds of imported stone before LM IB all again are at Knossos.

E. As inspiration for practical variations to existing Minoan forms

Warren (1969) indicates the appearance of a new vessel type in MM III, called the 'blossom bowl' (his Type 5), made of a relatively soft stone with a thick section, high shoulder, flat base and rather small interior capacity. Classically, it is of steatite, and decorated with incised lines imitating a stylised 6- or 8-petalled flower, hence the name. Other decorative embellishments also are found with this same general profile, either diagonally spiralled body ribs or horizontally fluted shoulder (his Type 9.A–B). A very few are left entirely undecorated, and are known as 'bird's nest bowls' (his Type 3). All actually have developed from the earlier, smaller undecorated Pre-Palatial 'bird's nest bowl'. They all have the same basic shape and size, but the various designs presumably indicate either different functions or different inherent meaning for the MM III Minoans. They are found throughout the island in MM III–LM I.

Certain features of this bowl type are hybridised with the imported spheroid jar form, to which the high-shouldered jar form may be considered a variant in Minoan eyes. The stone used is harder, and the vessel can have horizontal solid lug handles, a slightly raised base and flattened, widened collar rim. These vessels do not actually *imitate* the Egyptian form *per se,* but rather take elements from it that were, for some reason, required for its function. Not all features need be found on any one vessel but seem to have no direct indigenous source. They are no larger than 'usual' but are, without exception, no more than half the scale of the imported vessels and, in MM III–LM IA contexts, are all found in élite Knossos tombs but for a single instance recovered in its harbour town at Poros. Later, they spread to non-palatial sites, usually still associated with a large élite-type structure when the provenance is known, and mostly (for some reason) at Haghia Triadha. One very similar vessel, with matching lid, was recovered from the MB IIB 'Tomb of the Cisterns' at Ebla, and Sparks (1998; III, 54 #431a) has concluded may be is Minoan. One recently published vessel, of even smaller scale and recovered in a settlement area at Poros rather than a tomb, indicates a different decoration and a possibly different function than the others. Associated hybridised Minoan spheroid bowls are:

E.I. KSM (1957–61) C 8. Fragment, dolomitic marble, half preserved, shoulder fluting (Warren 1969; 74,75 Type 30.A). Warren dates this fragment to MM IIB, noting it is the earliest Minoan 'imitation' of the spheroid jar form. Recovered in a tholos tomb on Gypsades hill, presumably in the MM IIB–IIIA (lower) layer (see Hood 1958; 22–23 and Driessen and MacDonald 1997; 169).

E.2. KSM (1957–61) E 9 A. Fragment, probably antico rosso, no surface embellishment (Warren 1969; 75 Type 30.A). Warren dates this fragment to LM IA. Recovered in the same tholos tomb on Gypsades hill as the last, presumably in an LM IA context (see Hood 1958; 22–23 and Driessen and MacDonald 1997; 169).

E.3. HM L2146. Complete gabbro bowl, with wide flat collar, high shoulder, raised base, undercut interior. No surface decoration. Large circular hole on shoulder, incompletely executed (Warren 1969; 75 Type 30.A, P403, D222 [not D225 as stated]; see also Driessen and MacDonald 1997; 169). Found in Tomb III at Maro Spelio, generally dated MM II–LM I, but an LM II–IIIAI clay vessel also is reported.

E.4. Poros: HM L4956. Fragment, white marble, one-third preserved, with high shoulder, flat base, horizontal roll handle(s) on shoulder. Miniature scale. Entire body vertically ribbed (Dimopoulou 2000). Found in an MM IIIB–LM IA stratum in settlement area, on the Psykhogoudhakis plot (Dimopoulou 1993; 450–2), the only other vessel not directly found at Knossos, but from its 'harbour town' and very near Katsamba. This is smaller in scale, with different decoration and context type than the other examples, but is comparable in the use of hard stone and the addition of handles.

F. As artistic inspiration

Minoan vessels in Minoan stone, stylistically influenced by Egyptian iconography. The only example in a context within the chronological limits of this study is the famous bowl from Mycenae Grave Circle B, grave 0, depicting a duck floating on its back. The Egyptians produced no vessels of rock crystal after the Early Dynastic period (Aston 1994; 64–65 'translucent quartz crystal'), and the Mycenaeans none at all. Nor was it employed for vessels in the Near East. This must be a Minoan product.

The initially most obvious possible visual inspiration is the ivory cosmetic container type in the form of a duck floating on its back, but these vessels do not appear before sometime in the 18th Dynasty and so almost certainly post-date the Mycenae bowl. Two alabaster 'dishes' of 'later Thutmoside' date (Hayes 1959; 191 Fig. 106 upper left) are very similar to it except the birds are trussed; these too post-date the Mycenae vessel.

Egyptian 'duck' vessels, closed shapes but resembling the Mycenae bowl in general, first appear within Dynasty XII and are known into the Second Intermediate Period (Aston 1994; 141–42 #144), but they also represent trussed birds and are considerably removed from it. Open vessels of travertine, of a swimming duck with the head facing forward, are known from the early 18th Dynasty, the earliest dating to the reign of Amenhotep I (Aston 1994; 155, 189–90, from Gurob; see also Harageh example), and so just overlap with the end of LM IA. Ivory and wooden spoons with duck-head handles are known from the Middle Kingdom (Bourriau 1988; 145 #153), but the majority of these and other objects with duck-head terminal date to the New Kingdom (Hermann 1932); nonetheless, they seem the nearest inspiration. The inspiration for this vessel need not necessarily have developed from Egyptian duck-shaped vessels or spoons, but they are the most likely candidates. This unique vessel essentially is one Minoan (probably Knossian) artisan's own version of the general type.

F. 1. Mycenae: NAM 8638, open 'scoop-shaped' bowl with handle in the form of a duck's head and 'splayed' tail opposite, rock-crystal, recovered in Grave Circle B, Grave O, the associated burial dated to LH IB (= latter half of LM IA), and the vessel therefore manufactured and exported to Mycenae not later than late LM IA (Warren 1969; 104 Type 42.C; see also Dietz 1991; 128–30, 271 Fig. 82.36).

G. As diplomatic gifts or export abroad

See discussion above, in sections C, D and F. The specigic vessels involved are converted alabastron 1.e, jug D.3 and bowl F.1 at Mycenae, and chalice D.4 on Thera.[2] Minoan vessels produced in Minoan stones also are recovered at Mycenae in Shaft Graves IV and V (Warren 1969; 37 [NM 600, 584], 103 [NM389]), and at Akrotiri on Thera (Devetsi 2000, 125, 126, Fig. 1), some possibly but none more demonstrably earlier than LMIA in date.

COMMENTARY

The vast majority of imported vessel types generally have basic characteristics similar to contemporary Minoan forms: fairly large, sturdy and thick-walled. These are container vessels for utilitarian use, rather than the small shapes and delicate open forms imported during the Pre-Palatial to Old Palace period (Warren 1969; 110–112 Type 43.C?, E, F, G.4, G.6?). As such, they are practical vessels imported and employed by a society that has become more concerned with function than aesthetics, and reflect this change in circumstances. During the period under consideration, of so much construction and reconstruction at Knossos, the development of Egyptian stone vessel importation and use appears to follow a pattern related to post-quake social and material development.

Association with Egypt strikingly appears to avoid the early MM III period, an omission that cannot entirely be explained by the small number of known MM IIIA contexts. With one possible exception, known contexts with relevant material are no earlier than MM IIIB, and frequently not earlier than early LM IA, in date. It seems that the earliest period following the earthquake, when many new Minoan forms were introduced and developed, was little if at all affected by contact with Egyptian material.

The only possible exceptions are a spheroid bowl found in the Gypsades tomb, whose context (although possibly not the jar) extends back into MM IIB and the spheroid jar fragment in an ambiguous fill context. Use of imported spheroid jars themselves generally extends back into the Old Palace period or even earlier, as a few finds at Knossos are recovered in ambiguous or early contexts. However, the hybridised Minoan spheroid bowls should not date earlier than MM III, when the Minoan bowl type(s) from which they are partly derived were themselves introduced into the artisan's repertoire.

By MM IIIB, Egyptian stone vessel types found in greater quantity throughout the remainder of the New Palace period already have arrived at Knossos, including the alabastron and spheroid jar types, and the raw banded travertine stone of which at least one rhyton was produced. All are single fragments rather than intact vessels, so their actual *use* at Knossos may well extend back into MM IIIA and not necessarily be associated with their deposition, all in apparently occupation contexts near 'Hogarth's Houses.' Only the derivative Minoan spheroid bowl is found interred in a tomb, at nearby Gypsades. Presumably the rhyton was intended for ritual use by the living and, by the extreme end of this period, a high-shouldered jar also was included in a votive deposit at Katsamba. By the beginning of LM IA, then, a recognised pattern that continues to survive and even thrive in LM IB elsewhere on Crete already was well-established at Knossos.

It is not until sometime in LM IA, however, that we find the earliest dateable evidence for actual *reworking* of imported vessels in both variations of the phenomenon, with the sawn and drilled fragments at the Royal Road North and the converted alabastron at Mycenae. Both phenomena also are found in later periods. The second Gypsades spheroid bowl also is interred at this time, as is the duck bowl at Mycenae, the latter indicating the manufacture and export of 'status' *tour de force* vessels with an exotic flavour, presumably from Knossos. The artisan has increased his repertoire: whilst most MM III forms continue, the variety of Minoan stone vessels in general expands further in LM IA, so the apparent introduction of both phenomena here may be seen as an adjunct of his expanded range and greater technical expertise.

The context date range of many relevant vessels spans all or most of the MM III–LM IA period, so it is difficult to comment on their place in any social or technological development. These include the Stratigraphical Museum rim fragments, the Minoan spheroid bowl in the Mavro Spelio tomb, the two alabastra in the Isopata tomb? deposit, and the workers' waste fragments. Also too wide–ranging in date are the exported Mycenae tall jug and Thera chalice, although I see the latter as produced, exported and deposited later rather than earlier within the period, probably in LM IA. Nonetheless, these vessels reinforce the observation that the exotic – even if made at Knossos or at least on Crete – was considered an appropriate export commodity to a neighbouring élite by LM IA.

None of these Egyptian or related 'egyptianising' vessels are found within palace confines. Nor are any of the relevant tombs apparently associated with the palatial élite for, although these clearly are élite or at least wealthy tombs, they are not considered 'royal' burials. This *suggests* that the palatial élite were not involved in the importation and actual use of these vessels, but instead those living both immediately nearby (in the Royal Road and Stratigraphical Museum areas) and farther away (in the 'Hogarth's Houses' and Poros areas) from the palace, where fragments were recovered in debris or fill. Likely, however, the subsequent export of converted and other *tour de force* vessels would have involved the palatial élite. Warren's (1991) overall suggestion of a semi-independent Minoan (here only Knossian?) merchant class has some validity for this period, which might be viewed as one of origin and development of his New Palace period scenario.

The development in form and elaboration of Egyptian vessels imported to Crete and their Minoan relatives strongly suggests that these followed the general development of Knossian outlook on the world around them: initially utilitarian, latterly less so and with an aesthetic imagination that suggests the artisans were able to spend more time and effort, and to experiment with expensive imported stone, to produce individual vessels of technical quality. This may be either because they could expand their capability to doing so, or felt the need to embellish their vases further, or (more likely) their patrons or customers were now able to afford this extra luxury. This process continues and reaches its epitome in LM IB, but now is diffused throughout the rest of the island. Similar converted and imported stone vessels are recovered in LM IB contexts of élite structures at *Aghia Triadha, Archanes, *+Kato Zakro, *Malia and *Myrtos Pyrgos, as well as +Knossos and its environs where they are now found in palatial (+) as well as non-palatial élite contexts, many associated with ritual use (*). Others are associated with occupation and workshop debris, and imported stone vessels converted into Minoan forms are found in greater quantity – but now only on Crete itself. Export of converted vessels appears to be unknown in LM IB (=LH IIA), although unconverted alabastra are now found in quantity at Vapheio as well as Mycenae. The Minoans directly transfer their own goods, iconography and style to the Mainland in this period (Demakopoulou 1988; 52), one of profound interaction between the two cultures. Minoan influence now also is more diffused, evidence for it being recovered no longer only at Mycenae but at numerous other sites throughout the Mainland.

Thus, the development of New Palace period use of imported Egyptian vessels directly follows the general development of MM III–LM IA (and into LM IB) at Knossos, from the needs of practicality to a more self-assured urban society in communication with its cultural neighbours. Container vessels in two specifically chosen basic forms (alabastron and spheroid jar) already dominate the repertoire of imported vessels by MM IIIB, initially (one presumes) as transport vessels for specifically chosen(?) commodities such as unguents and/or as ballast, together with specifically chosen raw stone. The vessels (and their contents?) then were employed both as religious and funerary offerings. The spheroid jars also served as inspiration for practical variations of contemporary Minoan containers, probably again for a specific purpose or purposes. Only later, during LM IA so far as we can tell, did the imports also serve as artistic inspiration for the production of *tour de force* vessels, both for élite and unique Minoan forms (the duck bowl) and as the basis for reworking/conversion into Minoan types, often then exported probably as status symbols or for other political implications. Little evidence for Minoan *tour de force* vessels is known before LM IA. Whilst few contexts and vessels can be cited in MM III–LM IA, what can be isolated is evidence for the progressive and parallel nature of both social and technological development (at Knossos), and as a precursor to Minoan achievements of the succeeding LM IB period.

ACKNOWLEDGEMENTS

The present research is conducted and funded under the auspices of *The Synchronization of Civilizations in the Eastern Mediterranean in the Second Millennium B.C.* (SCIEM 2000) special research programme of the Österreichischen Akademie der Wissenschaften, whilst holding a Research Fellowship at the McDonald Institute for Archaeological Research, University of Cambridge. I am extremely grateful to both institutions for facilitating my work. I also warmly thank Sinclair Hood and Peter Warren for permission to describe or mention the unpublished stone vessels recovered in the 1957–1961 Royal Road (North) and Gypsades tomb excavations at Knossos, also Rachael Sparks for discussing Syro-Palestinian vessels with me, and permission to cite her unpublished dissertation and Lisa French for her help in dating the Mycenaean contexts

NOTES

1 I use the term 'spheroid jar' for the Egyptian imports, and 'spheroid bowl' for the related Minoan vessels.

2 See also Devetsi 2000; 125-27, Fig.2, Pl.32, an imported Egyptian vessel also recovered at Akrotiri, perhaps to be added to my Table 5.1. I had neglected to include it in this paper, and Devetsi's article helped to remind me of its existence.

REFERENCES

Aston, B. G., 1994, Ancient Egyptian Stone Vessels. Materials and Forms, *Studien zur Archäologie und Geschichte Altägyptens*, 5, Heidelberger Orientverlag, Heidelberg.

Betancourt, P. P., 1985, *The History of Minoan Pottery*, Princeton University Press, Princeton.

Bourriau, J., *Pharaohs and Mortals. Egyptian art in the Middle Kingdom*, Cambridge University Press, Cambridge.

Cadogan, G., 1983, Early Minoan and Middle Minoan Chronology, *American Journal of Archaeology*, 87, 507–18.

Catling, H. W., 1988, Archaeology in Greece, 1987–88, *Archaeological Reports*, 34, 3–85.

Cline, E., 1994, *Sailing the Wine-Dark Sea: International trade and the Late Bronze Age Aegean*, BAR International Series 591, Tempus Reparatum, Oxford.

Dietz, S., 1991, *The Argolid at the Transition to the Mycenaean Age: Studies in the Chronology and Cultural Development in the Shaft Grave Period*, The National Museum of Denmark, Department of Near Eastern and Classical Antiquities, Copenhagen.

Deniakopoulou, K., 1988, Mycenaean Pottery, in Demakopoulou, K. (ed.), *The Mycenaean World. Five Centuries of Early Greek Culture, 1600–1100 BC*, Greek Ministrry of Culture, Athens.

Devetsi, A., 2000, The 'imported' stone vases at Akrotiri, Thera: a mew approach to the material, *Annual of the British School in Athens*, 95, 121–139.

Dimopoulou, N., 1993, Poros – Katsambas, *Archaiologikon Deltion*, 48, B'3, *Chronika*, 450–59.

Dimopoulou, N., 2000, Tmimia kupellidiou, in Karetsou, A., Andreadaki-Viazaki, M. and

Papadakis, N. (eds.), *Kriti – Aigyptos, Politismikoi desmoi trion chileition, Katalogos,* Kapon Editions, Athens, 216.

Driessen, J. and MacDonald, C. F., 1997, *The Troubled Island. Minoan Crete before and after the Santorini eruption, Aegaeum* 17, Eupen, Kliemo SA.

Evans, A. J., 1914, The 'Tomb of the Double Axes' and Associated Group, and the Pillar Rooms and Ritual Vessels of the 'Little Palace' at Knossos, *Archaeologia* 65, 1–94.

Evans, A. J., 1921–1935, *The Palace of Minos at Knossos,* Macmillan, London. Reprinted 1964, Biblo and Tannen, London.

Evely, D, 1984, The Other Finds of Stone, Clay, Ivory, Faience, Lead, etc, in Popham, M. R., *The Minoan Unexplored Mansion at Knossos,* British School of Archaeology at Athens and Thames and Hudson, London, 223–59.

Halbherr, F., Stefani, E. and Banti, L., 1977, Haghia Triada nel periodo tardo palaziale, *Annuario delta Scuola Archeologica di Atene e delle Missioni Italiane in Oriente, 55,* 9–296.

Hawes, H. B. *et al, 1908, Gournia, Vasiliki and other Prehistoric Sites on the Isthmus of Hierapetra, Crete,* Amedean Exploration Society, Philadelphia.

Hayes, W. C., 1959, *The Scepter of Egypt II,* Metropolitan Museum of Art, New York.

Hermann, A., 1932, Das Motiv der Ente mit zurdckgewendetem Kopfe im ägyptischen Kunstgewerbe, *Zeitschrift für Ägyptische Sprache and Altertumskunde,* 68, 86–105.

Hood, M. S. F., 1958, Archaeology in Greece, 1957, *Archaeological Reports,* 4, 3–25,

Hood, M. S. F., 2000a, The Wall Paintings of Crete, in Sherratt, S. (ed.), *The Wall Paintings of Thera,* Athens, Thera Foundation-Petros M. Nomikos and The Thera Foundation, 21–32.

Hood, M. S. F., 2000b, Crete, Syria and Egypt, in Karetsou, A., Andreadaki-Viazaki, M. and Papadakis, N. (eds.), *Kriti – Aigyptos, Politismikoi desmoi trion chileition, Katalogos,* Kapon Editions, Athens, 21–23.

Karo, G., 1930–1933, *Die Schachtgräber von Mykenai,* Verlag F. Bruckman, Munich.

Levi, D., 1967–1968, L'abitato di Festos in localita Chalara, *Annuario delta Scuola Archeologica di Atene e delle Missioni Italiane in Oriente,* 45–46, 55–166.

Platon, L., 1999, New Evidence for the occupation at Zakros before the LM 1 Palace, in Betancourt, P.P., Karageorghis, V., Laffincur, R., and Niemcier, W. D. (eds.), *Meletemata. Studies in Aegean Archaeology presented to Malcolm H. Weiner as he enters his 65th year, Aegaeum* 20, Histoire de 1'art et archdologic de la Grece antique Liege,, III, 671–81.

Platon, N., *1972, Zakros. The Discovery of a Lost Palace of ancient Crete,* Charles Scribner's Sons, New York.

Sparks, R. T., 1998, *Stone Vessels in the Levant During the Second Millennium B. C. A Study of the Interaction Between Imported Forms and Local Workshops,* unpublished Ph.D. thesis, Sydney, University of Sydney.

Walberg, G., 1976, *Kamares: A Study of the Character of Palatial Middle Minoan Pottery,* Acta Universitatis Upsaliensis, Uppsala.

Walberg, G., 1983, *Provincial Middle Minoan Pottery,* Veriag Philipp von Zabern, Mainz am Rhein.

Walberg, G., 1992, *Middle Minoan III - A Time of Transition, Studies in Mediterranean Archaeology* 97, Paul Astrijms Forlag, Jonsered.

Warren, P. M., 1969, *Minoan Stone.Vases,* Cambridge University Press, Cambridge.

Warren, P. M., 1972, Royal Road South excavations, *Archaiologikon Deltion,* 27, B'2, 627–29.

Warren, P. M., 1981, Knossos: Stratigraphical Museum Excavations, 1978–1980, Part 1, *Archaeological Reports,* 27, 73–93.

Warren, P. M., 1989, Egyptian Stone Vessels from the City of Knossos. Contributions Towards Minoan Economic and Social Structure, *Ariadni. Epistimoniki Epetirida tis Philosophikis Scholis (Aphieroma ston Styliano Alexiou),* V, 1–9.

Warren, P. M., 1991, A Merchant Class in Bronze Age Crete? The Evidence of Egyptian Stone Vessels from the City of Knossos, in Gale, N.H. (ed.), *Bronze Age Trade in the Mediterranean, Studies in Mediterranean Archaeology* 90, Paul Aströms Förlag, Jonsered 295–301.

Yule, P., 1980, *Early Cretan Seals: A Study of Chronology, Marburger Studien zur Vor- und Frügeschichte* 4, Verlag Philipp von Zabern, Mainz am Rhein,.

Zervos, C., 1956, *L'Art de la Crete Neolithique et Minoenne*, Éditions Cahiers d'Art, Paris.

Chapter 6

Stone Vessel Workshops in the Levant: Luxury Products of a Cosmopolitan Age

Rachael Sparks

Abstract
During the Middle and Late Bronze Ages in the Levant several stone vessel workshops emerged producing luxury vessels for local consumption and wider distribution. This paper will focus on two examples of this phenomenon: an obsidian vessel workshop at Tell Atchana, and gypsum vessel workshops in the Jordan Valley. These will be used to evaluate the different mechanisms by which such industries operated, based on a study of the range and sources of the raw materials used, physical evidence for on-site production, and the distribution patterns of the finished products.

INTRODUCTION

Stone vessels were popular in antiquity as luxury containers for a variety of cosmetic preparations and perfumes. These kinds of objects are found in many parts of the Near East during the Third Millennium, when several industries began to produce a diverse repertoire of forms to meet growing demand. The market for such products was more restricted in the Levant, and imported examples were mostly confined to large, well developed centres such as Byblos and Ebla. It was not until the Second Millennium that this region appears to have intensively developed its own schools of production. This paper will focus on the products of two such schools; obsidian vessels from Atchana, and gypsum vessels from the Jordan Valley. These provide an opportunity to study the nature and extent of such workshops through manufacturing debris and discards, evaluation of resource availability, and a quantitative and stylistic analysis of the distribution of the finished products.

THE OBSIDIAN VESSEL WORKSHOP AT ATCHANA

In 1948, Woolley put down a number of soundings through the floor of the Level IV palace at Tell Atchana, in order to determine the nature of previous structures in this area. The third of these was placed in the northwest corner of courtyard 22, coming down onto part of a stone vessel workshop. No plan was ever published and the details of this remarkable locus are known only from Woolley's brief descriptions and the objects which were found within it (Woolley 1955, 109–110, 292–296). These indicate that Woolley discovered a single room with a brick paved floor and doors in the southeast and southwest walls, sealed by a thick deposit of ash and burnt debris. The floor ran up to substantial wall foundations which were attributed to level VII. The context is therefore thought to be contemporary with the level VII palace further to the east, and in view of the size of the associated wall, may well have been a part of it. This would date the structure to the last phase of the Middle Bronze Age, following the low dating for the site (Bourke 1991, 208–211; *contra* Na'aman 1976; see also the discussion in McClellan 1989). The extensive destruction layer that was found suggests that this area was destroyed in the same episode that brought the level VII palace to an end. A male skeleton found in the debris bears witness to the traumatic circumstances of this event (Woolley 1955, 293).

Little further information is provided about the architecture of this locus, and nothing is known about its relationship to adjacent structures or rooms. There is no reference to specialised features such as benches, storage bins, troughs or work pits that might mark this as a craft area. The only installation found was a large cooking vessel, which had been set into the centre of the pavement. As this was not associated with a hearth, it may have been used for storage; however its contents at the time of discovery could represent debris which had fallen in from above rather than anything deliberately placed inside. Despite the lack of architectural clues, the locus was able to be identified as a workshop on the basis of its other finds, including partially modified stone blocks, unfinished vessels and a few tools.

Raw material was represented by numerous obsidian fragments. These had come from several rectangular blocks which had been shattered by the heat of the fire which destroyed the room. Although these could not be lifted intact, Woolley was able to measure the pieces in situ, determining that these ranged in size from around 30 x 20 cm to 10 x 14 cm (AT/48/102; Woolley 1955, 292–293). The sides of each block had been polished. This group probably represents obsidian that had been pre-cut for easier transportation from the quarry. This kind of preparation at the source area is known in modern operations, where the size of the blanks are determined by prior arrangement with the purchaser (Wulff 1966, 161; Kohl 1975, 122).

The locus also produced a number of unfinished stone vessels (Woolley 1955, 293; Fig. 6.1). Three of these were made of obsidian, consisting of a shallow leaf-shaped bowl, a carinated bowl and a tripod mortar (AT/48/99–101; Fig. 6.1A). There was also a finished semi-circular obsidian bowl or pyxis with flat rim and internal dividing wall (AT/48/98; Fig. 6.1B). All these objects fit easily into the size range represented by the pre-cut blocks found with them in the room. A shallow, unfinished granite bowl was also recovered from the debris (AT/48/105).

This group represents several different stages of production and can suggest the sequence in which stone vessels were formed. For the obsidian bowls, the first step was to block out the desired shape, sometimes making use of the polished surfaces of the original block to provide a flat, ready finished rim or base area (e.g. AT/48/99, AT/48/101). The interior was then removed by drilling a series of adjacent holes using a

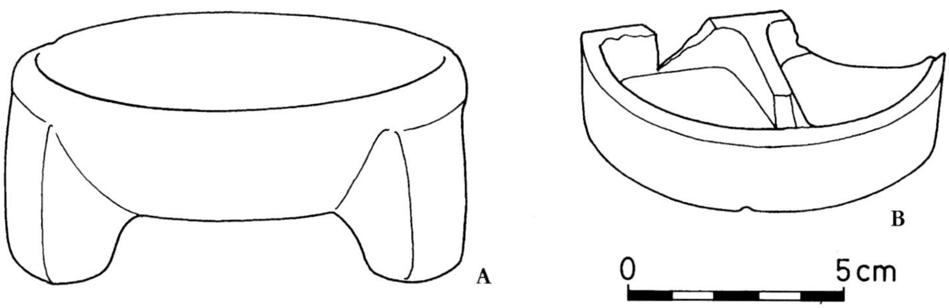

Figure 6.1: Obsidian vessels from the Atchana workshop: A. Tripod mortar, AT/48/100; B. bowl, AT/48/98 (after Woolley 1955, pl. LXXXIII).

solid-headed drill bit, then chipping out the remaining obsidian with a chisel. Once the interior was fully formed, the exterior could be properly shaped and the entire vessel polished. This sequence of events appears to have been reversed on the granite vessel, with the outside being shaped before removing the interior, which in this case has yet to be removed. One reason for this difference may lie in the different physical characteristics of the raw materials involved. As obsidian could easily fracture under percussive stress, it was probably more efficient to carry out the most risky procedures early in the sequence. Granite is more durable and less likely to break during manufacture, so the order in which these steps were carried out may have been less important.

Associated with the vessels and fragments were a series of tools. These included a black chert axe with hammer marks on the upper end, a granite weight, and a lump of red ochre. Woolley suggested that the ochre may have been used to outline vessel forms on the stone blocks before proceeding to cut out the shape (unpublished field record AT/48/103, University College London). However none of the surviving blocks are reported as carrying such marks. Several pieces of hematite were also discovered in the debris, which could have been used to give each vessel is final polished surface (Woolley 1955, 294). No metal tools were found, although solid drill bits, chisels or files would probably have been necessary parts of the artisans' tool kit. It is possible that these were stored in a nearby room, to be issued to the workmen at the start of work. Alternatively, they could have been removed from the scene by either craftsmen or looters, or recovered from the debris after the fire.

It would appear that the workshop was in operation at the time of the destruction of the city. This single room contained bowls at varying stages of completeness, but none of them appear to have been abandoned because of manufacturing flaws. Yet Woolley does not appear to have discovered any debitage. This could mean one of two things. It is possible that working debris existed, but because of its fragmentary condition it was mistaken for unworked raw material, or simply not reported. The second explanation is that there was a separate work area elsewhere in the complex where the vessels were actually carved. In that case, this context may have represented a store for raw material, equipment, and objects which were in progress. There was insufficient horizontal clearance of the architecture in this area to determine if this was the case; however this type of functional division has been noted in modern stone vessel workshops (Hester and Heizer 1981).

Other aspects of the operation are also open to conjecture. It is not clear whether the concentration on open obsidian vessels was typical of the overall output of the workshop. Craftsmen may have worked with other materials or on other shapes at different times, depending on the availability of raw materials or the commissions they had received. It is also unknown how this part of the city relates to the main palace complex to the east, or the temple area to the south. Some connection is to be expected, and it seems likely that this workshop operated under the control of the Atchana state or temple authorities. A high degree of organisation and wealth would have been essential in maintaining a craft requiring imported luxury raw materials such as obsidian.

Sources of supply and distribution of the finished product

Obsidian is not available in the immediate region of Atchana, and the raw material to support this industry must therefore have been brought to the site from some distance. No analysis was ever conducted on the fragments found in the workshop, so the location of this source is still unknown. The material was described as being of two distinct colours, black and dark green (Woolley 1955, 293). Potential source areas include the Aegean and various locations in Anatolia (for a selection of sourcing studies see Renfrew *et al.*, 1964; Cann *et al.* 1969; Cann and Renfrew 1972; Zarins 1989; Gratuze *et al.* 1993; Brennan 1996). In any case, the distance that this stone travelled must have been considerable, as the nearest obsidian deposits are several hundred kilometres away. This kind of trade in unworked or only slightly modified raw material is well documented in antiquity. Quantities of unworked lapis were recovered from the Early Bronze Age palace at Ebla and wasters of imported calcite from MMIII–LMI Knossos (Pinnock 1985, 88; Warren 1969, 125–126). Egyptian texts such as the Annals of Tuthmosis III, the Medinet Habu inscriptions of Ramesses III and the Papyrus Harris also mention blocks of raw lapis lazuli, alabaster, malachite and 'costly stone' (e.g.: Breasted 1906a, section 446; Breasted 1906b, sections 30, 231, 234, 245). Much of this trade focused on small, easily transported raw material for jewellery and amulets. Larger blocks for architecture and statuary could nonetheless travel a long way, provided that the material was sufficiently valuable to justify the effort required to transport it.

While it is clear that the raw materials for the obsidian vessel industry at Atchana were obtained via long-distance trade networks, the extent of trade in the finished product is less easy to establish. In general, obsidian vessels are rare in Bronze Age Levantine assemblages. At Atchana, only two further examples were discovered, a spouted rectangular bowl and a body sherd from a closed vessel (Woolley 1955, 294, AT/39/135, AT/48/41). These were found in level VII palace and temple areas. Neither corresponds with any of the shapes found in the workshop itself, and neither displays stylistic features to link them specifically to the Amuq Valley region. From the evidence of these two pieces alone, there would therefore have been little reason to suspect the existence of local obsidian vessel workshops.

Outside of Atchana, there is only one good candidate for an obsidian vessel which may have originated at that site. This is a tripod semi-circular bowl from Tell el-'Ajjul, whose basic body shape is reminiscent of the pyxis found within the workshop, as well as a series of similar bowls of Syrian manufacture in ivory and faience (Petrie 1934, pl. XXIII; cf Montet 1928, pl. CVI.725). However it has Egyptianising elements in the decoration, a feature not evident in the material from Atchana itself. Exchange of specialist craftsmen between royal courts shows the mechanism by which this kind of interaction could occur (Sasson 1968; Canby 1976, 33; Zaccagnini 1983). As comparable vessels are so poorly represented in contemporary assemblages, it would appear that the overall output of the Atchana operation was limited. This may have been because the workshop functioned for only a short time before the city was destroyed, or because it was producing an elite product to which access was restricted. In any event, the vessels that were produced appear to have been highly individual in form.

Further obsidian vessels found in the Near East are probably of Egyptian origin, including inscribed jars at Bogazköy, Byblos and Tell el-'Ajjul (Lilyquist 1995, 46, cat. A; Virolleaud 1922, pl. LXVII.1; Rockefeller Museum, 38.588). It has been suggested that Egyptian workshops obtained their obsidian primarily from Ethiopian rather than Anatolian sources, a view which has been supported by recent analyses of obsidian objects from Upper Egypt (Lucas and Harris 1962, 415; B. Aston *et al.* 2000, 46). If this proves to be the case, the origin of the raw material could be used to distinguish between vessels being produced by workshops in Egypt and traded into the Levant, and those being manufactured locally.

Anatolian workshops also produced a range of obsidian vessels (e.g.: Özten 1979, pl. III; Mellink 1988, 107), and it is possible that the original impetus for the Atchana industry came from this direction. However, the shapes of the vessels found at the site do not seem Anatolian in character. The tripod mortar, for example, mirrors a form more popular in basalt and which appears to have originated in the Levant. The semi-circular pyxis matches shapes more familiar in Syrian ivory and faience work, while the tripod pyxis from Tell el-'Ajjul combines both these elements in a single vessel. Examples such as these suggest that the primary stylistic influence on the Atchana artisans came from Syria itself, rather than from further north.

GYPSUM VESSEL WORKSHOPS OF THE SOUTHERN LEVANT

Gypsum is a soft sedimentary rock which was used in Bronze Age Palestine to make a variety of objects, including vessels, chariot fittings and weights. The industry initially developed in response to the growing trade in imported Egyptian calcite vessels to the region during the Middle Bronze Age. As calcite was not readily available, craftsmen began to produce copies of these imports using local supplies of gypsum. These two materials can appear visually similar, but differ in composition and hardness. Gypsum is considerably softer than calcite, allowing Palestinian craftsmen to carve out the interiors of their vessels using narrow-bladed chisels. Egyptian workshops made use of tubular and crescent-shaped drills for this purpose. As both techniques leave behind distinctive marks on the vessels themselves, these provide another means of distinguishing between the locally made and imported examples (Ben-Dor 1945, 96–97; Sparks 1998, 328, 356–60, 364–6).

Physical evidence for gypsum vessel production

Physical evidence for on-site manufacture of gypsum objects is rare in Palestine. Despite large numbers of gypsum vessels being discovered at Middle Bronze Age Jericho, the only direct evidence for local working of this material is found with a series of gypsum beads and roughed out and partially bored bead blanks from a much earlier tomb at the site (Kenyon and Holland 1983, 799, tomb A113). The same is true for other sites where gypsum vessels are found in the Second Millennium B.C., with the notable exception of Beth Shan in the north Jordan Valley.

Excavations at Beth Shan during the 1930's produced a series of gypsum vessel wasters (Ben-Dor 1945, A8, B1, C1, C4; Pennsylvania University Museum 32–15–221, 32–15–222, 32–15–244, 34–20–261, 34–20–262). These were mostly from juglets of similar type, and had been discarded because of various mistakes in their production, such as in removing too much from the interior or handle (Fig. 6.2A). They were discovered in various room and street contexts attributed to levels XA and XB (FitzGerald 1932, 147; Ben-Dor 1945, 97–101). Many of the loci originally attributed to level XB were subsequently reassigned to XA by Frances James (P.E. McGovern, pers. comm. 1992). Other unfinished forms were also found, including simple globular flasks with offset necks, from levels XB and IX (Fig. 6.2B). An unfinished lug-handled jar was also discovered in a much later level VI deposit (James 1966, Fig. 54.10, locus 1205; Fig. 6.2C). This had been shaped, but not hollowed out and may represent an unfinished, rather than a discarded example. The archaeological setting for these finds has been very poorly published, and it is not clear whether any of the unfinished vessels were found within an architecturally defined workshop space or in association with any tools or other debitage. However, unlike Atchana, all the shapes found are represented at the site in finished form (e.g.: Ben-Dor 1945, A10–13; James and McGovern 1993, Fig.

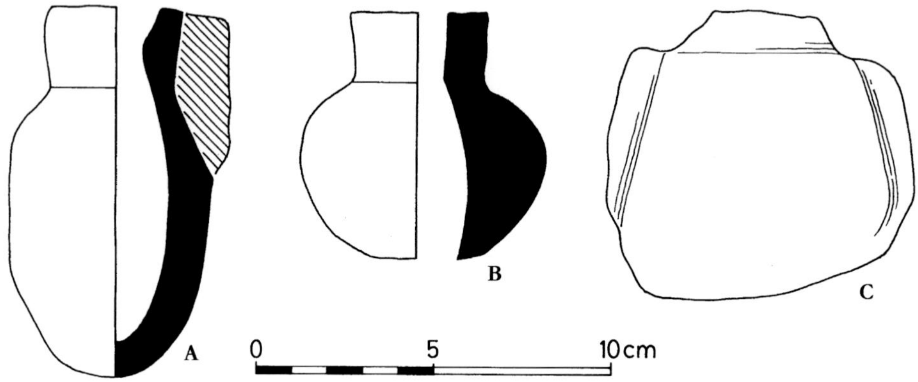

Figure 6.2: Unfinished vessels from Beth Shan: A. Level XA, Locus 1658 (Philadelphia University Museum 32-15-225), B. Level XB, below Loci 1666-73 (Philadelphia University Museum 32-15-244), C. Level VI, Locus 1205 (Philadelphia University Museum 29-107-327).

110.8–11; University of Pennsylvania Museum 34–20–261).

It is difficult to establish a beginning date for gypsum production at Beth Shan, for a number of reasons. Middle Bronze Age remains were initially exposed over a limited area and poorly published. There were also stratigraphic problems with this material, and the excavators' failure to recognise deposits such as burials cut down into earlier levels led to the attribution of several examples from strata XB or XA to the much earlier level XI (Fitzgerald 1934, 131). While the Middle Bronze Age strata have now been more thoroughly investigated by Mazar, only preliminary reports of this material have been published, and these do not mention additional finds of gypsum vessels (see Mazar 1990b, 108*; Mazar 1993; Wolff 1994, 488–489). Pinpointing the first stages of gypsum production is further complicated by controversy over the dating of levels XA and XB. Suggested dates for level XA have included LBI (Wright 1941, 485), the transitional MBIIC–LBI period (Oren 1973, 100), and MBIIC (Ben-Dor 1945, 100–101; Mazar 1990a, table 4).

The evidence from Beth Shan differs from that found at Atchana on a number of counts. It represents an industry that can be demonstrated to have been in operation over a long period of time, from the Middle Bronze through to the Iron Age. This industry was moderately standardised, and thus the shapes of the unfinished vessels found match not only each other but also finished examples from the site. These goods also seem to have been available to middle ranking individuals, occurring across a wider range of context types that includes domestic areas and burials.

The development of the gypsum vessel industry and stylistic evidence for additional centres of production

Unlike obsidian vessels, gypsum vessels are found in large quantities throughout the Levant. This enables the use of other techniques for identifying possible workshop groups, including stylistic analysis and examination of the distribution patterns of finished examples. Both these approaches suggest that there were multiple workshops in operation in the region throughout the course of the Bronze Age.

The earliest confirmed appearance of a gypsum vessel in the region occurs at Jericho in a burial dating to the late MBIIA or early MBIIB period (Kenyon 1960, Fig. 144.1, 353, 355). The industry may well have started at this site, as gypsum vessels form a significant part of tomb assemblages there throughout the rest of the Middle and early Late Bronze Age. In the earliest phase of this industry, gypsum vessels appear only sporadically elsewhere (Fig. 6.3). A second centre of production was to be found in the workshops at Beth Shan of this period, although their products are less well represented in the archaeological record.

In the following MBIIC and LBI periods, workshop production had clearly expanded and the distribution of gypsum vessels was more widespread (Fig. 6.4). Although the bulk of examples continue to appear at Jericho, major quantities are also now to be found at sites such as Beth Shan and Pella in the north Jordan Valley. Sporadic appearances of gypsum vessels outside of Palestine itself are possibly to be linked to this phase. A gypsum juglet of Jericho type was found at Ras Shamra (Caubet 1991, 240, pls I.4, VIII.9) and a decorated alabastron in a LMIA context at Thera (Warren 1979, 88, Fig. 4.1872). Warren suggested that the latter was Syro-Palestinian on the

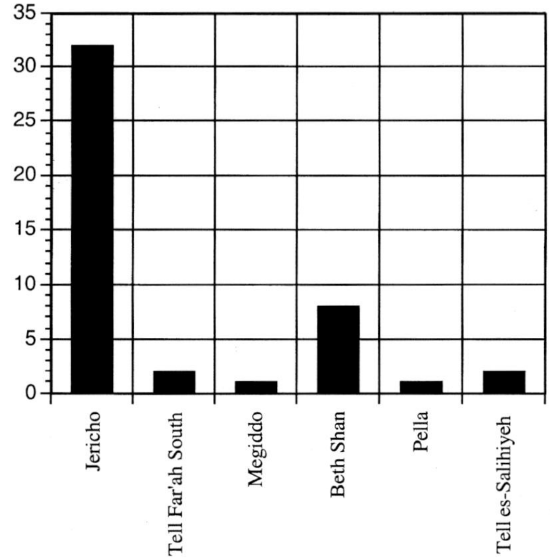

Figure 6.3: Gypsum vessel frequency in Palestine during the MBIIB period

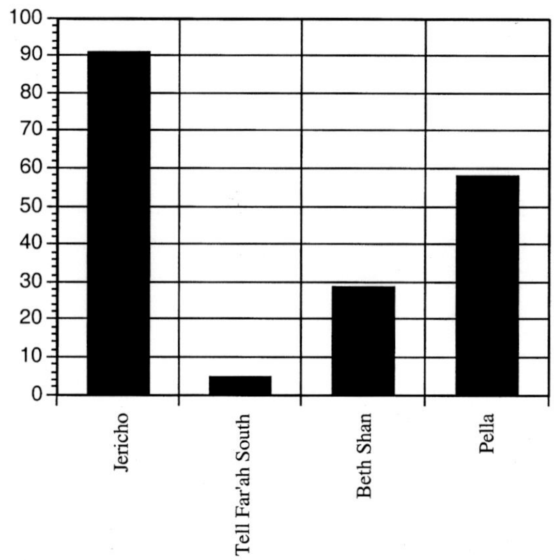

Figure 6.4: Gypsum vessel frequency in Palestine during MBIIC-LBI.

grounds of the general shape and material used; the unusual feature of incised decoration around the shoulder is closely paralleled in an example from Jericho tomb B35 (Kenyon 1960, 381; Nicholson Museum 52.640.119). Both vessel types closely match material from Jericho in shape and decoration, suggesting that they originated in workshops there. However such finds are extremely rare, and would seem to represent opportunistic or indirect exchange rather than an established trade in such goods (for a discussion of other Canaanite material on Thera during the LMIA period, see Warren 1979, 106–107, and note 2).

Distribution within the southern Levant is still restricted in this phase. Although the proportions of examples at single centres outside of Jericho increases, the overall dispersal of individual finds does not (see Figs. 6.3–4). One possible explanation for this phenomenon would be the development of direct trade links between Jericho and these other centres, with little subsequent redistribution of goods elsewhere. Alternatively, this period may see the development or expansion of vessel workshops outside of Jericho itself. The latter hypothesis gains considerable support from the manufacturing debris from Beth Shan mentioned earlier, and a stylistic examination of the main features of the gypsum vessel repertoires at Jericho, Beth Shan and Pella during this period. These allow us to propose the existance of distinct workshops centred on the north and south Jordan Valley areas.

Gypsum vessels at Jericho display features which appear across a range of shapes, and which may be seen as characteristic of a local workshop. This includes the tendency to define the junction between neck and body by means of an incised line, sometimes leading to an articulated, flattish upper shoulder. This is found on a number of types of alabastra and juglets (e.g.: Figs. 6.5A and 6.5C). A probable extension of this practice is the use of raised cordons, or ridges at the base of the neck, and occasionally just above the foot. These are frequently decorated with simple incised oblique lines or crosses, and are found on shapes such as the alabastron and goblet (Figs. 6.5D-G). Similar decoration also appears on juglet handles. Motifs include oblique parallel lines, vertical grooves, herringbone or chevron based designs (Kenyon 1960, Fig. 171.3; Kenyon 1965, figs 154.8, 171.17; 179.7–8, 17; Figs. 6.5A-C). It is rare to find gypsum juglets with comparable decoration elsewhere. Beth Shan produced only two similar examples, one of which featured a raised rope design down the centre of the handle that is different yet again (Ben-Dor 1945, 99, A12). Another shape that may have originated at Jericho is a type of bowl with four ram's-head shaped handles (Fig. 6.5H). This shape is found at the site in both gypsum and wood, and appears there during MBIIB; with the exception of a single example at Tell es-Salihiyeh, the few examples from dated contexts elsewhere appear to belong to the MBIIC period or later (Von Der Osten 1956, pl. 26.72, 34.81; Sparks 1998, cat. nos 135–6). Finally, there are a series of unique shapes not found in other gypsum repertoires, including a double-bodied juglet, an elongated bottle with unusual square plan-view, a tall goblet and a dipper juglet with pointed base (Kenyon 1960, Fig. 171.7; Ben-Dor 1945, 110, cat. 2; Kenyon 1965, Fig. 154.13, 171.17; Fig. 6.5G).

Pella is comparatively close to Beth Shan, and one might expect these two sites to display very similar stone vessel repertoires. Some overlap is evident with forms such as the drop-shaped alabastron, conical alabastra and cylindrical jars with angular shoulders (McNicoll *et al.* 1982, Fig. 106.11–13; Bourke *et al.* 1994, Fig. 7.5–6; reg. nos. 70370, 70445, cat. no. 920440). Similarly there are a few examples found at Pella which seem typical of the Jericho repertoire, representing either imports or local imitations of the style. These include two ram's-head handled bowls, a form also represented by a single example at Beth Shan and a conical alabastron with decorated cordon (Sparks

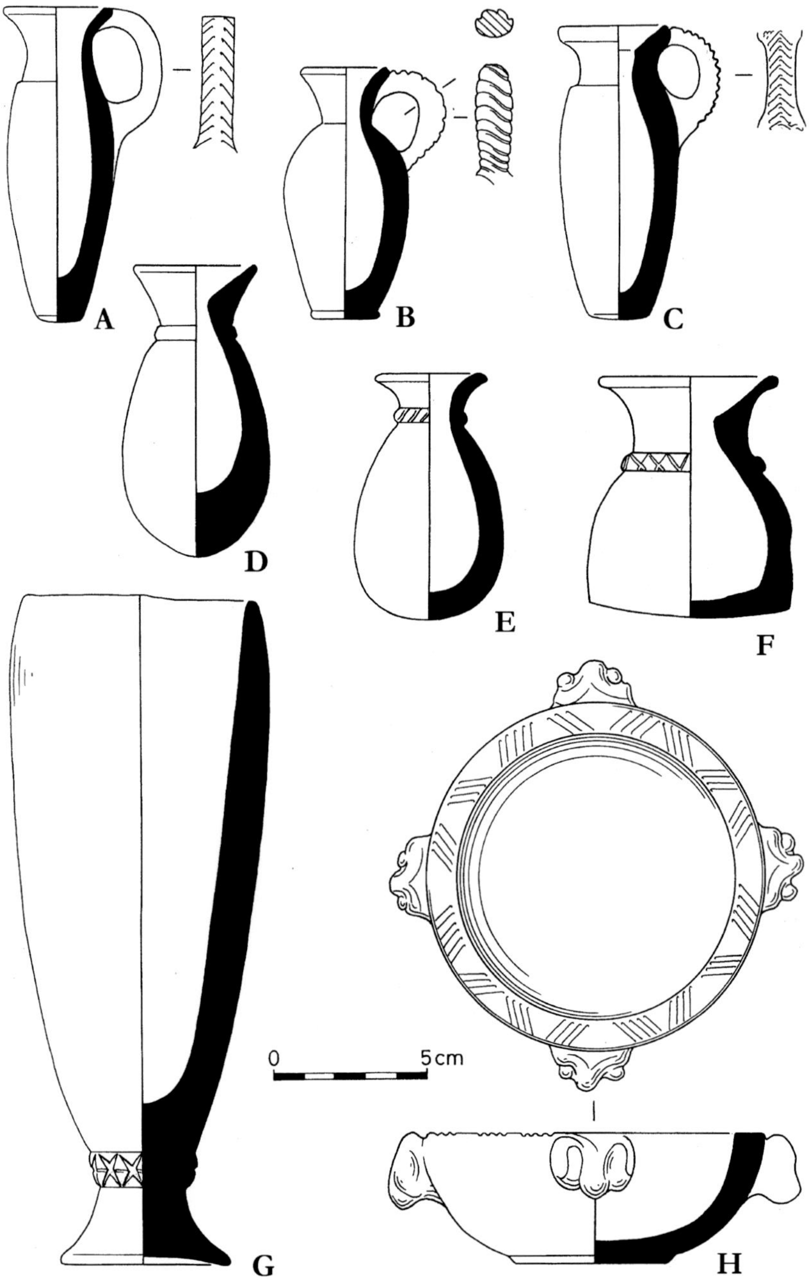

Figure 6.5: Characteristic products of gypsum vessel workshops at Jericho (all after Kenyon 1965, except as marked): A. Tomb J14, no.149; B. Tomb P17, no.25; C. Tomb M11, no.18; D. Tomb B51, no.165a; E. Tomb A136, no.50; F. Tomb B47, no.86 (Nicholson Museum 54.48.10); G. Tomb J14, no.87; H. Tomb J54, no.42.

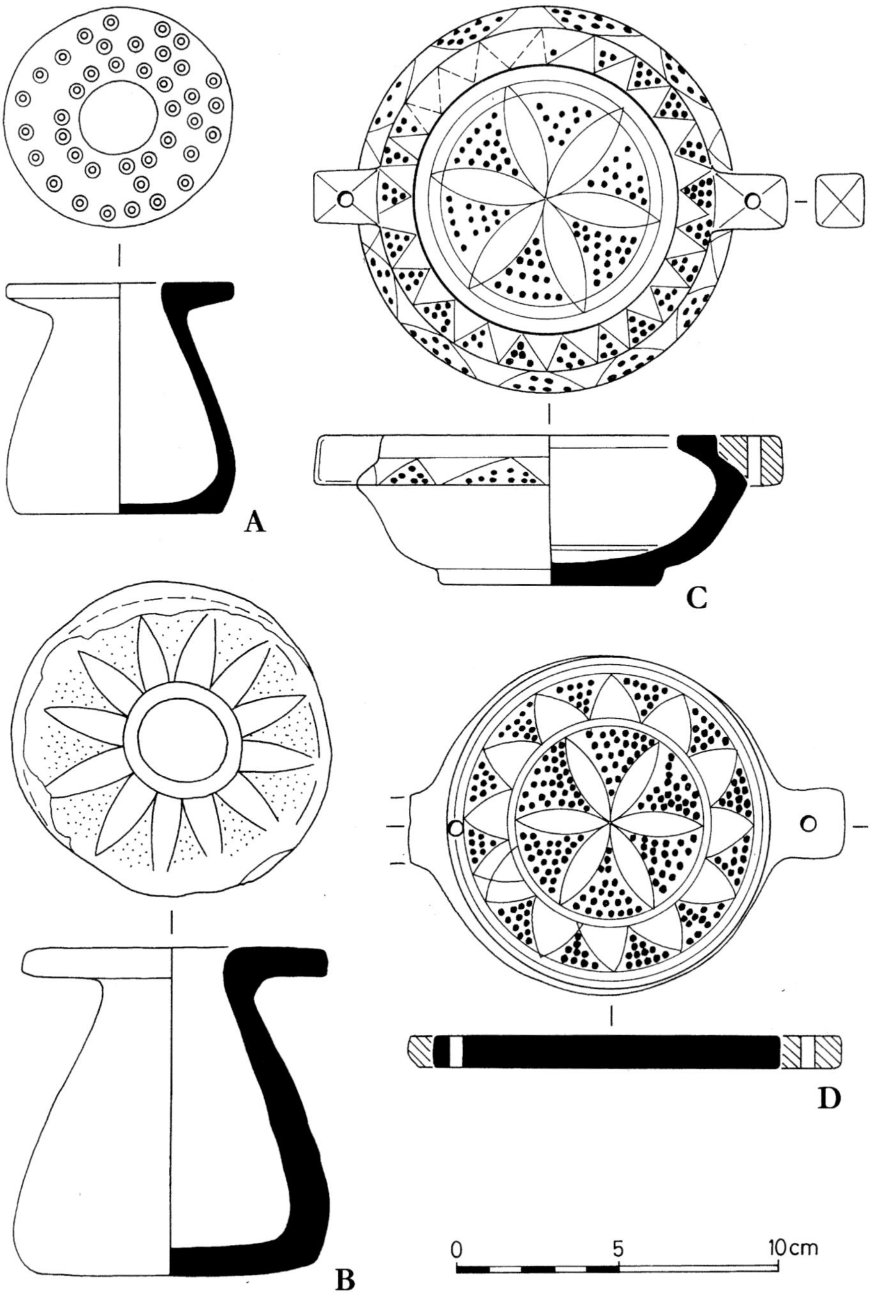

Figure 6.6: Characteristic products of gypsum vessel workshops at Pella (all after McNicoll et al. 1982, except as marked): A. Tomb 20, no.32187; B. Tomb 62 Locus 3E , no.70752, (Irbid Department of Antiquties 3815); C. Tomb 20, no.32212, pyxis body; D. Tomb 20, no.32213 with matching lid for (C).

1991, figs 2.5, 2.7; Fitzgerald 1931, pl. XL.3, 43; Smith 1973, pl. 57.131). Despite these similarities, it is notable that the bulk of the gypsum vessel repertoire at Pella during the MBIIC–LBI period appears to be largely distinct from that of either Beth Shan or Jericho (Sparks 2001).

Features which can be seen as characteristic of Pella workshops include shapes such as the conical alabastron with a broad, flat-topped rim, and a handled version of the type (McNicoll *et al.* 1992, pl. 61.2; McNicoll *et al.* 1982, pl. 114.13; Figs. 6.6A-B). Although the body shape is comparable to that seen in contemporary Palestinian alabastra, the movement from a funnel-shaped mouth to a flat rim is unique to Pella, as is the practice of decorating this upper surface with a series of incised and black inlaid designs. The motifs used include dotted circles and petals, elements well known in contemporary bone, wood and ivory products from the region (e.g.: Kenyon 1965, Fig. 173.5; Gachet 1987, pl. 5.42). A similar floral scheme is used to decorate a carinated pyxis with lid from Tomb 20, which was probably made in the same workshop (McNicholl *et al.* 1982, pl. 114.14; Fig. 6.6C–D). This form is also well known in bone and ivory (Mazar 1985, 12–13; Sparks 1991 note 40). Other gypsum vessels found at Pella during this period would appear to be unique to the site, such as a spouted jug and carinated bowl (Sparks 2001, figs 3.1–2).

At this point, it would be good to be able to demonstrate a similar workshop group from Beth Shan, based on stylistic evidence. Ironically, the one site which has produced clear indications of workshop activity displays few features in its gypsum assemblages which can be seen as particular to that site, and therefore characteristic of a 'Beth Shan' workshop. While one could perhaps argue that this lack of individuality points to close relations between the craftsmen of Beth Shan and other sites during the MBIIC–LBI period, it seems more likely that it indicates nothing more than the comparatively small size of the sample available. This makes it difficult to isolate general trends across the site as a whole, or to distinguish between vessels made locally and those that may have been imported from further south.

In LBII, the distribution pattern of gypsum vessels over Palestine changes yet again. The fortunes of Jericho go into decline, and consequently gypsum vessel production at this centre probably ceases (Bienkowski 1986, 136–137). Only a single vessel of Palestinian type can be attributed to the LBII phase at Jericho, compared to over 120 in the preceding periods (Garstang 1933, pl. XVI.8). Beth Shan and Pella continue to produce large numbers of examples, suggesting the continuation of north Jordan Valley workshops at this time, while the appearance of quantities of material at Deir 'Alla, Tell es-Sa'idiyeh and Tell Far'ah South point to new production or distribution points (Fig. 6.7). Distribution patterns show another phenomenon; gypsum vessels are now more widely dispersed outside the Jordan Valley, appearing along the Palestinian coast and inland hill country, and east into the Transjordanian hinterland. Some of this may be attributed to improved communications throughout Canaan after the region was absorbed into the Egyptian administrative network (Liebowitz 1986, 17).

In the early Iron Age, production appears to continue in the north Jordan Valley, with material focusing on the sites of Pella and Beth Shan. Workshops at the latter evidently continue operating well into the Iron Age, and gypsum vessels remain prominent in the local repertoire down until the end of level V. Outside of this area,

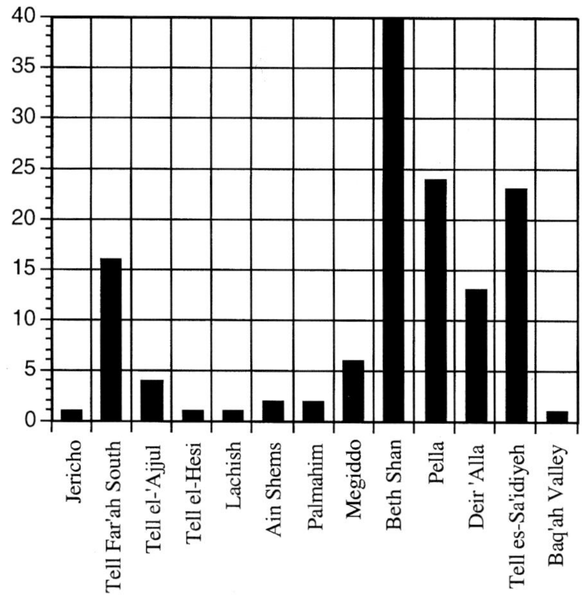

Figure 6.7: Gypsum vessel frequency in Palestine during the LBII period.

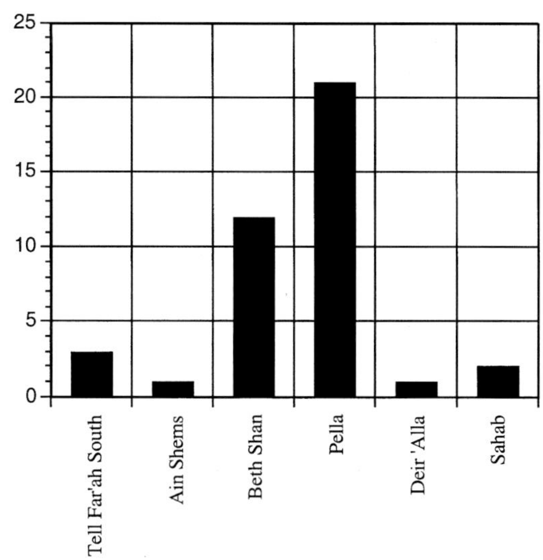

Figure 6.8: Gypsum vessel frequency in Palestine during Iron I.

distribution appears to be more sporadic, with only handfuls of examples appearing elsewhere (Fig. 6.8). This may be linked to the decline of Egyptian interest in Canaan, and a more general breakdown in communications and trade networks caused by population resettlements throughout the Eastern Mediterranean.

Distribution of gypsum sources in Palestine

The dominance of individual workshops across Palestine may have fluctuated over time, depending on economic and other conditions. In each case the owners would have been dependent on securing a supply of the necessary raw materials in order to ensure success. As various centres for these workshops have been proposed, it would be useful at this point to consider the distribution of potential gypsum sources and how these may have affected the development and location of particular workshops.

Gypsum beds of sufficient size for commercial exploitation have been reported at a number of sites throughout Israel and Jordan. A summary of this information is presented in Fig. 6.9. Sources closest to Beth Shan include those behind the Menahemia settlement, in the Wadi Bireh and near Jisr Mejami (Blake 1930, 27–28; Ben-Dor 1945, 96; Kleinmann 1976, 135–6). Gypsum also occurs in the Ghor, in the region between Pella and Deir 'Alla, although it is reported to be of lower quality (Bender 1973, Fig. 133; Blake 1930, 29; P. Macumber, pers. comm. 1992). Also within range of Deir 'Alla and Tell es-Sa'idiyeh are deposits which have been noted in the Wadi Zerqa and its offshoot, Wadi Huni (Ionides and Blake 1939, 64, 67, 69, 74; Bender 1974, 168). Further south, there are numerous deposits around the Dead Sea region and in the wadi systems running east, including Jebel Usdum, near Ras es-Zuweira, Zerqa Ma'in, Wadi Kerak, Wadi Mujib, Wadi Tafileh, Wadi Dra'a and Wadi al-Hasa (Blake 1930, 37; Ionides and Blake 1939, 118–119; Picard 1954, 13; Bender 1974, 168, Fig. 133; Kleinmann 1976, 137). Lankester Harding reported signs of ancient quarrying in the latter, although the exact date of this activity was not noted and may not have been identifiable (Ben-Dor, 1945, note 9). It is possible that either group of sources east of the Jordan were exploited by the craftsmen of Jericho. Finally, there are several sources in the Negev east and south of the Dead Sea, the most significant of which appears to be at Makhtesh Ramon (Braunfeld 1954, 75, Kleinmann 1976, 135). While this last group is the closest to Tell Far'ah South it is not certain that that these deposits had been discovered during the Bronze Age, as archaeological surveys of the region did not reveal any sites dating between MBII and Iron I (Rosen 1994). However the source reported near Timna may well have been known, as the area was being exploited at this time for copper and several sites are known in the region (Würzburger 1972, map 1; Rothenberg and Bachmann 1988, figs 2–4).

The variety of gypsum used for stone vessels during the Middle and Late Bronze Ages was usually faintly translucent and pure white in colour (Ben-Dor 1945, 94). This material was clearly preferred, but occasionally more strongly coloured gypsum was selected. Some vessels from Beth Shan were uniformly grey, a feature attributed to post-depositional burning by the excavators (Yadin and Geva 1986, 86). However in several cases the colour is quite consistent over the vessel and may be due to a naturally occurring mix with clay or bituminous substances (Ben Dor 1945, 95; Kleinmann 1976, 135; Yadin and Geva 1986, Fig. 36.1–2). A tazza from Tell Far'ah South is apparently made of a grey-veined gypsum (Jewish Museum New York 12–73.127, described by Lilyquist pers. comm. 1995), while similarly veined deposits have been reported (Ben-Dor 1945, 94). There are also rare instances of what appears to be brown gypsum (Kenyon 1965, Fig. 154.11; James and McGovern 1993, Fig. 113.4; Franken 1992, Fig. 4–18.15). This survey suggests that less pure varieties of gypsum were sometimes

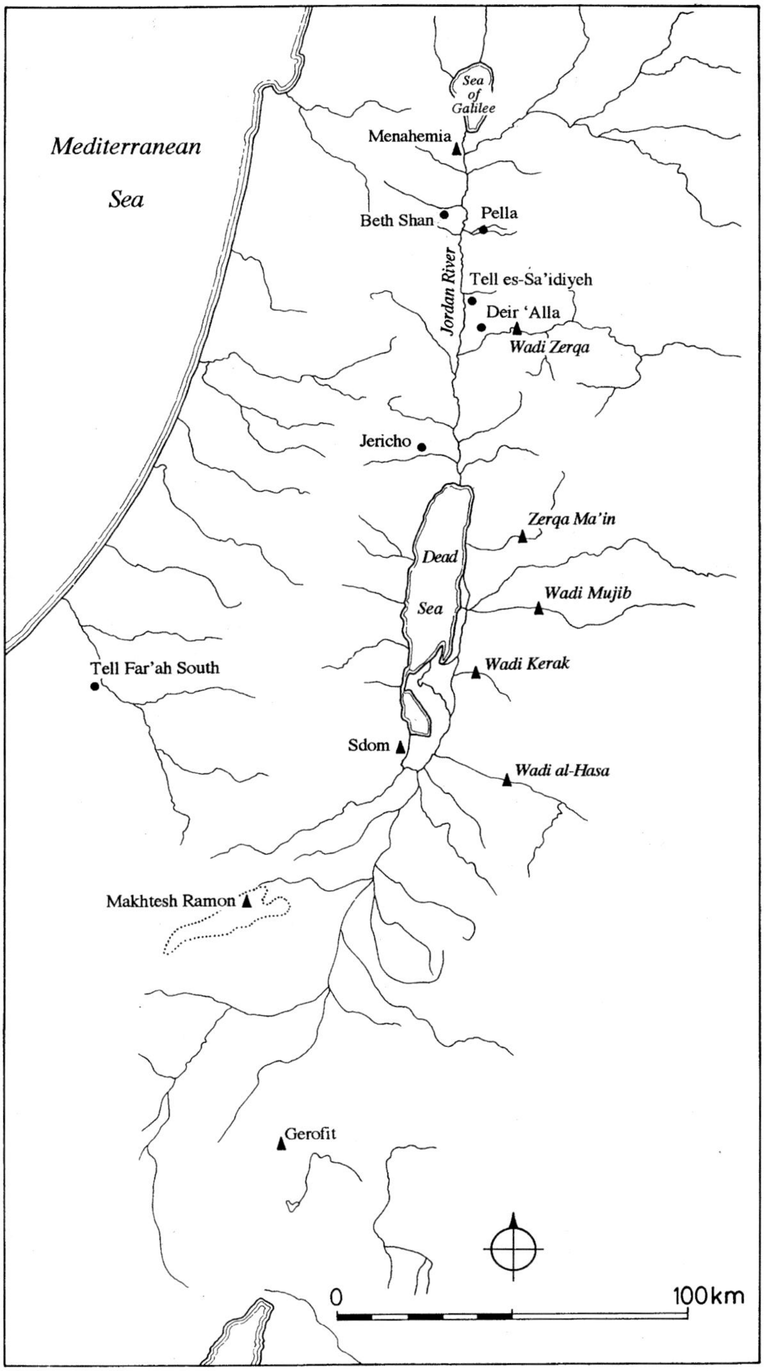

Figure 6.9: Map of gypsum sources in Israel and Jordan.

exploited in antiquity. This need not be surprising, as white and coloured gypsum often occur as layers in the same deposits, as is reported at Menahemia, Wadi Tafileh, Gesher and Grain Sabt (Ben-Dor 1945, 96; Kleinmann 1976 135–136).

The next step in determining the relationship between known sources of gypsum and Bronze Age workshops would be to compare representative samples from each of the source areas with the finished product. The main questions to be evaluated include whether these deposits match the physical characteristics of the vessels, whether the deposit was accessible to suspected producton centres, and whether there are indications that it had been worked in antiquity. The research that has been done on Palestinian sources has not yet been able to successfully address all these issues. A quantitative analysis was carried out on samples from the Menahemia quarries and a juglet from Beth Shan, which Ben-Dor claimed indicated a link between the two (Ben-Dor 1945, 96). However this link has been contested and further comparative data is necessary to determine whether the features isolated were peculiar to the Menahemia source (Ben-Dor 1945, 96; Kleinmann 1976, 136; Lilyquist 1996, 137 note 27). A more recent attempt to distinguish sources by comparison of thin-sections was also inconclusive (Kleinmann 1976).

The approach to this problem most likely of success may lie in alternative methods, such as isotopic analysis. Similar analyses have been successfully performed on Aegean sources of gypsum, where these were found to be of differing geological date (Gale *et al.* 1988). A similar diversity in the dates of deposits occurs in Palestine, where gypsum beds date variously to the Upper Miocene, Tertiary and Triassic periods (Picard 1954, 24–25; Braunfeld 1954, 75; Ionides and Blake 1939, 69, 74, 119). Such a study might indicate whether the proposed workshops at sites such as Beth Shan, Jericho, or Tell Far'ah South were obtaining their gypsum from different source areas. If this were the case, it would be a useful exercise to compare the workshop groupings suggested by stylistic analysis with the known origins of the raw material.

Although Palestinian workshops exploited regional gypsum resources, there was no attempt to widen their repertoire to include other local materials such as limestone. Like their preference for purer white varieties of gypsum, this single-mindedness probably reflects a continued desire to select materials with a strong visual similarity to imported Egyptian calcite vessels. These imports provided the chief inspiration for the development of this industry, and continued to influence the style and shapes of Palestinian versions throughout the course of the Bronze Age. The fact that this influence does not extend to manufacturing techniques implies that Egyptian influence came via trade goods, rather than through the arrival of Egyptian stone vessel craftsmen to the region.

The finished products were widely traded within the Jordan Valley, but appear less frequently beyond it. They do not appear to have taken much advantage of international trade networks, and it is probable that these vessels did not reach the levels of international popularity seen with the Egyptian products which they sought to emulate. They tend to represent small to medium-sized goods, aimed at middle ranking individuals rather than the ruling elite. Gypsum vessels therefore appear most frequently in funerary and domestic contexts, and only rarely in temple or palace areas.

GENERAL CONCLUSIONS

Stone vessel workshops took a number of different forms during the course of the Bronze Age. These case studies represent two separate and probably unrelated developments in the northern and southern Levant, although both regions sought to exploit a new found demand for luxury products in stone. At the top of the range are organisations such as that seen at Atchana, where the economic vitality and surplus generated by a successful state allows the production of individualised prestigious products using exotic materials obtained through long-distance trade networks. The vessels produced appear to be aimed at local elites, and may well have been sent further afield as part of greeting gifts used as a means of enhancing diplomatic ties (cf Moran 1992, EA 22, II.62–64, EA 25, II.60–61, EA 14, 30–45, 50–70). Such workshops probably existed at the palaces of other large cities of the period, such as Byblos and Ras Shamra. They would have operated in a cosmopolitan environment, where trade in foreign goods and artisans brought in fresh ideas and techniques. From such schools could emerge a more international style in arts and crafts, giving rise to a whole new range of Egyptianising and orientalising products, a trend that becomes particularly marked during the Late Bronze Age. The gypsum vessels of Palestine, however, represent a more readily accessible product. They were manufactured in larger quantities, were more standardised in form, and went on to be used as luxury containers for middle ranking individuals. These workshops made use of locally available resources, and the distribution of their products was more restricted in scope. These two studies also expose the unpredictability of the archaeological record. If Woolley had placed his palace sounding a few metres further away, he might never have uncovered the evidence for obsidian working at Atchana. If Kenyon had not discovered the extramural cemeteries at Jericho, the dominance of gypsum vessels at the site would never have been revealed. Similarly, if the unfinished vessels at Beth Shan had not been recovered, it might have been assumed that Beth Shan was a distribution centre rather than a production point for this material. We are bound entirely by what has been discovered, and can only use the evidence that is available to formulate ideas on the nature and extent of the stone vessel industry at any one given time. In order to minimalise the distorted picture that archaeology provides, it is best to use a holistic approach to the problem, combining data from every possible avenue of research. This study is only at a preliminary stage. The evidence from workshop debris and stylistic analysis combine to provide patterns which may be explained in a number of ways. The next step will be to add the results of sourcing studies to the mix, and see what additional light they can shed on the emerging picture.

ACKNOWLEDGEMENTS

The author would like to thank the following organisations for granting permission to study their material: the Pella Excavation Project, the Nicholson Museum, University of Sydney, the Australian Institute of Archaeology, Melbourne, the Jordanian Department of Antiquities, the Israel Antiquities Authority, and the Pennsylvania University Museum. Special thanks are also due to Phil Macumber, Christine Lilyquist and Patrick McGovern for helpful suggestions, and to Graham Reed for providing maps and object illustrations.

BIBLIOGRAPHY

Aston, B., Harrell, J. A., Shaw, I., 2000, Stone, in Nicholson, P. T., Shaw, I., *Ancient Egyptian materials and technology*, Cambridge University Press, Cambridge, 5–77.

Bender, F., 1974, *The geology of Jordan. Contributions to the regional geology of the earth*, Gebrüder Borntraeger, Berlin.

Ben-Dor, I., 1945, Palestinian alabaster vases, *Quarterly of the Department of Antiquities of Palestine, XI*, 93–112.

Bienkowski, P., 1986, *Jericho in the Late Bronze Age*, Aris and Phillips, Warminster.

Blake, G. S., 1930, *The mineral resources of Palestine and Transjordan*, Jerusalem.

Bourke, S. J., 1991, *The transition from the Middle to the Late Bronze Age in the northern Levant: The evidence from Tell Nebi Mend, Syria.* Unpublished doctoral dissertation, University College London.

Bourke, S. J. *et al.*, 1994, Preliminary report on the University of Sydney's fourteenth season of excavations at Pella (Tabaqat Fahl) in 1992, *Annual of the Department of Antiquities of Jordan*, XXXVIII, 81–126.

Bourke, S. J., Sparks, R. T., 1995, The DAJ excavations at Pella in Jordan in 1963/64, in Bourke, S. J., Descoeudres, J-P., *Trade, contact and the movement of peoples in the eastern Mediterranean. Studies in honour of J. Basil Hennessy*, Mediterranean Archaeology Supplement 3, Sydney, 149–167.

Braunfeld, A. S., 1954, Gypsum. I: general, *Israel Economic Forum*, 6.3, 74–75.

Breasted, J. H., 1906a, *Ancient records of Egypt volume 2*, University of Chicago Press, Chicago.

Breasted, J. H., 1906b, *Ancient records of Egypt volume 4*, University of Chicago Press, Chicago.

Brennan, P. V., 1996, Obsidian in the Bayburt-Erzurum area, eastern Anatolia, *in* Bunnens, G., *Cultural interaction in the ancient Near East*, Peeters Press, Louvain, 27–36.

Canby, J. V., 1976, The sculptors of the Hittite capital, *Oriens Antiquus*, 15, 33–42.

Cann, J. R. *et al.*, 1969, Obsidian analysis and the obsidian trade, in Brothwell, D., Higgs, E., *Science in archaeology*, Thames and Hudson, London, 578–591.

Cann, J. R., Renfrew, C., 1972, The characterisation of obsidian and its application to the Mediterranean region, *Proceedings of the Prehistoric Society*, 30, 111–113.

Caubet, A., 1991, Répertoire de la vaisselle de pierre, Ougarit 1929–1988, in Yon, M., *Arts et industries de la pierre*, éditions recherche sur les civilisations, Paris, 205–264.

FitzGerald, G. M., 1931, *Beth Shan excavations 1921–23. The Arab and Byzantine levels*, University of Pennsylvania Press, Philadelphia.

FitzGerald, G. M., 1932, Excavations at Beth-Shan in 1931, *Palestine Exploration Fund Quarterly Statement*, 138–148.

FitzGerald, G. M., 1934, "Excavations at Beth-Shan in 1933", *Palestine Exploration Fund Quarterly Statement*, 128–134.

Franken, H. J., 1992, *Excavations at Deir 'Alla, the LBA sanctuary*, Peeters Press, Louvain.

Gachet, J., 1987, Objets en os et en ivoire, in Yon, M., *Ras Shamra-Ougarit III: le centre de la ville*, éditions recherche sur les civilisations, Paris, 249–272.

Gale, N. H. *et al.*, 1988, The source of Mycenaean gypsum, *Journal of Archaeological Science*, 15.1, 57–72.

Garstang, J., 1933, Jericho: city and necropolis, *Liverpool Annals of Archaeology and Anthropology*, XX, 3–42.

Gratuze, B. *et al.*, 1993, Non-destructive analysis of obsidian artefacts using nuclear techniques: investigation of provenance of Near Eastern artifacts, *Archeometry*, 35.1, 11–21.

Hester, T. R. and Heizer, R. F., 1981, *Making stone vases: ethnoarchaeological studies at an alabaster workshop in upper Egypt*, Occasional Papers on the Near East, volume 1.2, Udena Publications, Malibu.

Ionides, M. G. and Blake, G. S., 1939, *Report on the water resources of Transjordan and their development. Incorporating a report on geology, soils and minerals and hydro-geological correlations*, Government of Transjordan, London.

James, F., 1966, *The Iron Age at Beth Shan: A study of levels VI–IV*, University Museum, University of Pennsylvania, Philadelphia.

James, F. W. and McGovern, P. E., 1993, *The Late Bronze Egyptian garrison at Beth Shan: a study of levels VII and VIII*, University Museum, University of Pennsylvania, Philadelphia.

Kenyon, K. M., 1960, *Excavations at Jericho I*, British School of Archaeology in Jerusalem, London.

Kenyon, K. M., 1965, *Excavations at Jericho II*, British School of Archaeology in Jerusalem, London.

Kenyon, K. M. and Holland, T. A., 1983, *Excavations at Jericho V*, British School of Archaeology in Jerusalem, London.

Kleinmann, B., 1976, "Mineralogical investigation of the gypsum vessels" *in* C. Clamer, *Late Bronze Age alabaster vessels found in Palestinian contexts with an emphasis on calcite and gypsum tazze.* Unpublished masters dissertation, Hebrew University of Jerusalem.

Kohl, P. L., 1975, *Seeds of upheaval.* Unpublished doctoral dissertation, Harvard University.

Liebowitz, H., 1986, Late Bronze II Ivory Work in Palestine: Evidence of a Cultural Highpoint, *Bulletin of the American School of Oriental Research* 265, 3–24.

Lilyquist, C., 1995, *Egyptian stone vessels: Khian Through Tuthmosis IV*, Metropolitan Museum of Art, New York.

Lilyquist, C., 1996, Stone vessels at Kamid el-Loz, Lebanon: Egyptian, Egyptianizing, or non-Egyptian? A question at sites from the Sudan to Iraq to the Greek mainland, in Hachmann, R., *Kamid el-Loz 16: 'Schatzhaus' – Studien*, R. Habelt, Bonn, 133–173.

Lucas, A.and Harris, J. R., 1962, *Ancient Egyptian materials and industries*, E. Arnold, London.

Mazar, A., 1985, *Excavations at Tell Qasile, Part 2*, Institute of Archaeology, Hebrew University of Jerusalem, Jerusalem.

Mazar, A., 1990a, *The Archaeology of the land of the bible, 10,000–586 B.C.E*, Doubleday, New York.

Mazar, A., 1990b,The excavations at Tel Beth-Shean in 1989–1990, *Eretz-Israel, 21*, 197–211 (Hebrew section), *108 (English summary).

Mazar, A., 1993, The excavations at Tel Beth-Shean in 1989–1990, *in* Biran, A., Aviram, J., *Biblical archaeology today*, 1990, Israel Exploration Society, Jerusalem, 606–619.

McClellan, T. L., 1989, The chronology and ceramic assemblages of Alalakh, in A. Leonard, A. and Williams, B. B., *Essays in ancient civilization presented to Helene J. Kantor*, Oriental Institute of the University of Chicago, Chicago.

McNicoll, A. W. *et al.*, 1982, *Pella in Jordan 1*, Australian National Gallery, Canberra.

McNicoll, A. W. *et al.*, 1992, *Pella in Jordan 2*, Mediterranean Archaeology, Sydney.

Mellink, M. J., 1988, Archaeology in Anatolia, *American Journal of Archaeology*, 92.1, 101–132.

Montet, P., 1928, *Byblos et l'Égypte*, Geunther, Paris.

Moran, W. L., 1992, *Tell el-Amarna tablets. The Amarna letters*, Johns Hopkins University Press, Baltimore.

Na'aman, N., 1976, A new look at the chronology of Alalakh VII, *Anatolian Studies*, XXVI, 129–144.

Oren, E., 1973, *The northern cemetery at Beth Shan*, Brill, Leiden.

Özten, A., 1979, Two stone plates from the Sarikaya palace at Acemhöyük, *Belleten*, XLIII.170, 385–388.

Petrie, W. M. F., 1934, *Ancient Gaza IV*, British School of Archaeology in Egypt, London.

Picard, L., 1954, History of mineral research in Israel, *Israel Economic Forum*, 6.3, 10–38.

Pinnock, F., 1985, About the Trade of Early Syrian Ebla, *MARI*, 4, 85–92.

Renfrew, C. *et al.*, 1964, Obsidian in the Aegean, *Annual of the British School at Athens*, 60, 225–247.

Rosen, S. A., 1994, *Map of Makhtesh Ramon (204)*, Archaeological Survey of Israel, Jerusalem.

Rothenberg, B. and Bachmann, H.-G., 1988, *The Egyptian mining temple at Timna*, Institute for Archaeo-Metallurgical Studies, London.

Sasson, J. M., 1968, Instances of mobility among Mari artisans, *Bulletin of the American Schools of Oriental Research*, 190, 46–54.

Smith, R. H., 1973, *Pella of the decapolis I*, College of Wooster, Wooster.

Sparks, R. T., 1991, A series of Middle Bronze Age bowls with ram's-head handles from the Jordan Valley, *Mediterranean Archaeology*, 4, 45–54.

Sparks, R. T., 1998, *Stone vessels in the Levant during the second millennium B.C.* Unpublished doctoral dissertation, University of Sydney.

Sparks, R. T., 2001, "The stone vessel repertoire of Pella during the Middle and Late Bronze Ages: Egyptian influence and local developments", in Bisheh, G., *Studies in the History and Archaeology of Jordan VII*, Department of Antiquities, Amman.

von der Osten, H. H., 1956, *Svenska Syrienexpeditionen 1952–1953. Vol. I. Die Grabung von Tell es-Salihiyeh*, CWK Gleerup, Lund.

Virolleaud, P., 1922, Découverte a Byblos d'un hypogée de la douziéme dynastie égyptienne, *Syria*, III, 235–290.

Warren, P., 1969, *Minoan Stone Vases*, Cambridge University Press, Cambridge.

Warren, P., 1979, The stone vessels from the Bronze Age settlement at Akrotiri, Thera, *Archaiologike Ephemeris*, 82–113.

Wolff, S. R., 1994, Archaeology in Israel, *American Journal of Archaeology*, 98.3, 481–519.

Woolley, C. L., 1955, *Alalakh - an account of the excavations at Tell Atchana in the Hatay, 1937–1949*, The Society of Antiquaries, London.

Wright, G. E., 1941, Review: the four canaanite temples of Beth Shan, *American Journal of Archaeology*, XLV, 483–485.

Wulff, H. E., 1966, *The traditional crafts of Persia*, M.I.T. Press, Cambridge.

Würzburger, U. S., 1972, State of Israel, natural resources, *Encyclopedia Judaica*, Jerusalem, 798–814.

Yadin, Y. and Geva, S., 1986, *Investigations at Beth Shean – the Early Iron Age strata*, Qedem 23, Institute of Archaeology, Hebrew University of Jerusalem, Jerusalem.

Zaccagnini, C., 1983, Patterns of mobility among ancient Near Eastern craftsmen, *Journal of the Near Eastern Society*, 42.4, 245–364.

Zarins, J., 1989, Ancient Egypt and the Red Sea trade: the case for obsidian in the predynastic and archaic periods, in Leonard, A. and Williams, B. B., *Essays in Ancient Civilization Presented to Helene J. Kantor*, Oriental Institute of the University of Chicago, Chicago, 339–368.

Chapter 7

The Provenance of Canaanite Amphorae found at Memphis and Amarna in the New Kingdom

Janine Bourriau, Laurence Smith and Margaret Serpico

Abstract

The focus of the paper is one element, that of fabric classification leading to provenance, within the collaborative Canaanite Amphorae project which covers all aspects of the transport amphorae from Syria-Palestine found in Egypt in the New Kingdom. Our data comes from current excavations by the Egypt Exploration Society at Memphis and Amarna supplemented by comparative samples from Israel, from sites excavated under the aegis of the Israel Antiquities Authority. Ceramic petrography has been used to refine the visual fabric classification; to create a concordance between sites and to suggest provenances for the resulting five fabric groups. Some use has been made of a pre-existing body of NAA data in the grouping of fabrics but further analysis by ICP/AES will be carried out in summer 2000 to test our results so far. Residue analysis carried out by Ben Stern and Carl Heron has confirmed the association of three of the fabric groups with particular commodities.

INTRODUCTION

This study is part of the Canaanite Amphorae Project, which examines the contexts, fabrics, contents and inscriptions on jars imported into New Kingdom Egypt. The project was begun in 1996 by Janine Bourriau and Margaret Serpico and arose directly out of their work on the Egypt Exploration Society's excavations at Memphis (Kom Rabi'a) and Amarna. It was recognized that these wheel thrown storage jars (Amiran 1969, 138–142; Grace 1956, 80–109; Parr 1973, 173–81), found throughout the Aegean, eastern Mediterranean, Egypt and Nubia (Serpico 1999), could offer important insights into trade networks and ancient economies. Research by Margaret Serpico into the New kingdom resin trade had already indicated that links could be made between specific fabrics and contents (Serpico 1996).

A broad line of shape development has already been established for the jars (Leonard 1996) but only site specific and largely unpublished investigations of fabrics have been carried out. With these points in mind, the aims of the project are sixfold:

1. to establish a visual classification system for the fabrics used for the jars, thus enabling their identification to be secured wherever they occur;
2. to determine their regions of manufacture;
3. to identify the commodities they carried, whether in primary or secondary use;
4. to relate products to specific fabrics;
5. to reconstruct trade networks by linking areas of amphorae manufacture to industrial locations supplying the commodities carried in them;
6. to relate these data to the contexts in which the jars are found in Egypt.

This paper's focus is on the fabric classification, which uses two levels of investigation: the basic one of visual description of fresh sherd sections examined under a microscope at 25x magnification, and thin section analysis of a standard series of sherds. The results obtained so far in relation to the question of the provenance of the jars will be presented. Chemical analysis is currently under way using ICP-AES techniques, but will not be complete in time for incorporation into this account.

ARCHAEOLOGICAL BACKGROUND

The study is based on amphorae from current excavations by the Egypt Exploration Society at Memphis (Bourriau 1990) and Amarna (Nicholson and Rose 1985, 139–40; Serpico 1996). These sites complement each other: Memphis was the capital of Egypt for most of the New Kingdom, Amarna only during the reign of Akhenaten, c. 1360–1330 B.C. Memphis provides a long stratified sequence from domestic housing spanning the whole of the New Kingdom, whereas Amarna has evidence from a variety of religious, royal and domestic contexts. This means that we can study the distribution of the fabrics both chronologically and spatially, utilising data from both sites. While not yet complete, this part of the work already shows, to take an example from the Memphis data, that the beginning of the 18th Dynasty marks a profound change: all except one fabric group (2) appears then for the first time. The relationship of Group 2 fabrics to the Middle Bronze Age jars present in the immediately preceding Middle Kingdom/Second Intermediate levels is evident but not yet clarified. When this finding is linked to the provenance we can now suggest for the fabric groups, it becomes clear that the sources for all the "new" fabrics lie in the northern part of the region. In modern terms this comprises northern Israel, Lebanon and Syria.

One of this study's starting points was the recognition that fabric had to be the primary criterion for classification and shape secondary. Fabric is defined as the potter's raw material plus any ingredients which may have been added or changes made in processing the material; plus alterations to colour, texture and hardness brought about by firing. At Memphis about 25% of the sherds found were shape diagnostic so that if only these had been recorded, a wholly distorted picture of the distribution would have arisen. For example, the presence of all the fabric groups from the beginning of the 18th Dynasty, albeit in very small quantities, would not have been evident. Our material also shows that shapes of rims, bases and handles are not exclusive to particular fabrics. Moreover, shapes do not change as quickly as in highly decorated fine wares, for example, for which there is evidence of trading in their own right.

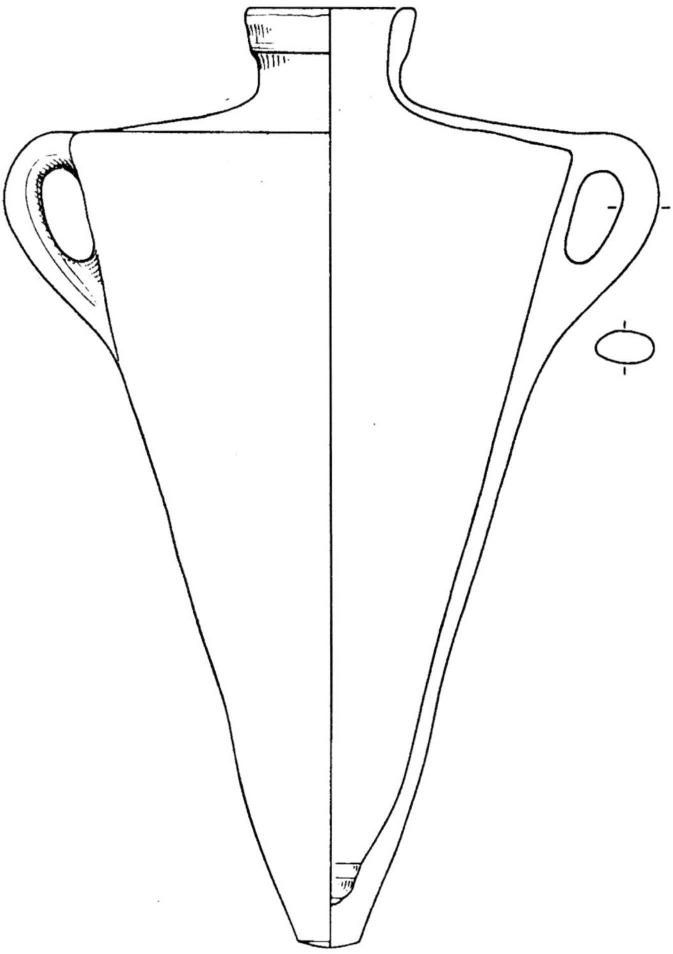

Figure 7.1: Canaanite Jar from Saqqara, Necropolis of Memphis (Tomb of Horemheb). By permission of Egypt Exploration Society.

Finally, fabric, as well as being the most significant property in distinguishing one jar from another, leads us directly to the question of where the jar was made.

METHODOLOGY

The method used for the basic description of fabric has been described in Bourriau and Nicholson (1992) and in more detail in Nordström and Bourriau (1993). The method is intended to provide a consistent and sufficiently high level of definition to relate both to the image seen through a 10x hand lens and to the image seen in thin section. Only a fraction of any assemblage can be thin sectioned or analysed chemically and the archaeologist must have a sound basis for the selection of samples for these analyses and then be able to feed back the information from the analyses to the whole assemblage. Our aim is to discover properties which are visible macroscopically yet correspond to or correlate with criteria identified as significant by petrological analysis.

The procedure adopted for the petrological study, which was carried out by Laurence Smith, was that outlined in Bourriau and Nicholson (1992) and Bourriau, Smith and Nicholson (2000). The first step was to substantiate, and where necessary, refine the existing classification, based on descriptions of sherd sections at 25x, and to confirm the concordance between the Memphis and Amarna fabric series. This 'Egyptian' series was then compared with samples from excavations in Israel. Bulk chemical analysis using NAA had already been undertaken on what was a largely different sample group of the same fabrics but together with the petrographic data it suggested the provisional grouping of fabrics set out below.

FABRIC GROUPS

The characteristic properties of each group are summarised, followed by the description of a representative specimen of each subgroup. The macroscopic description of the sherd break is given first, followed by the description of the specimen in thin-section. Constituents seen in the sherd break and in thin- section are, as far as possible, presented in the same order so that a direct comparison can be made between the two types of information. The quantitative data (mean percentages of inclusions) are also shown plotted as a barchart. Each specimen is illustrated by a photograph of the sherd break and of the thin section.

Abbreviations used in text

K-feldspar = potassium feldspar; IRF = igneous rock fragments; VRF = volcanic rock fragments; OAN = inclusions probably of organic origin, other than plant temper, mainly bone fragments. For abbreviations used in the bar charts (Figs. 7.2–7.7) see Appendix.

GROUP 1 FABRICS

Memphis P30, P11; Amarna III:10, IV:5; also identified from Tell Abu Hawam, Affule, Tel Dan, Megiddo, Tel Miqne, Tel Nami, Tel Ruian Survey, Safed, Wadi 'Aghara.

Sherd section description

High content of fine quartz, limestone (decomposed) and microfossils.
Dense, crumbly.
Colour (specimen 74): zones in section: outer reddish yellow 5YR6/6, inner light brown 7.5YR6/4. Surface pinkish white 7.5YR 8/2.

Thin section description

Colour (specimen 74) PPL: Mid brown core, mid reddish brown exterior zone; XPL: Dark brown core, mid orange-brown exterior zone.

Sorting: Moderate to well sorted.
Shape range of most inclusions: Quartz/Feldspars: 1.2–1.4, 2.2–2.4; Limestone: 1.3–1.5, 2.3–2.5.
Size range of most inclusions: Quartz/Feldspars: very fine-medium sand; rarely coarse sand.
Limestone: fine-coarse sand; rarely very coarse sand and granule.
Major inclusions: Abundant quartz and limestone, moderate percentage microfossils. Small amounts (generally under 2%) K-feldspars, chalk, marl clay pellets, shell fragments, red iron oxides, red clay pellets, volcanic rock fragments, mainly basalt, and replacement chert fragments.

SUB-GROUP 1.1.1

Abundant limestone and quartz, with a large difference in proportions, illustrated by specimen 33 (Amarna).

Sherd section description (Plate 7.1)

Inclusions: 'sand' (quartz and feldspar): fine [3]; limestone: fine [3], medium [1], coarse [1]; soft red-brown particles: fine [2]; black-rock particles: fine [1]; mica: fine [1]; other: clay pellets: medium [1].
Sorting: Good.
Porosity: Dense.
Hardness: Medium with decomposed limestone and elongated pores.
Vessel wall: 11 mm.
Section colour: zones light red (2.5YR 6/6), wide (8.0mm) grey core.
Outer surface: Pale yellow (5Y7/3).

Thin section description (Fig. 7.2A, Plate 7.22)

Colour PPL: Dark brown core and interior, mid brown exterior zone.
Colour XPL: Very dark brown core and interior, dark reddish-brown exterior zone.
Frequency of inclusions: mean 43.17%.
Sorting: Well sorted.
Approximate Shape Range: Quartz and feldspars 1.1–1.4 and 2.1–2.4, Limestone 1.2–1.5 and 2.2–2.5.
Main Additional Inclusions Present: Amphibole, pyroxene, calcite, shell, rutile, zircon, VRF.
Comments: K-feldspar includes orthoclase, microcline and perthite; pyroxenes include augite and aegirine-augite.

SUB-GROUP 1.1.2

Lower limestone content, abundant quartz with a small difference in proportions, illustrated by specimen 74 (Memphis).

Sherd section description (Plate 7.2)

Inclusions: 'sand' (quartz and feldspar): fine [2]; plant remains: fine [1]; limestone: fine [2], medium [1], coarse [1]; grey-white rock particles: fine [1], medium [2]; soft red-brown particles: fine [1], medium [1]; black-rock particles: fine [1]; shell microfossils: medium [1].
Sorting: Good.
Porosity: Dense.
Hardness: Medium with elongated pores.
Vessel wall: 6.0 mm.
Section colour: zones reddish yellow (5YR6/6), core light brown (7.5YR6/4).
Outer surface: Pinkish white (7.5YR8/2).

Thin section description (Fig. 7.2B, Plate 7.23)

Colour PPL: Mid brown core, mid reddish brown exterior zones.
Colour XPL: Dark brown core, mid orange-brown exterior zones.
Frequency of inclusions: mean 40.33%.
Sorting: Well sorted.
Approximate Shape Range: Quartz and feldspars 1.1–1.5 and 2.1–2.5, Limestone 1.2–1.5 and 2.2–2.5.
Main Additional Inclusions Present: Plagioclase, K-feldspars, muscovite mica, amphibole, pyroxene, opaques, calcite, shell, chalcedony, rutile, zircon, VRF, RF, OAN and voids from burnt-out plant remains.
Comments: K-feldspar includes orthoclase, microcline and perthite; amphiboles include basaltic hornblende and pargasitic hornblende; pyroxenes include augite.

SUB-GROUP 1.2.1

Low quartz and limestone content, with limestone more abundant than quartz, illustrated by specimen 60 (Memphis).

Sherd section description (Plate 7.3)

Inclusions: 'sand' (quartz and feldspar): fine [1], medium [2]; limestone: fine [2]; grey-white rock particles: fine [1], medium [2], coarse [1]; soft red-brown particles: fine [1]; shell microfossils: fine [1]; other: clay pellets: very coarse [1].
Sorting: Good.
Porosity: Medium.
Hardness: Medium with decomposed limestone particles.
Vessel wall: 4.5 mm.
Section colour: zones light red (2.5YR6/8), core pink (5YR7/4).
Outer surface: Light red like zones.

Thin Section description (Fig. 7.2C, Plate 7.24)

Colour PPL: Light brown core, light orange-brown exterior zones.
Colour XPL: Yellow-brown core, orange-brown exterior zones.

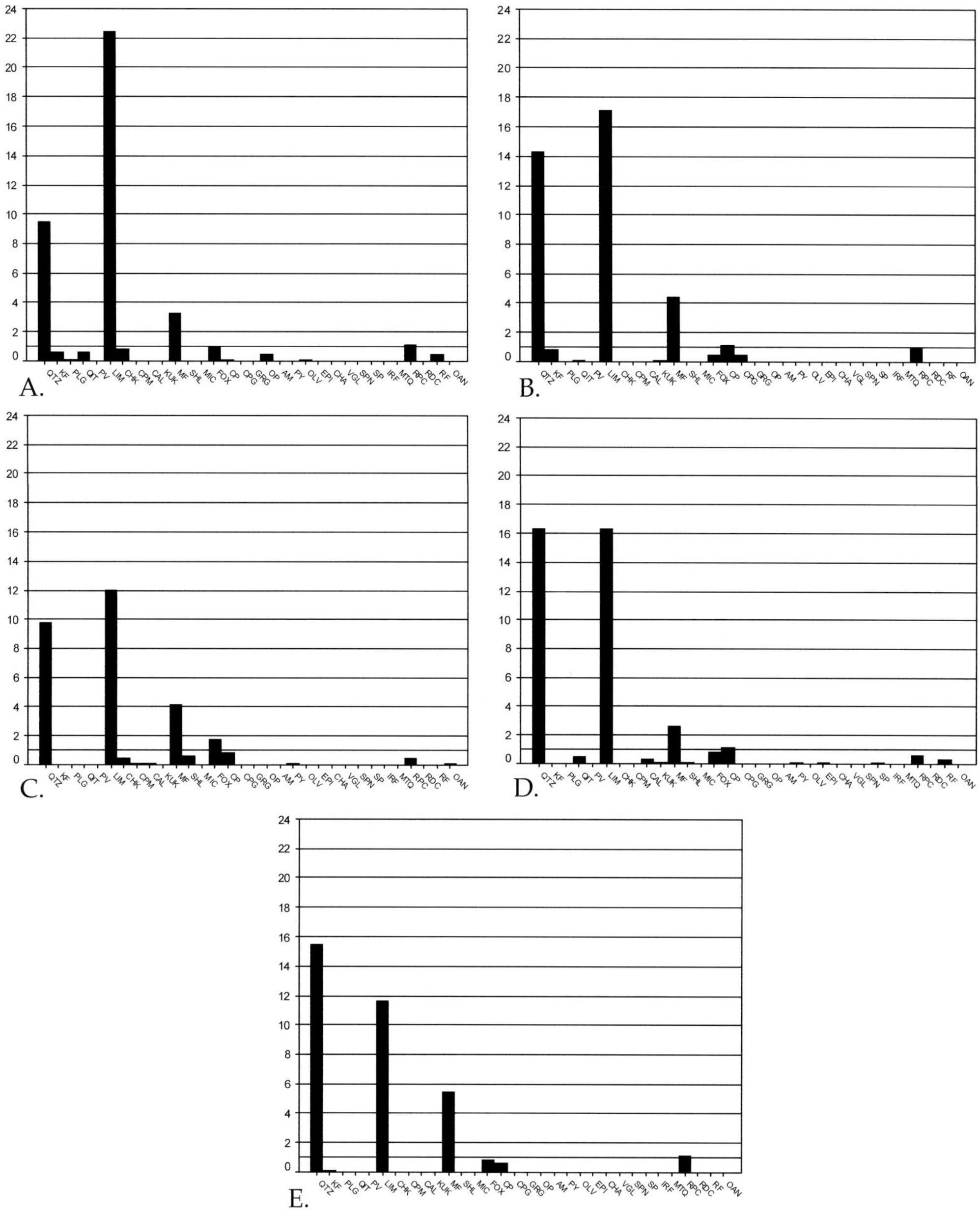

Figure 7.2: Mean percent of constituents for sherds of Group 1; A. Sherd 33, Fabric Sub-Group 1.1.1; B. Sherd 74, Fabric Sub-Group 1.1.2; C. Sherd 60, Fabric Sub-Group 1.2.1; D. Sherd 166, Fabric Sub-Group 1.2.2; E. Sherd 76, Fabric Sub-Group 1.2.3.

Plate 7.1 Sub Group 1.1.1

Plate 7.2 Sub Group 1.1.2

Plate 7.3 Sub Group 1.2.1

Plate 7.4 Sub Group 1.2.2

Plate 7.5 Sub Group 1.2.3

Plate 7.6 Sub Group 2.1.1

Plate 7.7 Sub Group 2.1.2

Plate 7.8 Sub Group 2.2.1

Plate 7.9 Sub Group 2.2.2

Plate 7.10 Sub Group 3.1.1

Plate 7.11 Sub Group 3.1.2

Plate 7.12 Sub Group 4.1.1

Plate 7.13 Sub Group 4.1.2

Plate 7.14 Sub Group 4.2.1

Plate 7.15 Sub Group 4.2.2

Plate 7.16 Sub Group 5.1.1

Plate 7.17 Sub Group 5.1.2

Plate 7.18 Sub Group 5.2.1

Plate 7.19 Sub Group 5.2.2

Plate 7.20 Rhas Shamra 210

Plate 7.21 Ras Shamra 212

Frequency of Inclusions: mean 31.0%.
Sorting: Well sorted.
Approximate Shape Range: Quartz and feldspars 1.1–1.5 and 2.1–2.5, Limestone 1.2–1.5 and 2.2–2.5.
Main Additional Inclusions Present: Plagioclase, K-feldspars, biotite and muscovite micas, amphibole, opaque minerals, olivine, chalcedony, IRF, quartzite, rutile, epidote, chlorite, ironstone fragment, kurkar fragments, claystone, OAN, RF.
Comments: K-feldspar includes microcline and perthite; amphiboles comprise clinoamphiboles, including basaltic hornblende and pargasitic hornblende; IRF include mainly basalt fragments and one inclusion textually on the border between basalt and dolerite.

SUB-GROUP 1.2.2

Low quartz and limestone content, with the amount of limestone similar to that of quartz, illustrated by specimen 166 (Amarna).

Sherd section description (Plate 7.4)

Inclusions: 'sand' (quartz and feldspar): very fine [1], fine [2], medium [1], coarse [1]; plant remains: fine [1]; limestone: fine [1], coarse [1]; soft red-brown particles: very fine [1], very coarse [1]; red-brown rock particles: fine [1]; black-rock particles: fine [1]; other: red-brown clay pellets: coarse [1].
Sorting: good.
Porosity: dense.
Hardness: medium with decomposed limestone particles.
Vessel wall: 11.0 mm.
Section colour: zones (3.0 mm wide) light reddish brown (5YR6/4), core grey.
Outer surface: as zones.
Inner surface: light reddish brown (5YR6/3).

Thin section description (Fig. 7.2D, Plate 7.25)

Colour PPL: Mid yellowish-brown core, light orange-brown exterior zones.
Colour XPL: Dark yellowish-brown core, mid orange-brown exterior zones.
Frequency of Inclusions: mean 42.17%.
Sorting: Well sorted.
Approximate Shape Range: Quartz and feldspars 1.1–1.5 and 2.1–2.5, Limestone 1.2–1.5 and 2.2–2.5.
Main Additional Inclusions Present: Plagioclase, K-feldspars, biotite and muscovite micas, amphibole, opaque minerals, chalcedony, VRF, rutile, zircon, clay pellets of the same material as the clay matrix, RF, OAN and voids from burnt-out plant remains.
Comments: K-feldspar includes microcline; amphiboles include pargasitic hornblende; pyroxenes include augite; IRF include mainly basalt fragments.

SUB-GROUP 1.2.3

Low quartz and limestone content, with quartz more abundant than limestone, illustrated by specimen 76 (Memphis).

Sherd section description (Plate 7.5)

Inclusions: 'sand' (quartz and feldspar): fine [3], medium [1]; limestone: fine [2], medium [1], coarse [1]; soft red-brown particles: fine [1]; black-rock particles: fine [1]; shell microfossils: fine [1].
Sorting: good.
Porosity: dense.
Hardness: hard with elongated pores.
Vessel wall: 8.0 mm.
Section colour: zones reddish yellow (7.5YR7/4), core light red (2.5YR6/8).
Outer surface: very pale brown (10YR8/4).

Thin section description (Fig. 7.2E, Plate 7.26)

Colour PPL: Light orange-brown core, light yellowish-brown exterior zones.
Colour XPL: Orange -brown core, yellow-brown exterior zones.
Frequency of Inclusions: mean 35.5%.
Sorting: Well sorted.
Approximate Shape Range: Quartz and feldspars 1.1–1.5 and 2.2–2.5, Limestone 1.2–1.5 and 2.2–2.5.
Main Additional Inclusions Present: Plagioclase, muscovite mica, pyroxene, opaque minerals, calcite, shell, chalcedony, VRF, quartzite, rutile, zircon, kurkar fragments.
Comments: K-feldspar includes orthoclase and microcline; pyroxenes include augite; amphiboles include hornblende; VRF include mainly basalt fragments.

GROUP 2 FABRICS

Memphis P31, P70; Amarna IV:6, V:5; also identified from Tell Abu Hawam, Affule, Aphek, Beth Dagan, Beth Shamesh, Pal Machim, Tel Nami.

Sherd section description

Dominated by high fine to very coarse limestone content (often decomposed); also fine to medium quartz, clay pellets, microfossils and elongate shell fragments and conspicuous chalk up to granule size. Occasional fine plant remains.
Medium porosity and hardness.
Colour (specimen 80): zones in section: outer reddish yellow 5YR6/6, core grey. Surface as outer zone.

Thin section description

Colour (specimen 80) PPL: Light brown throughout; XPL: Mid-brown core, orange-brown exterior zones.
Sorting: Poor to Moderate.

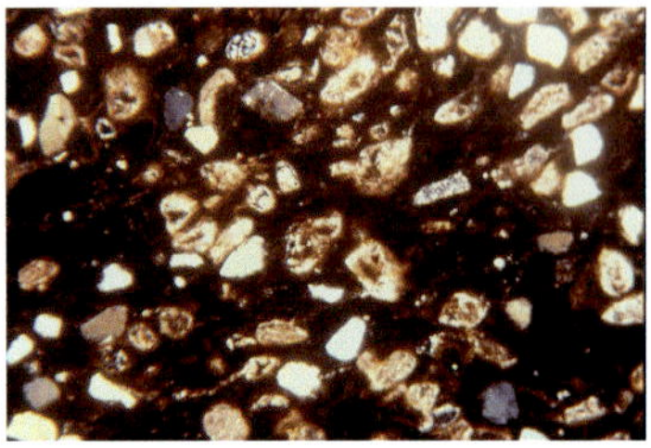

Plate 7.22 Sub Group 1.1.1

Plate 7.23 Sub Group 1.1.2

Plate 7.24 Sub Group 1.2.1

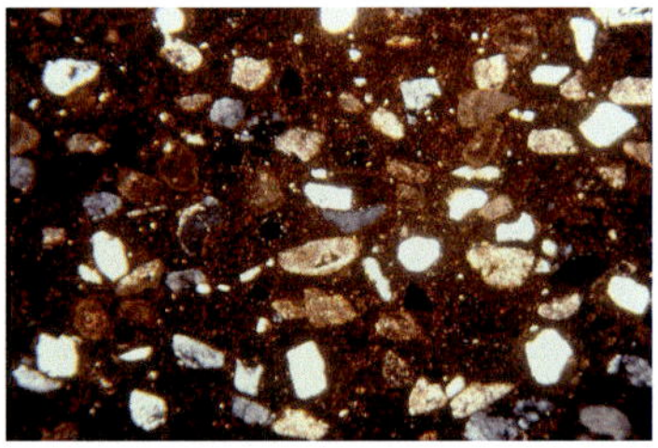

Plate 7.25 Sub Group 1.2.2

Plate 7.26 Sub Group 1.2.3

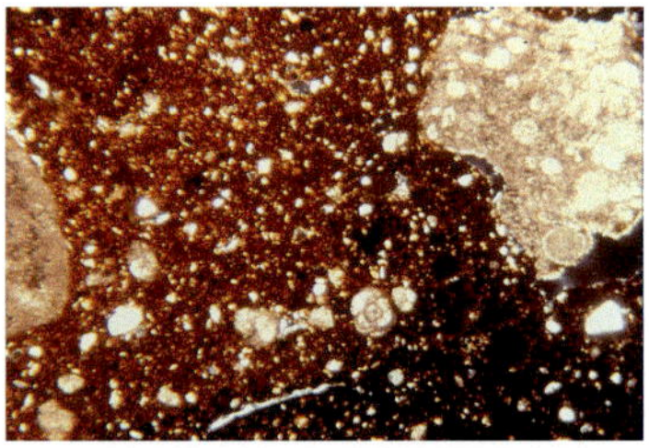

Plate 7.27 Sub Group 2.1.1

Plate 7.28 Sub Group 2.1.2

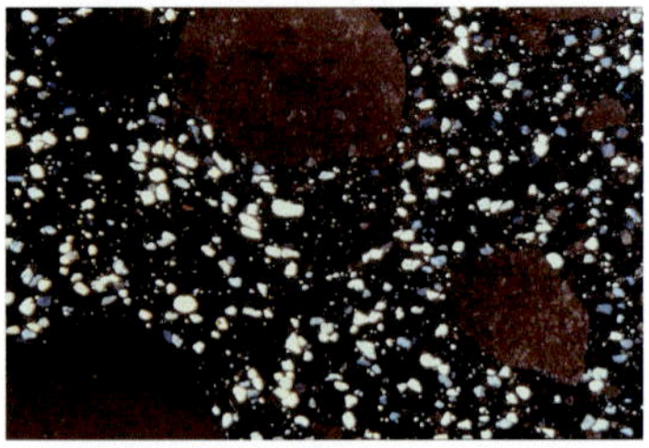

Plate 7.29 Sub Group 2.2.1

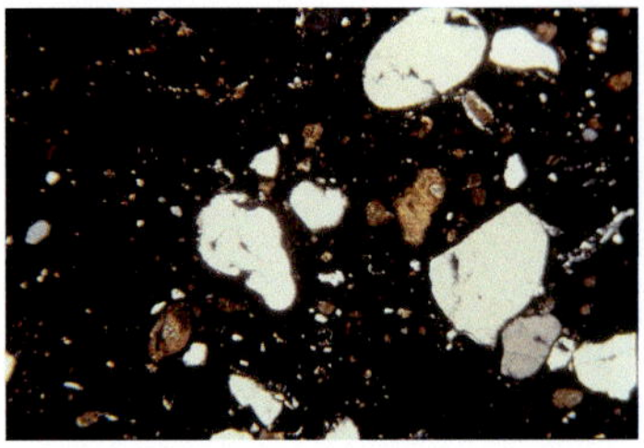

Plate 7.30 Sub Group 2.2.2

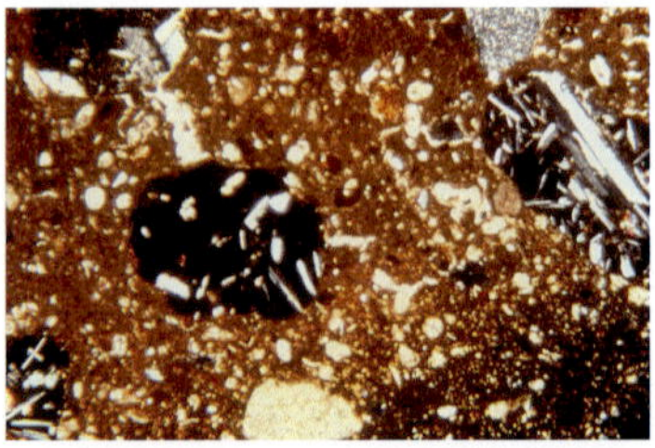

Plate 7.31 Sub Group 3.1.1

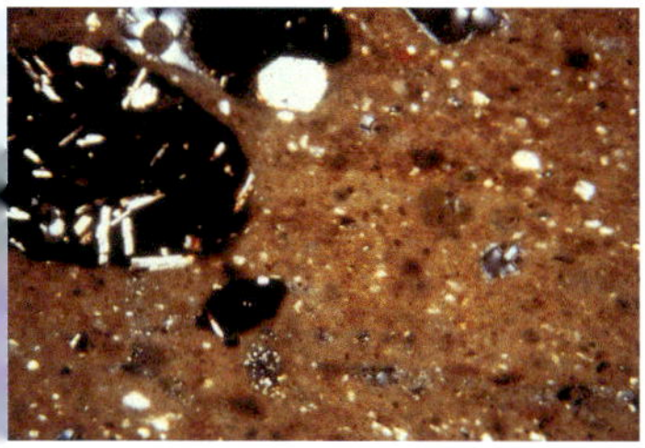

Plate 7.32 Sub Group 3.1.2

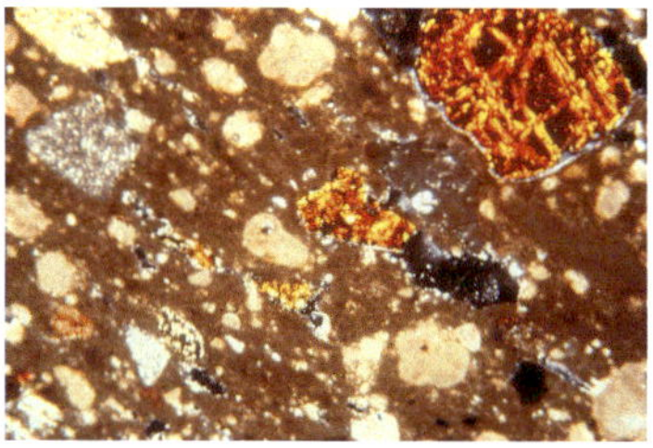

Plate 7.33 Sub Group 4.1.1

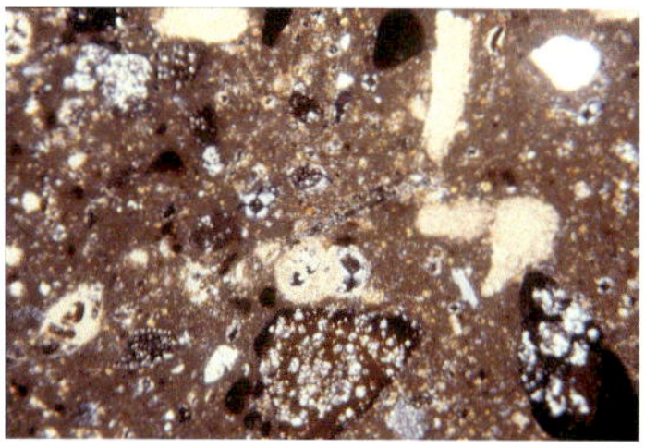

Plate 7.34 Sub Group 4.1.2

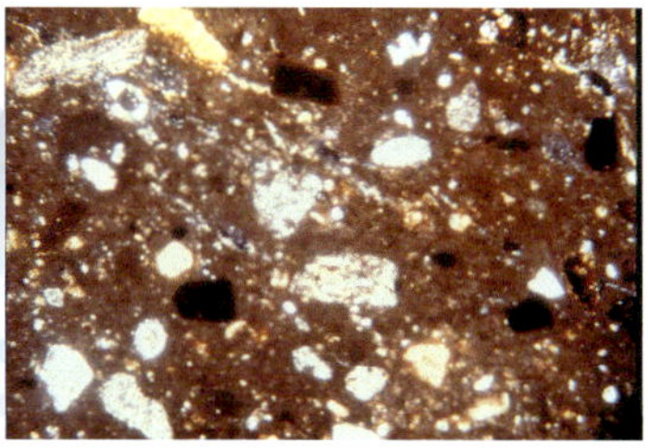

Plate 7.35 Sub Group 4.2.1

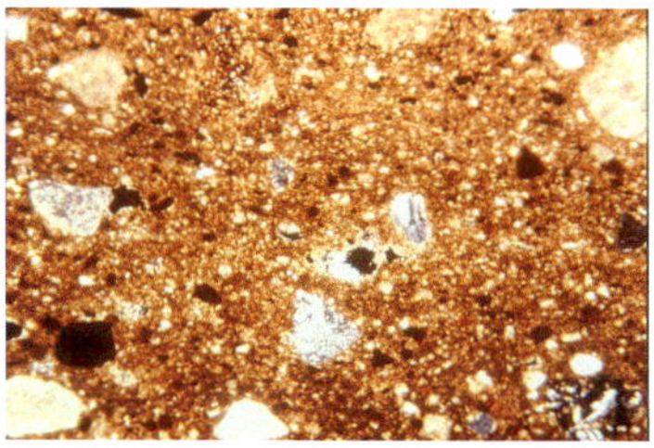

Plate 7.36 Sub Group 4.2.2

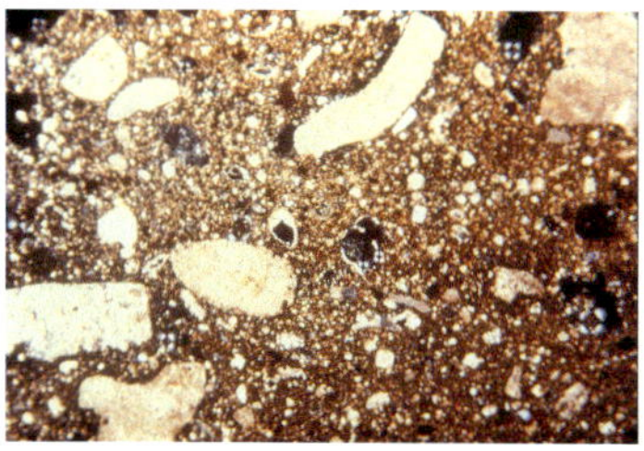

Plate 7.37 Sub Group 5.1.1

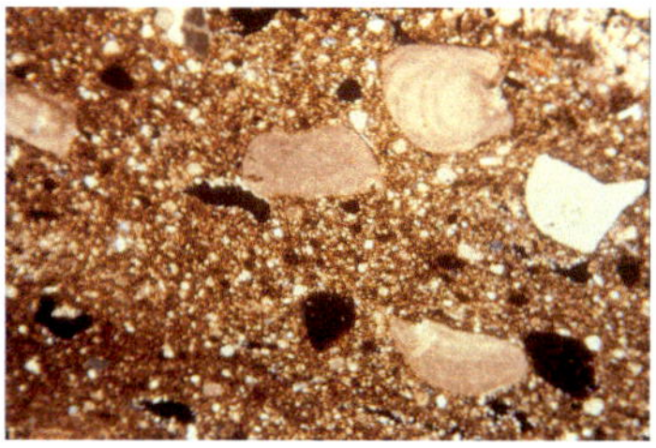

Plate 7.38 Sub Group 5.1.2

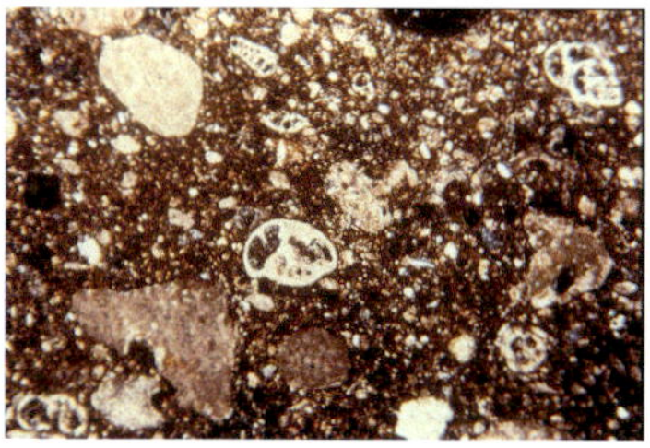

Plate 7.39 Sub Group 5.2.1

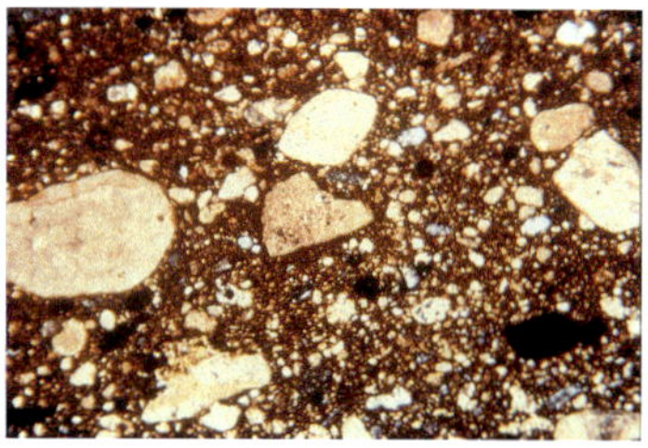

Plate 7.40 Sub Group 5.2.2

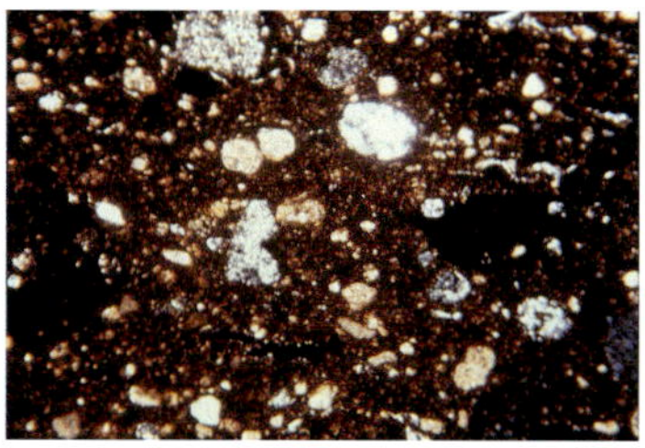

Plate 7.41 Ras Shamra 210

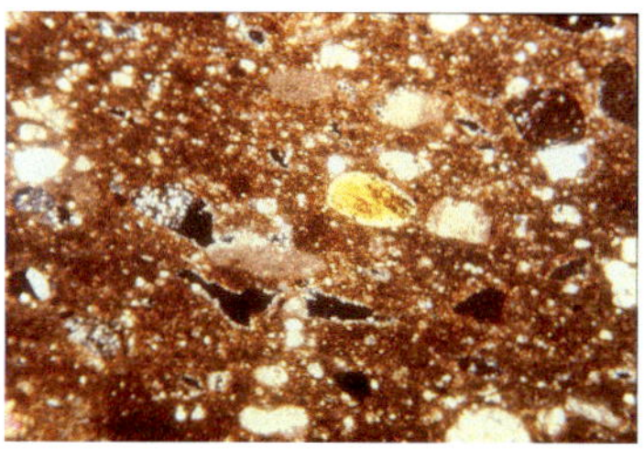

Plate 7.42 Ras Shamra 212

Shape range of most inclusions: Quartz/Feldspars: 1.3–1.4, 2.3–2.4; Limestone: 1.3–1.5, 2.3–2.5.
Size range of most inclusions: Quartz/Feldspar: very fine – coarse sand; Limestone: very fine – very coarse sand; rarely granule.
Major inclusions: Abundant quartz and limestone, moderate percentage microfossils. Chalk: rounded, medium to very coarse sand size, moderate to high percentage. Moderate percentage clay pellets. Volcanic rock fragments including basalt present in some specimens. Calcareous components include a small percentage of naricalcrete rock formed by dissolution of underlying chalk (Horowitz 1979, 168).

SUB-GROUP 2.1.1

Low quartz content, illustrated by specimen 146 (Memphis).

Sherd section description (Plate 7.6)

Inclusions: 'sand' (quartz and feldspar): very fine [2], fine [1]; limestone: fine [1], medium [1], coarse [1], very coarse [1]; soft red-brown particles: fine [1], medium [1]; shell microfossils: medium [1]; other: red-brown clay pellet: coarse [1]; kurkar: very coarse [1].
Sorting: poor.
Porosity: medium.
Hardness: medium with decomposed limestone particles.
Vessel wall: 7.0 mm.
Section colour: zones reddish brown (2.5YR5/4), core grey.
Outer surface: as zones.

Thin section description (Fig. 7.3A, Plate 7.27)

Colour PPL: Orange-brown core, light orange-brown exterior zones.
Colour XPL: Mid orange-brown core, orange-brown exterior zones.
Frequency of Inclusions: mean 25.83%.
Sorting: Poorly sorted.
Approximate Shape Range: Quartz and feldspars 1.1–1.5 and 2.2–2.5, Limestone 1.2–1.5 and 2.3–2.5.
Main Additional Inclusions Present: Plagioclase, K-feldspars, amphibole, pyroxene, opaque minerals, chert, red iron oxides, epidote, kurkar fragments.
Comments: K-feldspars include microcline and perthite; amphiboles include basaltic hornblende; pyroxenes include augite.

SUB-GROUP 2.1.2

Medium quartz content, illustrated by specimen 165 (Amarna).

Sherd section description (Plate 7.7)

Inclusions: 'sand' (quartz and feldspar): very fine [3], fine [1], medium [1], very coarse [1]; limestone: very fine [2], fine [1], medium [1], coarse [1]; soft red-brown

particles: fine [1]; red-brown rock particles: fine [1]; black-rock particles: very fine [1]; other: particles of unmixed matrix, coarse [1], very coarse [1].
Sorting: poor.
Porosity: dense.
Hardness: medium with decomposed limestone particles.
Vessel wall: 7.0 mm.
Section colour: zones brown (7.5YR5/4), core dark grey (7.5YR4/0).
Outer surface: pinkish grey (7.5YR6/2).

Thin section description (Fig. 7.3B; Plate 7.28)

Colour PPL: Brown core and exterior zone, light brown interior zones.
Colour XPL: Dark brown core and exterior zone, mid yellowish-brown exterior zones.
Frequency of Inclusions: mean 35.33%.
Sorting: Moderately sorted.
Approximate Shape Range: Quartz and feldspars 1.1–1.5 and 2.2–2.5, Limestone 1.2–1.5 and 2.2–2.5.
Main Additional Inclusions Present: Plagioclase, biotite and muscovite mica, amphibole, pyroxene, opaque minerals, shell, reddish-brown clay pellets, marl clay pellets, zircon, epidote, clinozoisite, VRF, OAN.
Comments: K-feldspars include microcline and perthite; amphiboles include basaltic and pargasitic hornblende; pyroxenes include augite.

SUB-GROUP 2.2.1

High quartz content, high chalk content, illustrated by specimen 80 (Memphis).

Sherd section description (Plate 7.8)

Inclusions: 'sand' (quartz and feldspar): fine [2], medium [1], coarse [1]; plant remains: fine [1]; limestone: fine [1], medium [1], coarse [1]; black-rock particles: fine [1]; mica: fine [1]; shell strips: fine [1], medium [1]; other: clay pellets: very coarse [1]; kurkar: medium [1], coarse [1].
Sorting: poor.
Porosity: medium.
Hardness: medium with elongated pores and decomposed limestone particles.
Vessel wall: 12.0 mm.
Section colour: zones reddish yellow (5YR6/6), grey core.
Outer surface: as zones.

Thin section description (Fig. 7.3C, Plate 7.29)

Colour PPL: Light brown throughout.
Colour XPL: Mid brown core, orange-brown exterior zones.
Frequency of Inclusions: mean 40.33%.
Sorting: Poorly sorted.
Approximate Shape Range: Quartz and feldspars 1.1–1.5 and 2.1–2.5, Limestone 1.2–1.5 and 2.2–2.5.

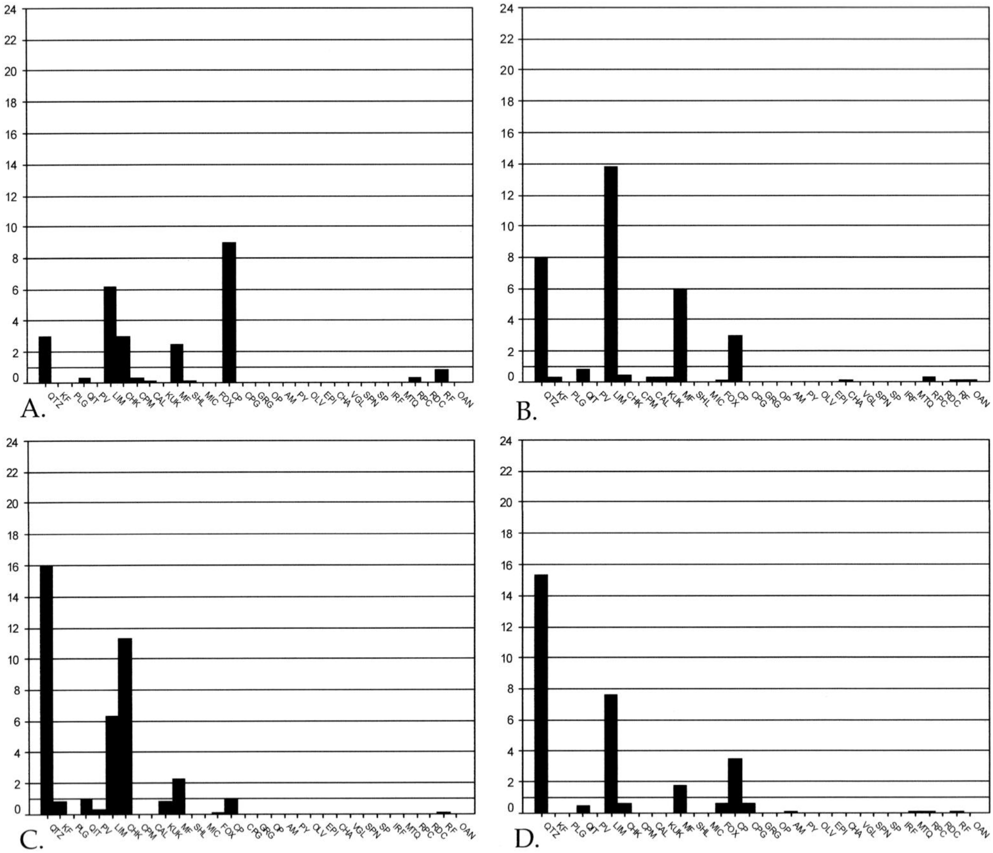

Figure 7.3: Mean percent of constituents for sherds of Group 2; A. Sherd 146, Fabric Sub-Group 2.1.1; B. Sherd 165, Fabric Group 2.1.2; C. Sherd 80, Fabric Sub-Group 2.2.1; D. Sherd 163, Fabric Sub-Group 2.2.2.

Main Additional Inclusions Present: Plagioclase, amphibole, pyroxene, opaques, shell, chert, chalcedony, zircon.

Comments: K-feldspars include microcline and perthite; amphiboles include basaltic and pargasitic hornblende; pyroxenes include augite.

SUB-GROUP 2.2.2

High subrounded to well rounded quartz content, low chalk content, illustrated by specimen 163 (Amarna).

Sherd section description (Plate 7.9)

Inclusions: 'sand' (quartz and feldspar): fine [1], medium [1], coarse [1]; plant remains : fine [1]; limestone: fine [1], medium [1], coarse [1], very coarse [1]; soft

red-brown particles: fine [1], medium [1], coarse [1]; black-rock particles: fine [1]; shell microfossils: fine [1], medium [1]; other: streaks of unmixed matrix.
Sorting: fair.
Porosity: dense.
Hardness: medium with decomposed limestone particles.
Vessel wall: 13.0 mm.
Section colour: narrow zones pinkish grey (7.5YR6/2), core grey.
Outer surface: as zones.

Thin section description (Fig. 7.3D, Plate 7.30)

Colour PPL: Very dark brown to black core and interior zone, dark brown exterior zone.
Colour XPL: Black core and interior zone, dark reddish brown exterior zone.
Frequency of Inclusions: mean 31.67%.
Sorting: Moderately well sorted.
Approximate Shape Range: Quartz and feldspars 1.2–1.5 and 2.1–2.5, Limestone 1.2–1.5 and 2.3–2.5.
Main Additional Inclusions Present: K-feldspars, biotite and muscovite mica, pyroxene, opaque minerals, rutile, epidote, ferruginous ooids, RF, voids from burnt-out plant remains.
Comments: K-feldspars include perthite; amphiboles include hornblende; pyroxenes include augite and aegirine-augite. There are present greyish clay pellets, which represent either claystone, or very altered basaltic fragments.

GROUP 3 FABRICS

Memphis P16; Amarna IV:4

Sherd section description

Dominated by black and red-brown rock particles (volcanic rock fragments) up to very coarse sand size. Abundant fine to medium limestone content (greatly decomposed); microfossils.
Dense and extremely hard.
Colour (specimen 62): zones in section: outer pink 7.5YR 7/4, core grey 5YR6/1.

Thin section description

Colour (specimen 62) PPL: Mid yellow-brown core and interior zone, light yellow-brown exterior zone; XPL: Dark yellow-brown core and interior zone, mid yellow-brown exterior zone.
Sorting: Poor to Moderate.
Shape range of most inclusions: Quartz/Feldspars: 1.2–1.3, 2.2–2.3; Limestone: 1.3–1.5, 2.3–2.5; Volcanic rock: 1.3–1.5, 2.3–2.5.
Size of most inclusions: Quartz/Feldspars : very fine-medium sand; Limestone: very fine-coarse sand; rarely very coarse sand; Volcanic rock: very fine-very coarse sand.
Major inclusions: Abundant igneous (volcanic) rock fragments, mainly olivine basalts,

followed by limestone, microfossils, quartz, replacement chert and clay pellets. Very low percentages of chalk inclusions, marl clay pellets, red iron oxides, olivine, non-volcanic lithic fragments, K-feldspars, plagioclase, calcite, mica, opaques, amphiboles, pyroxenes, chalcedonic quartz and serpentinite. Volcanic rock fragments include both fresh and altered material.

Comment: Six samples from Israel from Beth Shean, Kafr Nein, Tel Ruian and Affule were initially identified, using a hand lens, as belonging to this group but both NAA investigation and petrographic analysis showed them to belong to a different group although one whose major inclusion was also volcanic rock fragments, mainly basalts.

SUB-GROUP 3.1.1

Abundant volcanic rock fragments, >17%, illustrated by specimen 62 (Memphis).

Sherd section description (Plate 7.10)

Inclusions: 'sand' (quartz and feldspar): fine [1], medium [1], coarse [1], very coarse [1] ; limestone: fine [1], medium [1]; grey rock particles: fine [1], medium [1], coarse [1]; soft red-brown particles: fine [1]; red-brown rock particles: fine [1], medium [1], coarse [1]; black-rock particles: fine [1], medium [1], coarse [1]; shell microfossils: fine [1]; other: particles of unmixed matrix: very coarse [1].
Sorting: poor.
Porosity: dense.
Hardness: hard with elongated pores and decomposed limestone.
Vessel wall: 12.0 mm.
Section colour: zones pink (7.5YR7/4), core grey.
Outer surface: as zones.

Thin section description (Fig. 7.4A, Plate 7.31)

Colour PPL: Mid yellowish brown core and interior zone, light yellowish-brown exterior zone.
Colour XPL: Dark brown core, dark yellowish-brown exterior zone.
Frequency of Inclusions: mean 37.0%.
Sorting: Poorly sorted.
Approximate Shape Range: Quartz and feldspars 1.1–1.4 and 2.1–2.4, Limestone 1.2–1.5 and 2.2–2.5, Basalt fragments 1.2–1.5 and 2.2–2.5.
Main Additional Inclusions Present: Plagioclase, biotite mica, amphibole, shell, serpentinite, chalcedony, epidote, voids from burnt-out plant remains.
Comments: K-feldspars include perthite; amphiboles include basaltic hornblende.

SUB-GROUP 3.1.2

Low content of igneous (volcanic) rock fragments <12%, illustrated by specimen 63 (Memphis).

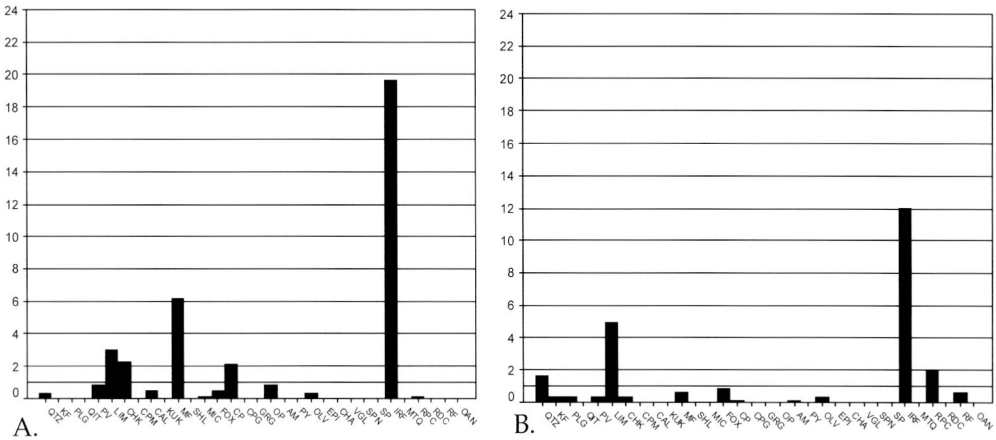

Figure 7.4: Mean percent of constituents for sherds of Group 3; A. Sherd 62, Fabric Sub-Group 3.1.1; B. Sherd 63, Fabric Sub-Group 3.1.2.

Sherd section description (Plate 7.11)

Inclusions: 'sand' (quartz and feldspar): very fine [1], fine [1], medium [1]; plant remains: fine [1]; limestone: very fine [1], fine [1], medium [1]; soft red-brown particles: very fine [1], fine [1]; red-brown rock particles: medium [1]; black-rock particles: very fine [1], fine [1], medium [1], coarse [1]; mica: fine [1].
Sorting: very poor.
Porosity: dense. Hardness: hard with decomposed limestone particles.
Vessel wall: 9.0 mm.
Section colour: zones pink (7.5YR7/4), core grey.
Outer surface: as zones.

Thin section description (Fig. 7.4B, Plate 7.32)

Colour PPL: Mid yellowish brown to mid olive-brown all through.
Colour XPL: Dark yellowish brown all through.
Frequency of Inclusions: mean 24.83%.
Sorting: Poorly sorted.
Approximate Shape Range: Quartz and feldspars 1.1–1.4 and 2.1–2.4, Limestone 1.2–1.5 and 2.2–2.5, Basalt fragments 1.2–1.5 and 2.2–2.5.
Main Additional Inclusions Present: Biotite and muscovite mica, pyroxene, opaque minerals, serpentinite, chalcedony, quartzite, rutile, epidote.
Comments: K-feldspars include orthoclase; amphiboles include basaltic hornblende; VRF are mainly basalt fragments. There is a proportion of altered basaltic inclusions.

GROUP 4 FABRICS

Memphis P40, P52; Amarna IV:1b, IV:7; also identified from; Akhziv; Aphek; Miqne; Ras Shamra.

Sherd section description

Conspicuous black and red-brown rock inclusions, but less abundant and much finer than Group 3; fine to coarse sand size quartz and limestone (decomposed); occasional fine plant remains and microfossils.

Dense, medium hard.

Colour (specimen 161): zones in section: outer pale red 10R 6/3, inner light reddish brown 5YR 6/4, core grey. Surface as inner zone.

Thin section description

Colour (specimen 161) PPL: Yellow-brown core, light orange-brown exterior zone; XPL: Mid yellow-brown core, orange-brown exterior.

Sorting: Poor to Moderate.

Shape range of most inclusions: Quartz/Feldspars: 1.2–1.4, 2.2–2.4; Limestone: 1.3–1.5, 2.3–2.5; Chert (Replacement) 1.2–1.4, 2.2–2.4.

Size range of most inclusions: Quartz/Feldspars: very fine – coarse sand; rarely very coarse sand; Limestone: very fine – very coarse sand; Chert (Replacement): very fine – coarse sand; rarely very coarse sand.

Major inclusions: Abundant limestone, replacement chert, igneous rock fragments, radiolarian chert and other lithic fragments; microfossils and serpentinite at moderate percentages. Small quantities (2% or less) quartz, plagioclase, K-feldspars, quartzite, red iron oxides, mica. Schist 0.5%. Igneous rock fragments include basalt and dolerite, also altered basaltic inclusions. The suite of inclusions is characteristic of an ophiolite complex and subgroups were based on variations in these constituents.

SUB-GROUP 4.1.1

High ophiolite content, with abundant serpentinite, illustrated by specimen 187 (Amarna).

Sherd section description (Plate 7.12)

Inclusions: 'sand' (quartz and feldspar): very fine [1], fine [1], medium [1]; limestone: very fine [1], fine [1], medium [1]; soft red-brown particles: fine [1], medium [1]; red-brown rock particles: fine [1], medium [1], coarse [1]; black-rock particles: fine [1], medium [1], coarse [1]; other: particles of unmixed matrix, fine [1], medium [1], coarse [1]; calcite with particles of fine and very fine quartz included in it, very coarse [1].

Sorting: fair.

Porosity: dense.

Hardness: hard with decomposed limestone particles.

Vessel wall: 7.0 mm

Section colour: zones pale yellow (5Y7/3), core light grey.

Outer surface: as zones.

Thin section description (Fig. 7.5A, Plate 7.33)

Colour PPL: Mid olive-brown all through.
Colour XPL: Dark yellowish brown all through.
Frequency of Inclusions: mean 38.5%.
Sorting: Moderately well sorted.
Approximate Shape Range: Quartz and feldspars 1.2–1.4 and 2.1–2.5, Limestone 1.2–1.5 and 2.2–2.5, Replacement chert 1.2–1.4 and 2.2–2.4.
Main Additional Inclusions Present: Quartz, plagioclase, opaque minerals, red iron oxides, radiolarian chert, spinel, grey clay pellets.
Comments: K-feldspars include orthoclase and perthite; amphiboles include basaltic hornblende; pyroxenes include augite; IRF are mainly basalt fragments, but include some coarser-grained inclusions likely to be dolerite. There are present a proportion of altered basaltic inclusions. The grey clay pellets are likely to be a very altered IRF fragments. Rare radiolarites occur isolated in the clay matrix.

SUB-GROUP 4.1.2

High ophiolite content with abundant radiolarian chert, illustrated by specimen 52 (Amarna).

Sherd section description (Plate 7.13)

Inclusions: 'sand' (quartz and feldspar): fine [1], medium [1], coarse [1]; limestone: fine [1], medium [1], coarse [1], very coarse [1]; soft red-brown particles: fine [1]; red-brown rock particles: fine [1], medium [1], coarse [1], very coarse [1]; black-rock particles: very fine [1], fine [1], medium [1], coarse [1]; shell microfossils: medium [1].
Sorting: fair.
Porosity: dense.
Hardness: hard with decomposed limestone particles.
Vessel wall: 5.0 mm.
Section colour: pinkish grey (7.5YR7/2).
Outer surface: as zones.

Thin section description (Fig. 7.5B, Plate 7.34)

Colour PPL: Light olive-brown all through.
Colour XPL: Dark yellowish-brown all through.
Frequency of Inclusions: mean 36.83%.
Sorting: Poorly sorted.
Approximate Shape Range: Quartz and feldspars 1.1–1.4 and 2.1–2.4, Limestone 1.2–1.5 and 2.2–2.5, Replacement chert 1.1–1.4 and 2.2–2.5.
Main Additional Inclusions Present: Plagioclase, amphibole, pyroxene, opaque minerals, shell, red iron oxides, chalcedony, rutile, spinel, RF, possible fragment of bone.
Comments: K-feldspars include orthoclase; amphiboles include basaltic hornblende; pyroxenes include augite; IRF are mainly basalt fragments, together with a

proportion of altered basaltic grains. There are present grey clay pellets, likely to be a very altered VRF; rare radiolarites occur isolated in the clay matrix.

SUB-GROUP 4.2.1

High ophiolite content, with abundant replacement chert and low radiolarian chert, illustrated by specimen 40 (Amarna).

Sherd section description (Plate 7.14)

Inclusions: 'sand' (quartz and feldspar): very fine [1], fine [1], medium [1], coarse [1]; plant remains: fine [1]; limestone: very fine [1], fine [1], medium [1], coarse [1]; grey-white rock particles: fine [1]; soft red-brown particles: fine [1]; red-brown rock particles: fine [1], medium [1], coarse [1]; black-rock particles: very fine [1], fine [1] ; other: particles of unmixed matrix: coarse [1].
Sorting: fair. Porosity: dense.
Hardness: medium with decomposed limestone particles
Vessel wall: 8.0 mm.
Section colour: zones light red (2.5YR6/4), core grey.
Outer surface: as zones.

Thin section description (Fig. 7.5C, Plate 7.35)

Colour PPL: Mid olive-brown core and interior zone, thin reddish-brown exterior zone.
Colour XPL: Dark olive-brown core and interior zone, thin dark reddish-brown exterior zone.
Frequency of Inclusions: mean 27.0%.
Sorting: Moderately well sorted.
Approximate Shape Range: Quartz and feldspars 1.1–1.4 and 2.1–2.5, Limestone 1.2–1.5 and 2.1–2.5, Replacement chert 1.2–1.5 and 2.2–2.5.
Main Additional Inclusions Present: Plagioclase, biotite mica, amphibole, pyroxene, olivine, rutile, epidote, and phosphatic grains.
Comments: K-feldspars include orthoclase and perthite; amphiboles include basaltic and pargasitic hornblende, riebeckite and a metamorphic clinoamphibole; pyroxenes include augite; IRF include basalt fragments, together with a proportion of altered basaltic grains, containing iddingsite or serpentine. Rare radiolarites occur isolated in the clay matrix.

SUB-GROUP 4.2.2

Low ophiolite content, with low replacement chert, illustrated by specimen 161 (Amarna).

Sherd section description (Plate 7.15)

Inclusions: 'sand' (quartz and feldspar): very fine [1], fine [1], medium [1], coarse [1]; plant remains: fine [1]; limestone: very fine [1], fine [1], medium [1], coarse [1], very coarse [1]; soft red-brown particles: fine [1], medium [1]; red-brown rock particles:

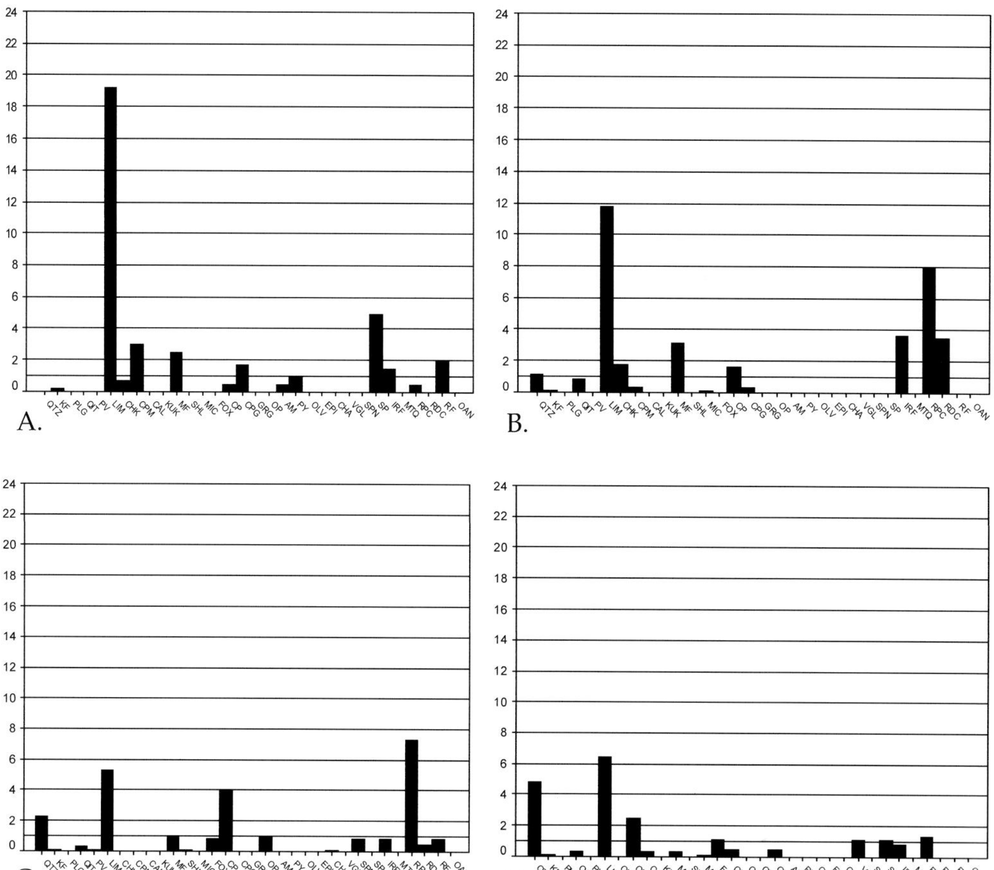

Figure 7.5: Mean percent of constituents for sherds of Group 4; A. Sherd 187, Fabric Sub-Group 4.1.1; B. Sherd 52, Fabric Sub-Group 4.1.2; C. Sherd 40, Fabric Sub-Group 4.2.1; D. Sherd 161 Fabric Sub-Group 4.2.2.

fine [1], medium [1], coarse [1]; black-rock particles: fine [1], medium [1]; mica: fine [1]; other: particles of unmixed matrix: medium [1]; calcite particles: with very fine and fine quartz included in it, medium [1], coarse [1].

Sorting: poor.

Porosity: dense.

Hardness: medium with decomposed limestone particles.

Vessel wall: 8.0mm.

Section colour: zones outer pale red (10R6/3), inner light reddish brown (5YR6/4), core grey.

Outer surface: as outer zone.

Thin section description (Fig. 7.5D, Plate 7.36)

Colour PPL: Yellowish-brown core, light yellowish-brown exterior zones.
Colour XPL: Mid yellowish brown core, yellowish brown exterior zones.
Frequency of Inclusions: mean 22.0%.
Sorting: Moderately well sorted.
Approximate Shape Range: Quartz and feldspars 1.1–1.5 and 2.2–2.5, Limestone 1.2–1.5 and 2.2–2.5, Replacement chert 1.3–1.5 and 2.2–2.4.
Main Additional Inclusions Present: Muscovite mica, pyroxene, olivine, chalcedony, spinel, devitrified volcanic glass, ironstone fragment, chalk fragment, cassiterite, shell, OAN, RF and voids from burnt-out plant remains.
Comments: K-feldspars include orthoclase and perthite; amphiboles include riebeckite; pyroxenes include augite; IRF comprise mainly basalt and dolerite fragments, together with a proportion of altered basaltic grains.

GROUP 5 FABRICS

Memphis P33, P51; Amarna IV:1a, IV:1c.

Sherd section description

Dominated by fine to very coarse limestone content (occasionally decomposed); fine to medium elongate shell fragments and microfossils. Low in fine to medium quartz; abundant fine to medium red-brown clay pellets.
Dense and hard.
Colour (specimen 160): zones in section: outer light reddish brown 5YR 6/4, wide grey core; surface: pink 5YR 7/4.

Thin section description

Colour (specimen 160) PPL: Yellow-brown core, light orange-brown exterior zone. XPL: Mid yellow-brown core, mid orange-brown exterior zone.
Sorting: Poor to Moderate
Shape range of most inclusions: Quartz/Feldspars: 1.1–1.4, 2.1–2.4; Limestone: 1.3–1.5, 2.3–2.5.
Size range of most inclusions: Quartz/Feldspars: very fine – coarse sand; Limestone: very fine – very coarse sand.
Major inclusions: Abundant limestone and microfossils, both up to very coarse sand size, and reddish-brown or brown clay pellets. Low (3% or less) quartz, red iron oxides, opaques, chert, shell and plant voids.

SUB-GROUP 5.1.1

High limestone content, with low microfossil content, illustrated by specimen 42 (Amarna).

Sherd section description (Plate 7.16)

Inclusions: 'sand' (quartz and feldspar): fine [1], medium [1]; plant remains : fine [1];

limestone: very fine [1], fine [1], medium [1], coarse [1], very coarse [1]; soft red-brown particles: fine [1], medium [1], coarse [1]; black-rock particles: fine [1], medium [1]; shell strips: medium [1]; shell microfossils: fine [1], medium [1].
Sorting: poor.
Porosity: dense.
Hardness: crumbly with decomposed limestone particles.
Vessel wall: 8.0 mm.
Section colour: zones pinkish grey (7.5YR7/2), core dark grey.
Outer surface: as zones.

Thin section description (Fig. 7.6A; Plate 7.37)

Colour PPL: Mid yellowish brown core, light brown interior zone, mid brown exterior zone.
Colour XPL: Dark yellowish brown core, light yellowish-brown interior zone, dark reddish-brown exterior zone.
Frequency of Inclusions: mean 26.17%.
Sorting: Moderately well sorted.
Approximate Shape Range: Quartz and feldspars 1.1–1.4 and 2.1–2.5, Limestone 1.1–1.5 and 2.1–2.5.
Main Additional Inclusions Present: Plagioclase, amphibole, pyroxene, chalcedony, quartzite, zircon, OAN, shell.
Comments: K-feldspars include orthoclase and perthite; amphiboles include basaltic hornblende; pyroxenes include augite.

SUB-GROUP 5.1.2

Lower limestone content, with low microfossil content, illustrated by specimen 160 (Amarna).

Sherd section description (Plate 7.17)

Inclusions: 'sand' (quartz and feldspar): very fine [1], fine [1]; limestone: very fine [1], fine [1], medium [1], coarse [1], very coarse [1]; grey-white rock particles: fine [1], medium [1], coarse [1]; soft red-brown particles: fine [1]; black-rock particles: fine [1]; shell microfossils: medium [1], coarse [1].
Sorting: poor.
Porosity: dense.
Hardness: hard.
Vessel wall: 10.0 mm.
Section colour: zones light reddish brown (5YR6/4), wide grey core.
Outer surface: pink (5YR7/4).

Thin section description (Fig. 7.6B; Plate 7.38)

Colour PPL: Mid brown core, light yellowish brown interior zone, light orange-brown exterior zone.

Colour XPL: Mid brown core, mid yellowish brown interior zone, mid orange-brown exterior zone.
Frequency of Inclusions: mean 23.33%.
Sorting: Poorly sorted.
Approximate Shape Range: Quartz and feldspars 1.1–1.5 and 2.1–2.5, Limestone 1.1–1.5 and 2.2–2.5.
Main Additional Inclusions Present: Plagioclase, K-Feldspars, biotite mica, pyroxene, amphibole, opaque minerals, serpentine, chalcedony, chalk fragment, shell, RF.
Comments: K-feldspars include orthoclase and perthite; amphiboles include pargasitic hornblende; pyroxenes include augite. Calcitic component includes rare rhombs, indicating the presence of dolomite.

SUB-GROUP 5.2.1

High limestone content, with high microfossil content, illustrated by specimen 43 (Amarna).

Sherd section description (Plate 7.18)

Inclusions: 'sand' (quartz and feldspar): very fine [1], fine [1]; plant remains: fine [1]; limestone: very fine [1], fine [1], medium [1], coarse [1]; soft red-brown particles: fine [1], medium [1]; red-brown rock particles: medium [1]; black-rock particles: fine [1]; mica: fine [1]; shell strips: medium [1]; shell microfossils: fine [1], medium [1], coarse [1].
Sorting: poor.
Porosity: dense.
Hardness: medium hard with decomposed limestone particles.
Vessel wall: 4.0 mm
Section colour: zones reddish brown (2.5YR 5/4), core dark grey
Outer surface: as zones.

Thin section description (Fig. 7.6C; Plate 7.39)

Colour PPL: Dark brown all through.
Colour XPL: Very dark brown core and interior, thin mid brown exterior zone.
Frequency of Inclusions: mean 34.17%.
Sorting: Moderately sorted.
Approximate Shape Range: Quartz and feldspars 1.1–1.4 and 2.1–2.4, Limestone 1.2–1.5 and 2.2–2.5.
Main Additional Inclusions Present: K-Feldspar, biotite mica, amphibole, pyroxene, calcite, chalcedony, quartzite, epidote, voids from burnt-out plant remains.
Comments: K-feldspars include orthoclase; amphiboles include basaltic hornblende; pyroxenes include augite. Calcitic component includes rare rhombs indicating the presence of dolomite.

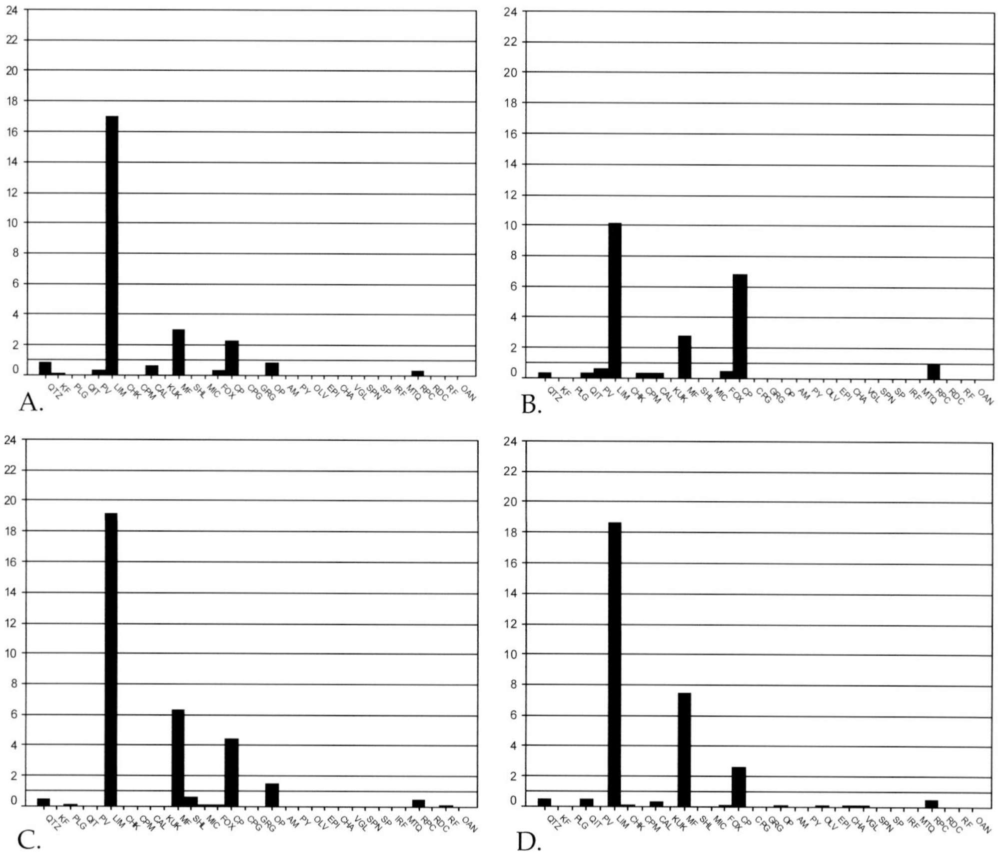

Figure 7.6: Mean percent of constituents for sherds of Group 5; A. Sherd 42, Fabric Sub-Group 5.1.1; B. Sherd 160, Fabric Sub-Group 5.1.2; C. Sherd 43, Fabric Sub-Group 5.2.1; D. Sherd 149, Fabric Sub-Group 5.2.2.

SUB-GROUP 5.2.2

High limestone content, with high microfossil content, illustrated by specimen 149 (Memphis).

Sherd section description (Plate 7.19)

Inclusions: 'sand' (quartz and feldspar): very fine [1], fine [1], medium [1]; limestone: very fine [2], fine [1], medium [1], coarse [1], very coarse [1]; soft red-brown particles: fine [1], medium [1], coarse [1]; red-brown rock particles: medium [1]; black-rock particles: very fine [1], fine [1]; shell microfossils: medium [1].
Sorting: poor.
Porosity: dense.
Hardness: hard.

Vessel wall: 7.0 mm.
Section colour: zones light reddish brown (5YR6/3), core pinkish grey (7.5YR6/2).
Outer surface: as zones.

Thin section description (Fig. 7.6D, Plate 7.40)

Colour PPL: Mid brown core. Light orange brown interior zone, mid brown exterior zone.
Colour XPL: Mid brown core, mid yellowish brown exterior zones.
Frequency of Inclusions: mean 31.67%.
Sorting: Moderately well sorted.
Approximate Shape Range: Quartz and feldspars 1.1–1.5 and 2.2–2.5, Limestone 1.2–1.5 and 2.2–2.5.
Main Additional Inclusions Present: Plagioclase, K-Feldspar, biotite and muscovite mica, amphibole, pyroxene, shell, rutile, epidote, devitrified volcanic glass, kurkar fragments.
Comments: K-feldspars include orthoclase; amphiboles include pargasitic hornblende; pyroxenes include augite.

RAS SHAMRA GROUP FOR COMPARISON

Sherd section description

Dominated by fine to medium black and red-brown rock particles and occasional very coarse ones; moderate fine to medium quartz and limestone (decomposed); few medium microfossils.
Dense, hard.
Colour (specimen 211): zones in section: outer pink 7.5YR 7/4, wide grey core; surface: as outer zone.

Thin section description

Colour (specimen 211) PPL: Mid yellow-brown core and exterior zone, light yellow-brown interior zone. XPL: Mid yellow-brown core and exterior zone, yellow-brown interior zone.
Sorting: Poor to Moderate
Shape range of most inclusions: Quartz/Feldspars: 1.2–1.4, 2.2–2.4; Limestone: 1.3–1.5, 2.3–2.5; Chert (replacement): 1.2–1.4, 2.2–2.4.
Size range of most inclusions: Quartz/Feldspars: very fine – coarse sand; Limestone: very fine-very coarse sand.

SPECIMEN 210 MATCH FOR SUBGROUP 4.2.1

Sherd section description (Plate 7.20)

Inclusions: 'sand' (quartz and feldspar): fine [1], medium [1]; limestone: fine [1], medium [1]; red-brown rock particles: fine [1], medium [1], very coarse [1]; black-rock particles: fine [1], medium [1]; shell strips: medium [1].

Sorting: fair.
Porosity: dense.
Hardness: medium hard with decomposed limestone and elongated pores.
Vessel wall: 5.0 mm.
Section colour: Outer zones pink (7.5YR7/4), core grey.
Outer surface: as zones.

Thin section description (Fig. 7.7A, Plate 7.41)

Colour PPL: Brown core, mid yellowish brown intermediate zones, thin yellowish-brown zone on exterior only.
Colour XPL: Dark brown core, yellowish brown intermediate zones, thin light yellowish-brown zone on exterior only.
Frequency of Inclusions: mean 32.25%.
Sorting: Moderately sorted.
Approximate Shape Range: Quartz and feldspars 1.1–1.4 and 2.2–2.5, Limestone 1.3–1.5 and 2.3–2.5, Replacement chert 1.1–1.5 and 2.2–2.5.
Main Additional Inclusions Present: Biotite mica, amphibole, opaque minerals, chalcedony, calcareous clay pellets.
Comments: Amphiboles include basaltic hornblende. Rare radiolarites occur as isolated inclusions within the clay matrix. Grey clay pellets are present and are likely to be highly altered volcanic rock fragments.

SPECIMEN 212 MATCH FOR SUBGROUP 4.2.1

Sherd section description (Plate 7.21)

Inclusions: 'sand' (quartz and feldspar): fine [1], medium [1], coarse [1]; limestone: fine [1], medium [1], coarse [1], very coarse [1]; soft red-brown particles: fine [1]; red-brown rock particles: fine [1], medium [1], coarse [1]; black-rock particles: fine [1]; shell microfossils: medium [1], very coarse [1].
Sorting: poor.
Porosity: dense.
Hardness: medium hard with decomposed limestone particles.
Vessel wall: 6.0 mm.
Section colour: Outer zones light brown (7.5YR 6/4), core grey.
Outer surface: as zones.

Thin section description (Fig. 7.7B, Plate 7.42)

Colour PPL: Mid yellowish-brown core, light yellowish-brown exterior zones.
Colour XPL: Dark yellowish-brown core, mid yellowish-brown exterior zones.
Frequency of Inclusions: mean 26.5%.
Sorting: Moderately sorted.
Approximate Shape Range: Quartz and feldspars 1.2–1.5 and 2.2–2.5, Limestone 1.2–1.5 and 2.2–2.5, Replacement chert 1.2–1.5 and 2.1–2.5.
Main Additional Inclusions Present: K-Feldspar, amphibole, rutile, reddish-brown clay pellets, possible devitrified glass, OAN.

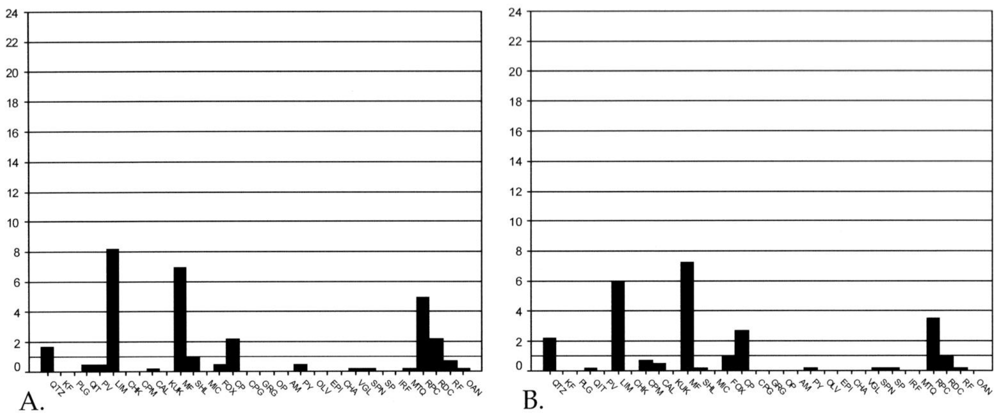

Figure 7.7: Mean percent of constituents for sherds from Ras Shamra, for comparison with Fabric Group 4; A.Sherd 210; B. Sherd 212.

Comments: K-Feldspar includes orthoclase and perthite; amphiboles include pargasitic hornblende. Rare radiolarites occur as isolated inclusions within the clay matrix.

PROVENANCE STUDY

The provenance study followed. Our sections were first compared with the collection held by Yuval Goren in the Department of Archaeology at Tel Aviv University. Sections representative of local geology and comparable with our series were selected and recorded by Laurence Smith using the methods applied to the Egyptian examples. We included two workshop sites considerably later than the Bronze Age, because of evidence for continuity of resource exploitation (Y. W. Goren, pers. comm., 1998). We benefited greatly at this point from discussions with Yuval Goren and he will collaborate in the writing of this part of the final report. For comparison with group 4, examples of local pottery from Ras Shamra (Schaeffer excavations) were obtained from the Louvre.

Geological background

Four elements of Levantine geology are relevant (Fig. 7.8), discussed further in the provenance section. The first comprises deposits along the coast of Israel and its immediate hinterland, including later Pleistocene to early Holocene coastal sands and formations such as the kurkar and the hamra units (Sivan 1996; Sivan, Gvirtzman and Sass 1999; Neev, Bakler and Emery 1987; Emery and Neev 1960). In contrast, the coast through Lebanon and Syria contains few significant Quaternary deposits, those of the Cretaceous and Tertiary predominating except south of Tartus and between Baniyas and Latakiya (Beydoun 1977, 321–322).

The second element comprises deposits of limestone, dolomite, chalk, chert and marls of Cenomanian to Turonian, Senonian to Paleocene and Eocene periods. They

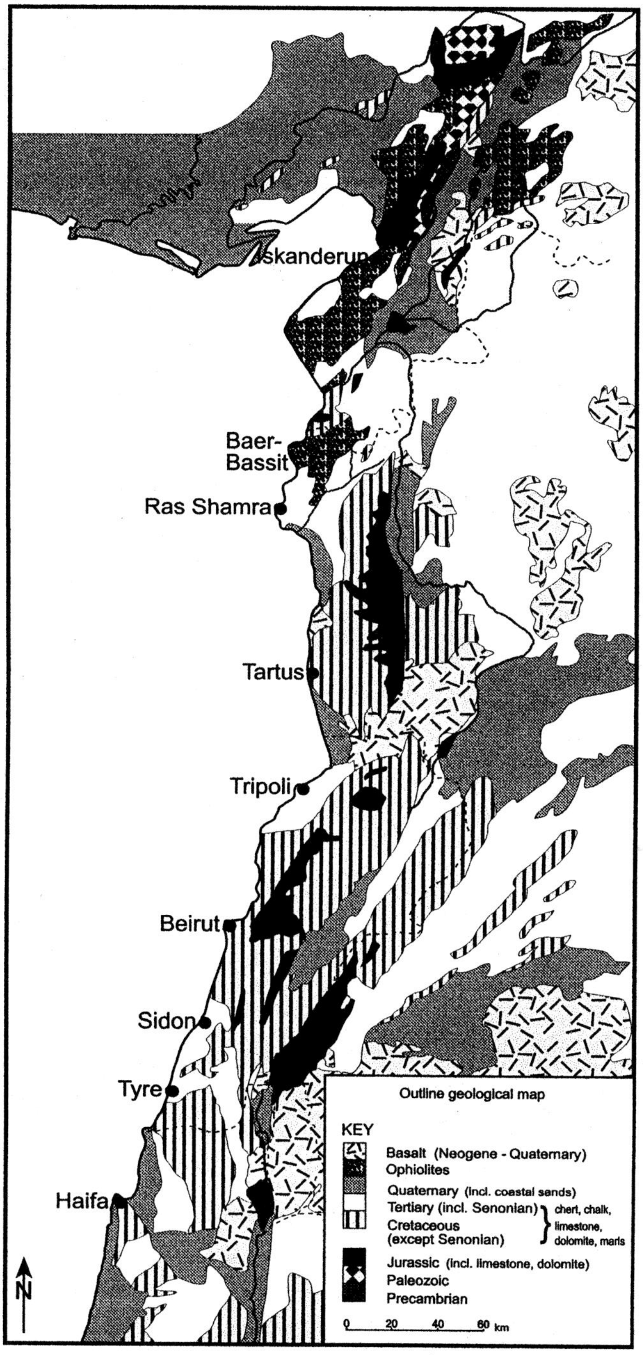

Figure 7.8: Simplified geological map of the main research area (after Beydoun, 1977 and IFP-CNEXO, 1974)

are widespread inland in southern and central Israel and extend from north and south of the Jezreel Valley to the coast. The Cenomanian to Paleocene and Eocene formations extend into southern Lebanon, the latter rarer north of Beirut (Bartov 1994; Picard and Golani 1992).

Thirdly, there are outcrops of basalts, particularly those of Miocene and Pliocene age, most prominent around the Jezreel Valley, extending into Galilee and Syria, but present in southern Israel, Jordan and Turkey (Bartov 1994; Beydoun 1977; Pinar-Erdem and Ilhan 1977).

The fourth element comprises the ophiolite complexes, uplifted portions of oceanic crust containing a basic sequence of igneous rocks, together with metamorphic and sedimentary rocks. Igneous rocks contribute basalts and dolerite; metamorphic contribute schist; and sedimentary contribute replacement and radiolarian chert (Nockolds, Knox and Chinner 1978, 138, 294–296). Major areas of ophiolite complexes include west and south-west Cyprus, central and southern Turkey and north-western Syria (Whitechurch, Juteau and Montigny 1984; Juteau 1980; Parrot 1980).

PROVENANCE

Group 1

The size and roundness of the quartz and the presence of accessory minerals plagioclase, zircon and pyroxene and of limestone and microfossils, including the coraline algae *Amphiroa* (Sivan 1996, 76), indicate that these constituents may be derived from the 'Coastal Sand', a relatively high-quartz sand extending from coastal Sinai to the Israel/Lebanon border (Horowitz 1979, 25). In these fabrics, the Coastal Sand is combined with a small proportion of chert, chalk and basalt, indicating sources inland from the coastal zone itself. The most probable area is the seaward portion of the Jezreel Valley to the north-east and south-east of the Carmel Ridge, where the Kishon river, draining from an area of basalts inland near Affule, from Eocene formations to the north and south, and from the Carmel Ridge Cenomanian and Senonian formations, passes into the region of Coastal Sand (Fig. 7.9). This provenance may be related to a Bronze Age port, Tell Abu Hawam, north east of Haifa. We also found good matches with samples from nearby Tell Nami (samples provided by Michal Artzy).

Group 2

This fabric group is less homogeneous than Group 1. Some examples overlap with Group 1, by containing Coastal Sand with calcareous inclusions and a small quantity of volcanic rock fragments, including basalt. The degree of similarity is insufficient to restrict the source of Group 2 to the same area as the former, having a lower percentage of basaltic fragments and sometimes a highly silty and ferruginous matrix perhaps from a terra rossa soil. This occurs with Cenomanian and Eocene formations, from which the calcareous components, including chalk and nari, and chert could be derived. These are present in the Shephela region of central Israel, but also further north into Lebanon, so either source is possible (Fig. 7.10).

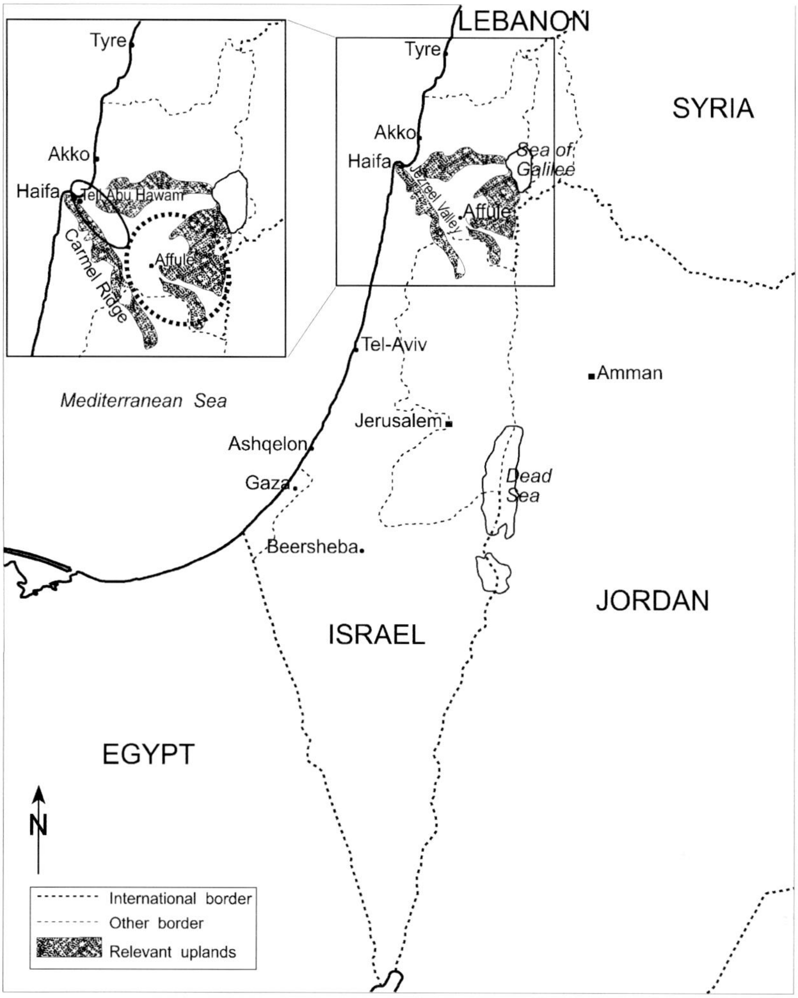

Figure 7.9: Southern Levant region, showing source areas for Fabric Group 1 (solid oval) and Fabric Group 3 (dotted circle) on the basis of geology.

Group 3

Basalt, limestone, chalk and chert, plus the absence of Coastal Sand, indicate an inland, not a coastal, origin. The limestone and chalk inclusions are well rounded, so appear to have been transported, but their size (medium to coarse sand) shows they are not far from their source. This appears to indicate incorporation of a wadi sand and basalt as temper. The most likely origin is the central Jezreel Valley into eastern Galilee (see Fig. 7.9). However, a similar combination of constituents exists elsewhere, around the north Lebanon coast, from which no comparative material was available at the time of writing.

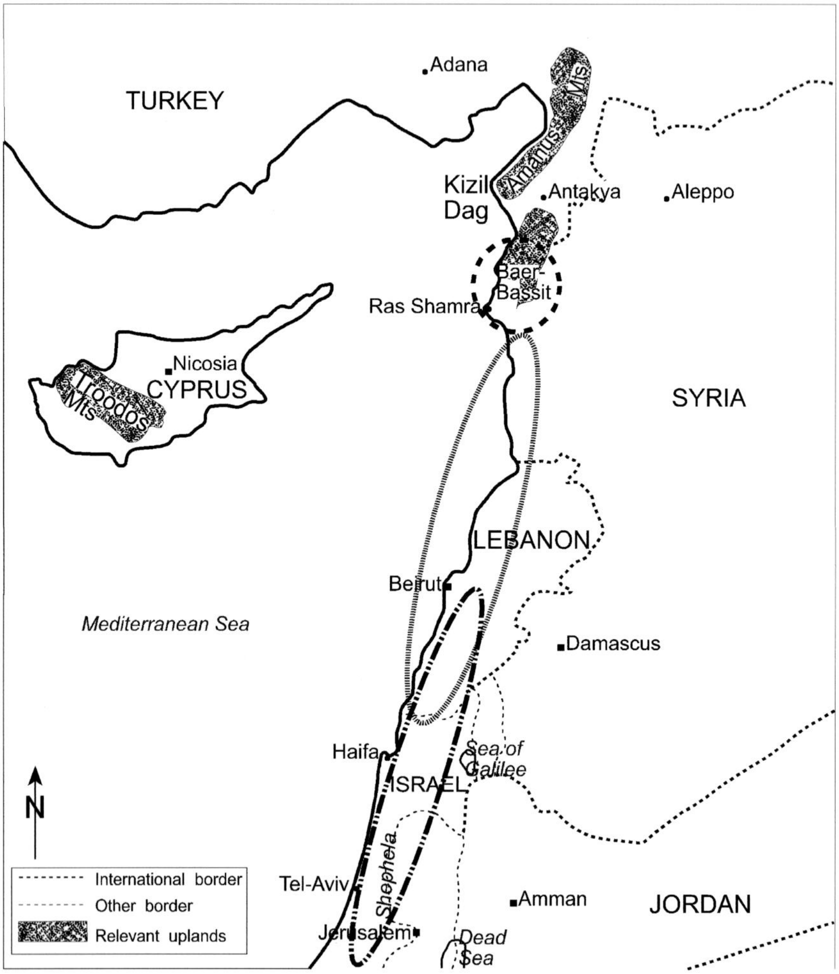

Figure 7.10: Northern Levant region, showing source areas for Fabric Group 2 (dashed and dotted oval), Fabric Group 4 (dashed circle) and Fabric Group 5 (dotted oval) on the basis of geology.

Group 4

The igneous rock fragments, including those showing alteration to iddingsite or serpentinite, together with quartzite, schist and both types of chert, indicate a source where components of an ophiolite complex exist. The ophiolite-related constituents occur with pelagic and shallow water carbonates, including planktonic and benthic foramenifera. This combination is consistent with a possible origin in the Antalya or Diyarbakir areas of Turkey, the Baër-Bassit area of north-west Syria or the area of the Mamonia and the Troodos complexes of west and south-west Cyprus (A. Robertson, pers. comm. 1999). On archaeological grounds, an origin in Turkey seems the least likely. A short visit to Cyprus was made in June 2000 and ten samples were collected. They are still under study but preliminary results suggest Cyprus was not a manufacturing centre for Group 4 amphorae. However, local pottery from the coastal site of Ras Shamra (Ugarit), immediately south-west of the Baër-Bassit contains the same ophiolite-related suite of minerals and lithic fragments with components indicating a coastal provenance. Further support comes from a similar set of constituents in thin sections of two Amarna Tablets, known to have been sent from Ugarit (Y. W. Goren, pers. comm., 1999).

Group 5

No closely-circumscribed source area has yet been identified for this fabric group. The chert includes opaques, which often occur as a replacement after dolomite, indicating a source in a Cenomanian chert region. Preliminary examination of the constituents, by N. Porat and L. Grossowicz of the Geological Survey of Israel, showed that the microfossils include shallow-water algae and planktonic foramenifera, indicating that they are derived from near the coast. The high content of carbonates rather than quartz is characteristic of the region north of Israel. Thus the source may lie in coastal Lebanon or Syria (see Fig. 7.10). Y. W. Goren (pers. comm.) has recently identified similar fossils in thin sections of Amarna Tablets send from Amurru, suggesting a provenance within Lebanon, north of Byblos.

CONCLUSIONS

It should be emphasised that the five fabric groups discussed in this paper represent only those fabrics found at Memphis and Amarna. In the planning of the Canaanite Amphorae project it was considered essential that core samples should come from current excavations in which the investigators themselves were involved, to ensure access to all the information needed. It is certain that the jars occur in many more fabrics, and even in Egypt itself during the New Kingdom, jars in other fabrics are probably to be found. Identifying fabrics is difficult and we have tried to present our results so that comparisons between sites can be more easily made.

A word of warning: samples collected in Israel were selected by identifying the relevant fabrics among the large collection of Late Bronze Age storage jars in the storerooms of the Israel Antiquities Authority in Jerusalem (courtesy of Baruch Brendel). Since the collection was not made comprehensively for the whole region, the absence of our fabric groups at particular sites in Israel may not be significant. The study of the amphorae in Israel, Lebanon, Syria, Turkey or the Aegean requires a different research plan from ours. Those from Israel and the Aegean are currently under study by Michael Sugerman.

We can suggest regions of origin, on geological grounds, for all five fabric groups. For four of them, we know the commodities the jars carried, from study of inscribed jars (Serpico 1996) and from residue analysis (Serpico and White 1998; Stern *et al.* 2000). It is clear that Groups 1 and 2 are linked to the trade in pistachia resin, while 4 and 5 are associated with oil. Excavations at Ras Shamra have provided evidence of olive oil production in commercial quantities in the Late Bronze Age. Patterns of trade are beginning to emerge. Hitherto it had been assumed that southern Palestine was the source of Canaanite Jars in Egypt in the Late Bronze as it was during the Middle Bronze Age, but all our fabric groups, except Group 2, originate in the north of the region. It is interesting to note that in the Late Bronze Age potter's workshop found at Lachish (Magrill and Middleton 1997) where amphorae as well as other vessel types were made, a local clay was utilised which did not match any of our fabric groups.

FURTHER STUDY

Analysis by ICP-AES of our fabric standards is planned, to confirm our findings, since a single type of analysis carries insufficient weight to support the complex trade and historical implications of our results. A monograph is being written which will incorporate all the archaeological data, including a shape typology for each group, the full publication of the petrographic and chemical analyses and the results of the residue analyses.

ACKNOWLEDGEMENTS

The Egyptian data derive from sherds from Memphis and Amarna exported by permission of the Supreme Council for Antiquities, Egypt. Some 40 comparative samples were collected in Israel by permission of the relevant excavators and the Israel Antiquities Authority; 6 samples from Ras Shamra were collected by permission of the Department of Near Eastern Antiquities, the Louvre, Paris; and 10 samples were collected in Cyprus, by permission of the Director of Antiquities and Ian Todd and Alison South, the excavators of Kalavassos. We should like to thank all those involved in granting these permissions, for without them the project would not have been possible. Funding was provided by the Egypt Exploration Society, McDonald Institute for Archaeological Research, Cambridge, Wainwright Fund in the University of Oxford, British Academy and Society of Antiquaries of London. Collaborators on residue analysis are C. Heron and B. Stern of the University of Bradford, supported by a NERC grant; and on the petrological study, Yuval Goren of the University of Tel Aviv and Judith Bunbury of the Department of Earth Sciences, University of Cambridge. The photographs are the work of the authors and Gwilym Owen of the Department of Archaeology, University of Cambridge.

APPENDIX: ABBREVIATIONS USED IN BAR CHARTS

QTZ	Quartz
KF	Total K-(alkali) Feldspar
PLG	Plagioclase
QIT	Quartzite
PV	Voids from burnt-out plant remains
LIM	Limestone
CHK	Chalk
CPM	Marl clay pellets
CAL	Calcite

KUK	Kurkar
MF	Microfossils
SHL	Elongated shell fragments
MIC	Total mica (including biotite and muscovite)
FOX	Red iron oxides
CP	Clay pellets
CPG	Grey clay pellets, including highly altered igneous rock fragments
GRG	Grog
OP	Opaque minerals
AM	Total amphibole (mainly hornblende)
PY	Total pyroxene (mainly augite)
OLV	Olivine
EPI	Epidote group minerals
CHA	Chalcedonic quartz
VGL	Volcanic glass
SPN	Spinel
SP	Serpentine
IRF	Total igneous rock fragments (mainly basalt, but including rare dolerite)
MTQ	Sheared metamorphic quartzite
RPC	Replacement chert
RDC	Radiolarian chert
RF	Total other rock fragments
OAN	Organic inclusions other than voids from plant temper (mainly bone fragments)

REFERENCES

Amiran, R., 1969, *Ancient pottery of the Holy Land*, Mascada Press, Jerusalem.

Bartov, Y., 1994, *Geological photomap of Israel and adjacent areas*. 1:750,000, 2nd edition, Geological Survey of Israel, Jerusalem.

Beydoun, Z. R., 1977, The Levantine countries: the geology of Syria and Lebanon (maritime regions), in Nairn, A. E. M., Kanes, W. H., and Stehli, F. G. (eds), *The ocean basins and margins. Vol. 4A The eastern Mediterranean*, Plenum Press, New York and London, 319–353.

Bourriau, J. D., 1990, Canaanite Jars from New Kingdom Deposits at Memphis, Kom Rabi'a. *Eretz-Israel*, 21, 18–26.

Bourriau J. D., and Nicholson, P. T., 1992, Marl clay pottery fabrics of the New Kingdom from Memphis, Saqqara and Amarna, *Journal of Egyptian Archaeology*, 78, 29–91.

Bourriau, J. D., Smith, L. M. V., and Nicholson, P. T., 2000, *New Kingdom pottery fabrics: Nile clay and mixed Nile/marl clay fabrics from Memphis and Amarna*, Egypt Exploration Society, London.

Emery, K. O., and Neev, D., 1960, Mediterranean beaches of Israel. *Bulletin of the Geological Survey of Israel*, 26, 1–24.

Grace, V. R., 1956, The Canaanite Jar, in Weinburg, S.S. (ed.) *The Aegean and the Near East: Studies Presented to Hetty Goldman on the occasion of her Seventy-Fifth Birthday*, J. J. Augustin, New York, 80–109.

Horowitz, A., 1979, *The Quaternary of Israel*, Academic Press, New York and London.

I.F.P.-C.N.E.X.O., 1974, *Carte géologique et structurale des bassins Tertiaires du domaine Méditerranéen. 1st edition 1:250,000*, L'Institut français du Pétrole, Le Centre National pour l'Exploitation des Océans, L'Institut National d'Astronomie et de Géophysique.

Juteau, T., 1980, Ophiolites of Turkey. *Ofioliti*, Special Issue: Tethyan ophiolites 2. Eastern area, 199–237.

Leonard, A. Jr., 1996, 'Canaanite Jars' and the Late Bronze Age Aegeo-Levantine Wine Trade, in McGovern, P. E., Fleming, S. J., and Katz, S. H. (eds), *The Origins and Ancient History of Wine*, Gordon and Breach Publishers, Amsterdam, 233–254.

Magrill, P. and Middleton, A., 1997, A Canaanite potter's workshop in Palestine in Freestone, I., and Gaimster, D. (eds), *Pottery in the Making*, BMP, London, 68–73.

Neev, D., Bakler, N., and Emery, K. O., 1987, *Mediterranean coasts of Israel and Sinai: Holocene tectonism from geology, geophysics and archaeology*, Taylor and Francis, New York.

Nicholson, P. T., and Rose, P., 1985, Pottery Fabrics and Ware Groups at el-Amarna, in Kemp, B. (ed), *Amarna Reports II*, Egypt Exploration Society, London, 133–174.

Nordström, H-Å., and Bourriau, J.,1993, Ceramic technology: clays and fabrics, in Arnold D., and.Bourriau J., (eds), *An introduction to ancient Egyptian pottery*, von Zabern, Mainz, 143–190.

Parr, P. J., 1973, The origin of the Canaanite jar, in Strong, D. E, (ed), *Archaeological theory and practice*, Seminar Press, London, 173–81.

Parrot, J.-F., 1980, The Baër-Bassit (northwestern Syria) ophiolitic area, *Ofioliti* Special Issue: Tethyan ophiolites 2. Eastern area, 279–295.

Picard, L. Y., and Golani, U., 1992, *Geological map: northern sheet. 1:250,000*, Geological Survey of Israel, Jerusalem.

Pinar-Erdem, N. and Ilhan, E., 1977, Outlines of the stratigraphy and tectonics of Turkey, with notes on the geology of Cyprus, in Nairn, A. E. M., Kanes, W. H., and Stehli, F. G. (eds), *The ocean basins and margins. Vol. 4A The eastern Mediterranean*, 277–318.

Serpico, M., 1996, *Mediterranean resins in New Kingdom Egypt: A multidisciplinary approach to trade and usage*, Unpublished Ph.D. Thesis, University College, London.

Serpico, M., 1999, New Kingdom Canaanite Amphorae Fragments from Buhen, in Leahy, A., and.Tait, J. (eds), *Studies on ancient Egypt in honour of H. S. Smith*, Egypt Exploration Society, London, 267–272.

Serpico, M., 2000, Resins, amber and bitumen, in Nicholson, P. T., and Shaw, I. (eds), *Ancient Egyptian materials and technology*, CUP, Cambridge, 430–474.

Serpico, M., and White, R., 1998, Chemical Analysis of Coniferous Resins from Ancient Egypt Using Gas Chromatography/Mass Spectrometry (GC/MS), in Eyre, C. J. (ed), *Proceedings of the Seventh International Congress of Egyptologists*, Peeters, Leuven, 1037–1048.

Serpico, M., and White, R., 2000, Oil, fat and wax, in.Nicholson, P. T., and Shaw, I. (eds), *Ancient Egyptian materials and technology*, CUP, Cambridge, 390–429.

Sivan, D., 1996, *Paleogeography of the Galilee coastal plain during the Quaternary*. Report GSI/18/96, Geological Survey of Israel, Jerusalem.

Sivan, D., Gvirtzman, G., and Sass, E., 1999, Quaternary stratigraphy and paleogeography of the Galilee coastal plain, Israel, *Quaternary Research*, 51, 280–294.

Stern, B., Heron, C., Serpico, M., and Bourriau, J., 2000, A comparison of methods for establishing fatty acid concentration gradients through potsherds: a case study using Late Bronze Age Canaanite amphorae, *Archaeometry*, 42 (2), 399–414.

Whitechurch, H., Juteau, T. and Montigny, R., 1984, Role of the Eastern Mediterranean ophiolites (Turkey, Syria, Cyprus) in the history of the Neo-Tethys, in Dixon, J. E., and Robertson, A. H. F. (eds), *The geological evolution of the Eastern Mediterranean*, Geological Society Special Publication 17, published for the Geological Society, Blackwell Scientific Publications, Oxford, 301–317.

Chapter 8

Glass and faience at Amarna: different methods of both supply for production, and subsequent distribution

Andrew Shortland, Paul Nicholson and Caroline Jackson

Abstract

The ancient city of Amarna, situated beside the Nile in Middle Egypt, dates from a short period during the mid to late 14th century B.C. Amongst the remains of religious, civil and domestic buildings, several excavators have uncovered the sites of a number of glass and glazing workshops. At these workshops, sometimes situated in 'slum' areas right in the centre of the residential city, glassworking and faience making appears to be going on side by side probably by the same craftsmen. However, the organisation of supply and distribution of these two very similar (to the modern eye) materials is probably very different. The evidence for a royal monopoly in glassmaking is assessed and compared to the pattern of local production seen in faience. The distribution of glass and faience at Amarna is then considered, showing that glass was rare and linked to royalty, whereas faience common and widespread. The implications of this for the organisation of the industries and their place within Egyptian social and economic systems is then assessed and a reason for the location of the workshops proposed.

INTRODUCTION

Amarna is the modern name for the 18th Dynasty capital founded by Amenophis IV, later known as Akhenaten (1353–1337 B.C.). The city was inhabited for only a relatively short period of time, and subsequently robbed of almost everything of value during the 19th and 20th Dynasties (Kemp, 1989). The majority of the site was not built on again and therefore anything found in a stratified context within the City can be firmly dated to the reign of Akhenaten or soon after. The nature of Amarna was first revealed to modern eyes by excavations in the late 19th and early 20th centuries. W. M. F. Petrie embarked on the first scientific evaluation of the site in 1891, uncovering the royal palace and what he believed to be several glass and glazing factories (Petrie, 1894). From 1904 to 1914 the Deutsche Orient-Gesellschaft gained the concession and under the direction of Ludwig von Borchardt began a clearance of the residential part of the

Main City (Borchardt, 1907–14; Borchardt, 1980). After the 1914–18 War, the concession was transferred to the Egypt Exploration Society (EES), who held it between 1921 and 1937 (Peet and Woolley, 1923; Pendlebury, 1951; Frankfort and Pendlebury, 1933). They succeeded in making further clearances of the Main City and also surveyed and excavated some of the outlying areas, including the Workmen's Village. In 1937 work ceased until it was resumed again in 1977, when the EES began a systematic study of the site under the direction of Barry Kemp.

This paper considers the evidence for glass and faience making at Amarna. It endeavours to show that although there is good evidence to believe that glassworking and faience making was going on simultaneously in the same facilities, the supply of raw materials for the manufacture of these materials was probably quite different. It also shows that the end users of the products also fell into different socio-economic groups. To this end, the location of the various workshops is briefly described along with the evidence that indicates that the facilities were common. The evidence for glassmaking is then discussed, along with the distribution of glass finds at Amarna. This is then contrasted with the situation that existed for faience, where a different pattern of raw material acquisition existed, leading to a different distribution of the finished product. The conclusion then considers the wider social and economic implications of these patterns.

LAYOUT OF AMARNA

The City of Amarna is in Middle Egypt on the eastern bank of the Nile, occupying a low level plain fringed by cliffs. Various excavators have given names to parts of the City which are still widely used and are shown in Fig. 8.1. The area known as the Central City was amongst the first part of the City to be constructed. It lies at the heart of Amarna and includes most of the religious and administrative buildings. To the south of the Central City lies the Main City, in which the majority of the population resided and in which, due to the evidence here for industrial activity, the main focus for this study lies. Here there are a number of larger houses, built along the main thoroughfares, leaving a fair amount of space between their compound walls. The gaps between these buildings have been filled in with smaller houses, called 'slums' by some of the excavators (Frankfort and Pendlebury, 1933). Although adopting the same basic internal plan of the majority of the houses, houses in these 'slum' areas tend to be flimsily built and sharing common walls. They often lean right up against the compound walls of the larger houses and are connected by small narrow alleys to the main thoroughfares and, due to their thin walls, they are often very poorly preserved. Thus, there is a complex intermingling of the poorest and richest of the houses in some areas of the town, most notably the northern part of the Main City (Kemp, 1989).

To the south of the Main City is the South Suburb and further temple complexes (Maru-Aten and Kom el-Nana). Surrounding the City are various outlying areas, including the North Tombs and the South Tombs in the cliffs to the east of the City and the Royal tombs (Aldred, 1988, 28). The Workmen's Village in which the workers resided who were building these tombs has also been excavated and lies on the plain close to the tombs (Kemp, 1989, 273; Frankfort and Pendlebury, 1933).

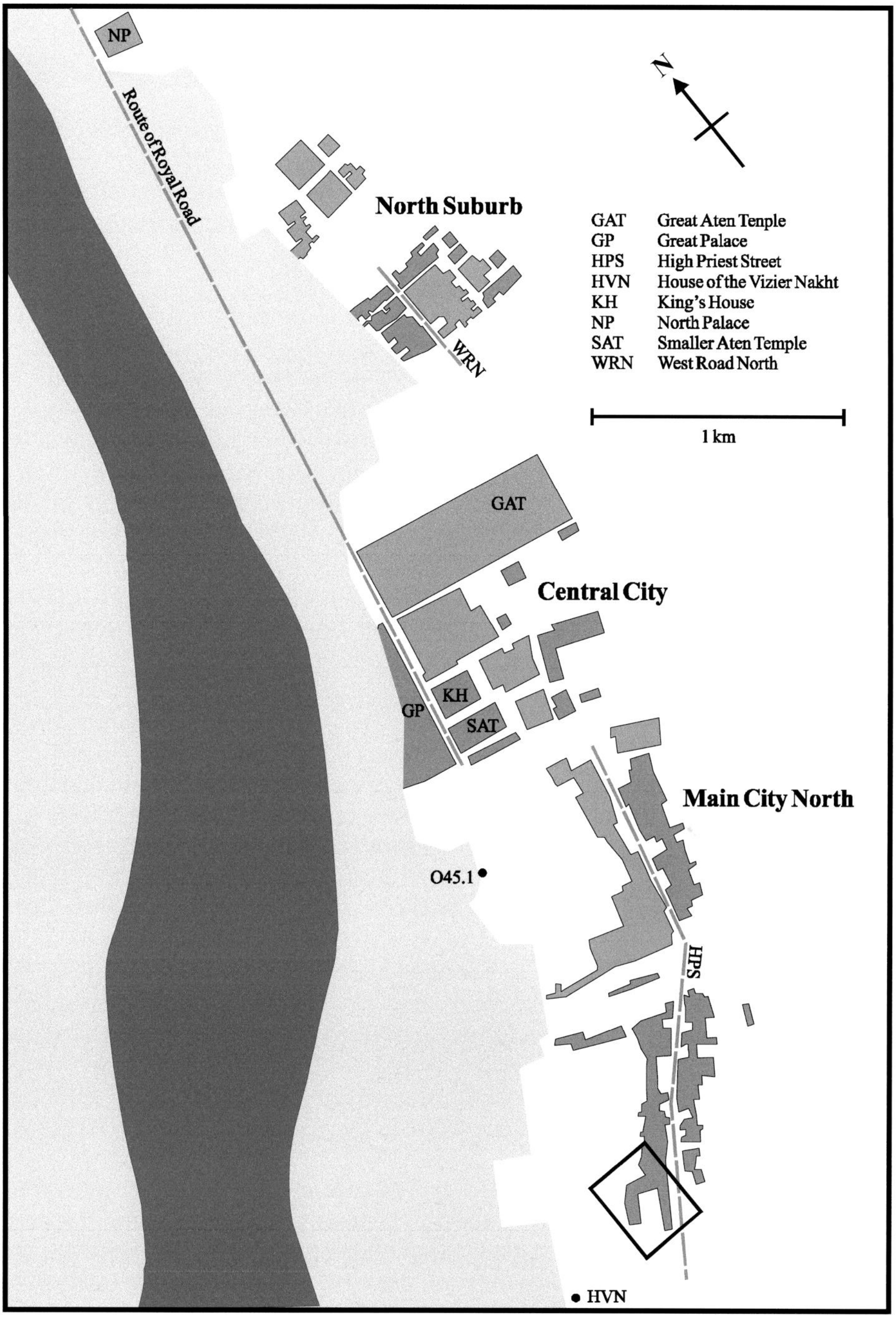

Figure 8.1: Layout of the main residential areas of Amarna beside the Nile (horizontal hatching) and cultivated area (dotted), after Kemp (1989, fig.89).

MANUFACTURE OF VITREOUS MATERIALS AT AMARNA

The first excavations of Amarna by Petrie revealed evidence of glassworking and glassmaking (Petrie, 1894, 25–7). This evidence was mostly drawn from extensive waste heaps of fragments and included frit and fritting pans, cylindrical crucibles with glass adhering and glass rods and spills. High concentrations of faience were found in the same areas as the glassworking debris. From the scatter of all this debris, Petrie reports that he found 'three or four glass factories, and two large glazing works and though the actual workrooms had almost vanished, the waste heaps were full of fragments that shewed [*sic*] the methods employed' (Petrie, 1894, 25).

Regrettably, Petrie omitted to note the location of some of the areas of the city where was excavating. Although the location of major sites like the temples and palaces are noted on his site map (Petrie 1894, plate XXXV), Petrie fails to state where his five or six workshops were, the only hint being the suggestive word 'moulds' that was written on his map about a quarter of a mile south-southeast of the palace. This site is in grid square O45, in the Main City, some distance from the areas that have been contiguously excavated and right up against the area of modern cultivation. Paul Nicholson subsequently excavated this site (Nicholson, 1995a, 1995b and 1998) and uncovered two kilns 2m in diameter, described as thick walled and highly vitrified, with a sacrificial lining and associated with a large amount black slag. A third smaller kiln was found, apparently associated with the other two, and of a type recognised by Nicholson as a pottery kiln. Associated with the site were frit, melted glass, glass rods, fragments of cylindrical vessels for the melting of glass and faience moulds. All of this suggests strongly that this site, designated O45.1, was used for the manufacture of both vitreous materials and pottery.

Petrie gives no clue as to the location of the other glass factories, the location and nature of which have been completely lost. However a survey of the distribution of the baked clay faience moulds and other debris (Shortland, 2000a) has revealed clusters of debris that sometimes centre on kilns or other features, which may be the sites of the workshops that Petrie describes. At two of these clusters there are a number of kilns associated with moulds and glass and faience debris, while at a third there is also tentative evidence for the working of precious metals. Two further sites have moulds and faience without the presence of large amounts of glass and are associated with sculptors' workshops. Three further areas are possible, two in the Central City and one in the Northern Suburb, but the information for each is poor. All of these workshops are sited in poorer areas of the heart of the city, within the flimsily built 'slum' houses described above and often built right against the perimeter wall of larger, richer properties.

The excavations by Nicholson at O45.1 and the survey of other debris over the Main City area therefore show that many of the different industries (stone, pottery, precious metals and particularly glass and faience) were taking place very close together. Indeed in most cases the same buildings were being used for a combination of industries and these were usually sited in the heart of the residential area of the city.

GLASS AND GLASSMAKING – A ROYAL MONOPOLY?

The many glass rods that have been found at Amarna and Malkata (the palace of Amenhotep III on the west beank of modern Luxor) definitely indicate that glass*working* is occurring in these workshops. The presence at Amarna of cylindrical vessels with drips of glass indicate glass melting is also being carried out, perhaps associated once again with glassworking. Finds of fritting pans and frits was originally taken as being evidence for glass*making* (Turner, 1954; Petrie, 1894), but recent work (Shortland and Tite, 1998) has shown that these frits are almost certainly not associated with glassmaking. It is instead more likely that they are an intermediate stage in the manufacture of what Lucas (1962: 163-4) calls Variant D vitreous faience. Therefore, the question of glassmaking has to be considered in more detail.

A number of authors have suggested that glass may be the subject of a royal monopoly, at least during the early stages of its history (Nolte, 1968; Nicholson, 1995b, and others). The evidence for this is as follows:

1. Textual evidence
    a) presence of cartouches associated with glass in reliefs
    b) high value given to glass in Egypt
    c) evidence that glass was given to the King in tribute
2. large glassmaking workshops at Qantir
3. distribution of finds of glass at Amarna

Textual evidence

In the relief depicting the dedication of the spoils of the campaigns of Tuthmosis III in the Festival Hall at Karnak, wherever the words for glass appear, they seem to be followed by the royal cartouche, a distinction not granted to any of the other materials on the relief. This shows a special royal interest in the material. This relief and other written records indicate that glass was a valuable material and therefore likely to be in great demand by the court. It would seem reasonable therefore that such a valuable material would be closely controlled. The Amarna Letters show that the King was very keen to obtain glass from satellite states in the Levant and Syria. Glass was therefore being sent to Egypt from abroad, probably in ingot form.

Glassmaking at Qantir

Rehren and his colleagues (Rehren and Pusch, 1997, and Rehren *et al*, this volume) excavating at the Ramesside glass workshops at Qantir have revealed additional information which sheds new light on the problem of glassmaking. Unlike at Amarna, examples of intermediate stages in glassmaking and raw glass, refractory materials and diagnostic slags have been found at Qantir (and are suggested by Rehren and Pusch to be indicators of glassmaking). However, Qantir has none of the glassworking debris found at Amarna and Malkata. Rehren therefore suggests that different sites are used for glass*making* (Qantir) and for glass*working* (Amarna and Malkata), with glassmaking sites perhaps specialising in one colour (red at Qantir). It is unclear whether there was also a glassmaking site at Amarna. Of various locations for faience

Table 8.1: Glass recovered in the excavation of 260 houses at Amarna (Shortland, 2000c)

type	blue	blue-green	green	grey-blue	yellow	white	red	violet	polychrome	colourless	not stated	*total*
scarabs	0	0	0	0	0	0	0	0	0	0	0	*0*
jewellery	20	0	3	1	2	9	4	0	14	2	16	*71*
inlays	1	0	0	0	0	1	0	0	0	0	1	*3*
tiles	5	0	0	0	0	0	1	0	1	0	0	*7*
vessels	5	0	0	0	0	0	0	0	11	0	3	*19*
plant models	0	0	0	0	0	0	0	0	0	0	0	*0*
animal models	0	0	0	0	0	0	0	1	1	0	0	*2*
figurines	0	0	0	0	0	0	0	0	0	0	0	*0*
others	1	0	0	0	1	1	0	0	2	0	0	*5*
manufacturing debris	8	0	1	1	4	0	2	1	1	0	0	*18*
not stated	1	0	1	0	0	0	0	0	3	1	0	*6*
total	*41*	*0*	*5*	*2*	*7*	*11*	*7*	*2*	*33*	*3*	*20*	*131*

and glass workshops discussed above, only the site of O45.1 seems possible in terms of the scale of the kilns, but conclusive proof has yet to be found. However, it is clear that cobalt blue glass is being made of Egyptian materials and therefore almost certainly somewhere in Egypt (Shortland, 2000b). It has also been shown that the cobalt blue glass ingots recovered from the Ulu Burun shipwreck are the right size to have been made in the cylindrical vessels ('ingot moulds') found at Amarna (Nicholson *et al.*, 1997). The balance of evidence therefore suggests that there is a glassmaking facility somewhere at Amarna, either at O45.1 or as yet undiscovered. The only other alternative is that the cylindrical vessels are a uniform shape and size and used at all the glass workshops and therefore the glassmaking is going on elsewhere in Egypt.

Distribution of glass

Table 8.1 lists the results of the survey of glass recovered in the excavation of 260 houses at Amarna (Shortland, 2000c, see Appendix for methodology). The overall finds of glass number only 131, with very small items of jewellery being the most common type of object at around 54%. This percentage is made up of few amulets and very few finger rings, but a high percentage of ear-rings and beads. The other major usage of glass is glass vessels, fragments of 19 being found, eleven being polychrome. A significant proportion of all the glass objects but especially the glass vessel fragments were found in, or around the houses defined in Shortland (2000a) as being connected with glass working. Perhaps half the vessel fragments can be accounted for in this way. This leaves a total of five or six pieces of polychrome glass vessel in the survey that

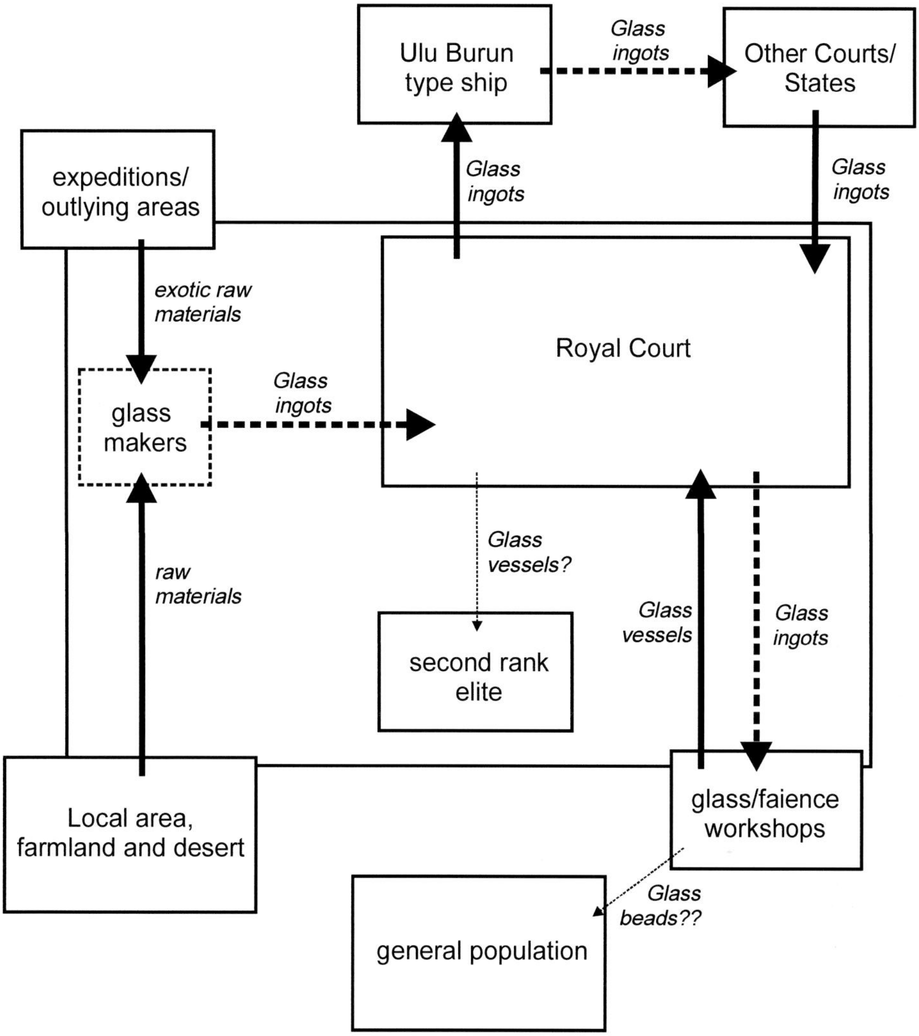

Figure 8.2: Production and distribution of glass.

were found in non-workshop houses. It is instructive to compare this with the finds that Petrie made in the Palace dumps. Petrie (1894) did not keep, or even count, all the pieces of glass vessel he found in his excavation of the dumps of broken material thrown out by the Palace complex. He counted only the polychrome fragments that 'had a distinct pattern', that is to say that small pieces where it was not clear whether they were patterned by dragging, spiralling or blotching, etc, were not counted. Despite this, Petrie lists 282 fragments of glass vessel that he found. The finds from the Palace dumps therefore outnumber the finds in 260 houses of the Main City by around 50 to one when the glassworking areas are taken into account or 25 to one overall. This is again good evidence that tight control was exercised in the distribution of glass and that the glass is largely associated with royal usage. The small objects of jewellery found around Amarna may well have been produced from spare glass left over from the manufacture of the vessels.

Summary

Glassmaking seems to have been restricted to a small number of large Qantir-type manufacturing facilities in Egypt and abroad, each specialising in a small subset of the possible colours. These facilities were tightly controlled by the King and state and glass from them was exchanged between courts as gifts or tributes. The production of glass vessels was achieved in faience and glass workshops like those at Amarna. They were perhaps supplied with an appropriate weight of coloured glass ingots obtained by the King from both Egypt and abroad. These would then be worked up into vessels in their workshops and the vessels sent back to the commissioning body. They may have been paid with the scrap glass or by some other method. Figure 8.2 shows a summary for the production and distribution of glass.

FAIENCE MAKING – A PRIVATE MARKET?

There is no evidence that faience was treated as a royal monopoly. While glass was only introduced into Egypt some 100 years or so before Amarna, faience had been produced since the Predynastic Period. Its production was therefore well and truly embedded in Egyptian society. The debris of the faience production at Amarna includes baked clay moulds with unfired faience paste still stuck in them (Friedman, 1998, 166; Shortland, 2000a, Fig. 2). This makes it very clear that faience *making* is definitely occurring at Amarna. This manufacture is working from the original raw materials, probably sand and plant ash with a little colorant, usually copper. The first two of these raw materials are readily available around Amarna and no special organisation would be required to obtain them (Stocks, 1997). Copper and bronze scrap would be available from metal workshops and require slightly more effort to acquire, but relatively little copper would be required. Rarer more exotic colorants like lead antimonate and cobalt may have been obtained by barter or, perhaps more likely, through 'official' channels. For example, the lead source for the lead antimonate may lie at Gebel Zeit (Shortland *et al.*, 2000) and it is known that the Temple of Amun at Karnak was granted rights to extract galena from this site for kohl (Castel *et al.*, 1988). Galena for the production of lead antimonate was therefore available through the temples. Similarly gifts/taxes were being given to the King by the governor of the Western Oases (Giddy, 1987), where the alums used in the production of cobalt colorant are found. Probably cobalt colorant could therefore also have been obtained through the temples or state bodies.

The manufacture of blue 'ordinary faience' (Lucas and Harris, 1962) required very little in the way of specialist materials, and the surviving kilns at Amarna show that a wide variety of shape and design of kiln could have been used. The 1500 years of history of faience production would also suggest that the techniques employed were probably widely known.

Table 8.2: Faience found in the excavation of 260 houses at Amarna (Shortland, 2000c)

type	blue	blue-green	green	grey-blue	yellow	white	red	violet	polychrome	colourless	not stated	*total*
scarabs	118	14	27	3	1	1	0	0	0	0	1	*165*
jewellery	599	31	87	11	19	13	8	9	88	0	9	*874*
inlays	69	1	4	2	2	3	5	1	70	0	6	*163*
tiles	29	1	6	0	0	0	0	1	31	0	1	*69*
vessels	20	1	5	0	2	1	0	0	14	0	1	*44*
plant models	58	0	2	1	1	1	1	0	10	0	1	*75*
animal models	68	3	1	1	0	0	2	0	3	0	0	*78*
figurines	25	0	1	0	0	0	0	0	4	0	0	*30*
not stated	7	0	2	1	1	0	1	0	4	0	0	*16*
others	42	1	1	1	2	1	1	1	3	0	0	*53*
total	*1035*	*52*	*136*	*20*	*28*	*20*	*18*	*12*	*227*	*0*	*19*	*1567*

Distribution of faience

A total of 1567 pieces of faience were recorded in the survey of the 260 houses (see appendix) and can be classified as shown in Table 8.2. Faience was used to construct a wide range of objects, with over half the objects being jewellery (56%), followed by roughly equal numbers of scarabs and inlays (both about 10%), and lesser numbers of tiles and items modelled in the round, like animals plants and figures. Rarer examples of other items, playing pieces, writing instruments and so forth were also found. Of the jewellery, which included beads, rings, amulets and ear-rings, a very high proportion of the finds were finger rings. Bezels and shanks from these objects accounted for a total of 44% of the jewellery (388 objects), or one in four of all the pieces of faience from Amarna.

Faience can therefore be seen to be a common material at Amarna. Over 90% of the houses in the survey contained at least one piece of faience. This shows that faience objects were a regular part of everyday life amongst the population and also, since so many of the smaller items were found left on the site, that they were probably not of great value.

Control of faience workers

Little direct evidence is available for the organisation of faience workers, but what little is known suggests they were organised into teams controlled by a chief faience worker (*ḥry irw ḥsbḏ*), ultimately under an overseer of faience workers (*imy-r irw ḥsbḏ*) (Nicholson, 1998). These are state titles and suggest control directly by the administration. A further stela (20th Dynasty, National Museum of Scotland

A.1956.153) gives the title of a man called Rekhamun as 'faience maker of Amun' (*irw ḫsbḏ n'Imn*) showing that temples also controlled their own faience workers. This might suggest that tight administrative control was also exercised on the production of faience and certainly the faience workers seem to come under the administrative structure. However, the widespread distribution of the material argues that the picture is more complicated than this.

Some authors (Kemp, 1989; Janssen, 1975) have shown that while certain parts of the Egyptian economy were controlled by an administrative, redistributive system (for example, as seen in the glass described above), there was also a 'relatively dynamic private sector' (Kemp 1989). Records from Deir el-Medina show that although the artisans in this village had titles suggesting that they were in the employ of the state, they were free to engage in private commissions as well. Everything had a value expressed in *dbn* (weight of copper) and purchases were made by exchanging goods of equal value. For example, Ostracon Deir el-Medina 73 details the exchange of a coffin (value 25.5 *dbn*) for two goats, one pig, two sycamore logs and a weight of copper scrap (Janssen, 1975). For richer households, the preliminaries for these exchanges was carried out by an employed servant, who held the title of *šwty*, translated as either 'trader' or 'commercial agent' (Kemp, 1989). All *šwty* were employed by an institution, temple or family, there do not seem to have been independent merchants deriving their income from the profits of trade, indeed there is no word for 'profit' in the Egyptian vocabulary.

The wide distribution of the faience and its implied low value suggests that it was, in part, the subject of private exchanges and barters. Texts from Deir el-Medina and elsewhere give every reason to believe that this would be the case. The presence of *šwty* and an active private sector also explains the distribution of the faience workshop sites at Amarna. The main demand for larger more expensive pieces would obviously come from richer people who lived in the more spacious houses. These people employed *šwty* to carry out there purchases and there would be a significant advantage to have your workshop as close as possible to the demand, ie. right up against the wall of the larger houses. Being in the centre of the main residential areas would also enable exchange with those who could not afford to employ specialist traders. Figure 8.3 shows a summary for the production and distribution of faience.

It is very probable that the faience makers were also making items for the court and the temples, but it is difficult to pick out which items might have been made for the private market and which for either state display or redistribution – perhaps all items were made for both. The faience finger rings are particularly interesting. They are so common that it is not unreasonable to suggest that perhaps most people at Amarna might have owned one or more. Could they represent a 'must have' fashion that lead to everyone purchasing one, or were they handed out to all in a window of appearances type ceremony? Either is possible, but without further textual or pictorial evidence it is impossible to be sure.

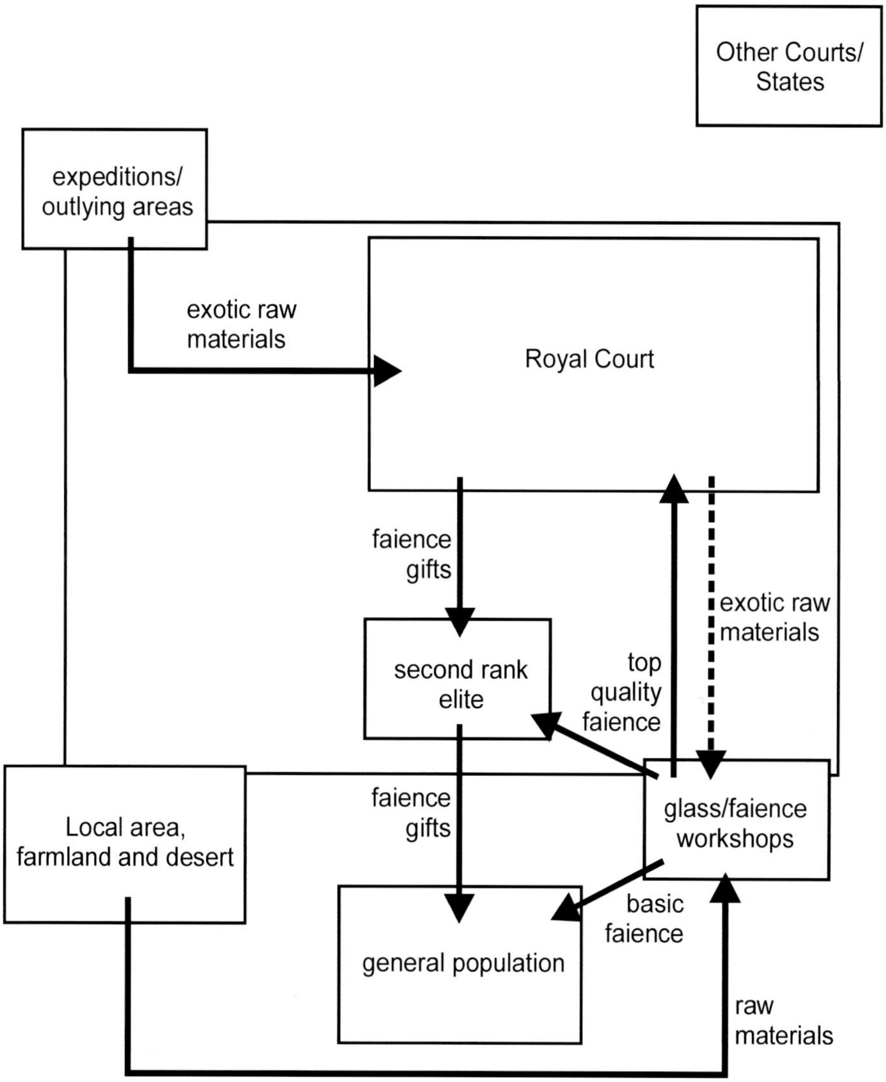

Figure 8.3: Production and distribution of faience.

CONCLUSION

The organisation at Amarna of supply for the production of faience and glass and its subsequent distribution is different for each material. Glass, a new technology, was the subject of tight royal control amounting almost to a monopoly. It was produced in large specialised facilities probably under some degree of secrecy, and exchanged between courts principally in the form of ingots, but also of glass vessels. The glass and faience workshops at Amarna were probably supplied with coloured glass ingots, which they worked into glass vessels and returned to the court. A small amount of perhaps left over glass was used to manufacture other items like beads and earrings.

In contrast, faience was an old technology firmly embedded within Egyptian society. Its raw materials were common and (for the most part) easily obtained without the need of complex organisational structures. Faience is a common material and of relatively low value. Some objects, for example faience finger rings are still so common on the site that it is perhaps possible that most adults might have owned one or more. After objects of fired clay, it is the most common man-made material surviving on the site. It is probable that a relatively large and active private market existed in faience, free of the restrictions of value and supply of glass and similar to that seen in many different objects at Deir el-Medina. The active private market also explains why so many of the faience workshops are right in the centre of the residential area of the city, since there is a big advantage to being close to the customers for their products.

ACKNOWLEDGEMENTS

The authors are most grateful the Egyptian Supreme Council for Antiquities for permitting work at Amarna. PN's work at this site has been funded by the Wainwright Fund for Near Eastern Archaeology, to whom he is indebted, and conducted under the auspices of the Egypt Exploration Society and Mr Barry Kemp, director of the Amarna project. PN and CJ's work was supported by Leverhulme Grant F/407/J and they would like to thank them for their support in this work.

APPENDIX – SURVEY METHODOLOGY

Previous work (Shortland, 2000c) includes a survey of the finds of faience and glass throughout part of the Main City of Amarna. The data for the survey was derived from the finds lists of 260 houses excavated by the Deutsche Orient-Gesellschaft between 1904–14 and published by (Borchardt, 1980). From this list, a database was produced, which classified each object by type (jewellery, figurine, tile, etc.), colour, and house where it was found (Shortland, 2000c). The difficulty in conducting such a survey and interpreting the results are numerous and must be taken into account. The houses in Amarna were abandoned by their owners, who were probably in no hurry to leave and subsequently had plenty of time to remove any goods that were of interest to them, leaving only the remnants for the excavator. Further losses occur through subsequent occupation, weathering, treasure hunting and so forth, severely depleting the remains. The evidence is therefore heavily skewed towards broken or unwanted items, a process that has been termed 'negative selection' (Patch, 1998), with an occasional utilitarian vessel and very rare, not necessarily typical, but frequently cited first class examples from hoards or mislaid small items. However, if the limitations are taken into account, the information derived from a survey at Amarna could give a very valuable insight into possible distribution patterns.

REFERENCES

Aldred, C., 1988, *Akhenaten, King of Egypt*, London.

Borchardt, L., 1907–14, Ausgrabungen in Tell el-Amarna, *Mitteilungen der Deutschen Orient-Gesellschaft*, 34, 46, 50, 52, 55.

Borchardt, L. and Ricke., H., 1980, *Die Wohnhäuser in Tell el-Amarna*, Deutschen Orient-Gesellschaft, Berlin.

Castel, G., Pouit, G. and Soukiassian, G., 1988, Les mines de galene pharaonique du Gebel Zeit (Egypte), dans le Miocène du rift de la Mer Rouge, *Chronique de la Recherche Minière*, 492, 19–32.

Frankfort, H. and Pendlebury, J. D. S., 1933, *The City of Akhenaten II*, Egypt Exploration Society, London.

Friedman, F. D. (ed.) 1998, *Gifts of the Nile – Ancient Egyptian faience*, Thames and Hudson, London.

Giddy, L. L., 1987, *Egyptian Oases*, Aris and Phillips, London.

Janssen, J. J., 1975, *Commodity prices from the Ramesside Period: An economic study of the village of necropolis workmen at Thebes*, Leiden.

Kemp, B. J., 1989, *Ancient Egypt: Anatomy of a civilization*, Routledge, London.

Lucas, A. and Harris, J. R., 1962, *Ancient Egyptian materials and industries*, 4th Edition, London.

Nicholson, P. T., 1995a, Glass-making and glass-working at Amarna: some new work, *Glass Technology*, 37.1, 11–19.

Nicholson, P. T., 1995b, Recent excavations at an ancient Egyptian glassworks: Tell el-Amarna 1993, *Glass Technology*, 36.4, 125–128.

Nicholson, P. T., 1998, Materials and technology, in Friedman, F. D. (ed.), *Gifts of the Nile – Ancient Egyptian faience*, Thames and Hudson, London.

Nicholson, P. T., Jackson, C. M. and Trott, K. M., 1997, The Ulu Burun glass ingots, cylindrical vessels and Egyptian glass, *Journal of Egyptian Archaeology*, 83.

Nolte, B., 1968, *Die Glassgefässe im alten Ägypten*, Verlag Bruno Hessling, Berlin.

Patch, D. C., 1998, By necessity of design: faience use in ancient Egypt, in Friedman, F. D. (ed.), *Gifts of the Nile – Ancient Egyptian faience*, Thames and Hudson, London.

Peet, T. E. and Woolley, C. I., 1923, *The city of Akhenaten I: excavations of 1921 and 1922 at el-Amarna*, Egypt Exploration Society, London.

Pendlebury, J. D. S., 1951, *The city of Akhenaten III*, Egypt Exploration Society, London.

Petrie, W. M. F., 1894, *Tell el-Amarna*, London.

Rehren, T. and Pusch, E., 1997, New Kingdom glass melting crucibles from Qantir-Piramesses, Nile Delta, *Journal of Egyptian Archaeology*, 83.

Shortland, A. J., 2000a, The number, extent and distribution of the vitreous materials workshops at Amarna, *Oxford Journal of Archaeology*, 19.2, 115–34.

Shortland, A. J., 2000b, Raw materials of glass from Amarna and implications for the origins of Egyptian glass, *Archaeometry*, 42.1, 141–53.

Shortland, A. J., 2000c, *Vitreous materials at Amarna: the production of glass and faience in 18th dynasty Egypt*, British Archaeological Reports International Series, S827, Archaeopress, Oxford.

Shortland, A. J., Nicholson, P. T. and Jackson, C. M., 2000, Lead isotopic analysis of 18th dynasty Egyptian eyepaints and lead antimonate colorants, *Archaeometry*, 42.1, 153–159.

Shortland, A. J. and Tite, M. S., 1998, The interdependence of glass and vitreous faience at Amarna, in McCray, P. and Kingery, W. D. (eds.), *The Prehistory and History of Glassmaking Technology*, The American Ceramics Society, USA, 251–268.

Stocks, D. A., 1997, Derivation of ancient Egyptian faience core and glaze materials, *Antiquity*, 71, 179–82.

Turner, W. E. S., 1954, Studies in ancient glasses and glass making processes. Part I. Crucibles and melting temperatures employed in Ancient Egypt at about 1370 BC, *Journal of the Society of Glass Technology*, 38.

Chapter 9

Gold and Granulation: Exploring the Social Implications of a Prestige Technology in the Bronze Age Mediterranean

Thea Politis

Abstract
The variability present in the early developments of the technology of granulation, the most difficult gold-smithing technique to perfect and master, can be used to define and explore socio-economic relationships within the archaeological systems of the Early to Late Bronze Age Mediterranean. This paper will discuss and outline the integration of the theoretical and the practical, developing a middle range theory specific to assessing the significance of the early history of this prestige technology. Furthermore, concepts such as technological innovation, style, and cross-craft interaction will be considered along with the intersection of local and international meaning and practice with illustrations drawn from material from the Hyksos stronghold of Tell el 'Ajjul in Southern Palestine, and Thera and the Greek Mainland in the Aegean.

INTRODUCTION

Granulation is generally understood as a smithing technique that fuses small granules of gold to a metallic substrate using heat, most often with the aide of a copper agent. This technology allows the artisan to create highly light reflective and intricate patterns. The Etruscans produced perhaps the finest and most famous examples of the technique, but the metallurgical effect has a long history of development dating back to the third millennium B.C. Granulation belongs to a class of techniques, that are especially important in the earliest development of granulation. The characteristic that they all have in common is the maximization of the light reflective qualities of gold using raised spheres. Consequently, discussion will include a number of techniques that may not even require heat, but do aim to qualitatively produce the same light reflective patterning.

Most studies of granulation address either style or technology, and rarely combine the two elements. Both types of studies are extremely useful in that they provide a rich corpus of discussions of details of the objects and parallels from various sites to draw upon; yet, if the questions posed to the archaeological data are more social in nature, emphasis must shift away from descriptions of the objects as the *main* objective of the study. Instead, it is necessary to emphasize the people creating and using the objects rather than the process or end products. The objects still must be carefully examined, but in a forensic manner as evidence of past actions. There are, of course, examples of more socially oriented studies done with jewelry/goldwork (e.g. Clark, 1986; Hoffman, 1997; Renfrew, 1986) but few deal carefully with technological and material evidence. Notable exceptions to this statement come from the disciplines of Pre-Columbian archaeology (e.g. Hosler, 1994; Lechtman, 1984) and European Prehistory (e.g. Eluère, 1989; Echt & Thiele, 1995). With patience, prestige materials can be quite successfully used as a vehicle with which to bridge the gap between the 'things' dug up by field archaeologists and the issues and questions raised by archaeological theorists. Other types of material culture such as pottery, lithics, and less noble metals certainly do not have an exclusive ability to shed light on such issues as agency, innovation, or political economy. Furthermore, prestige goods and technologies are often *more* useful for studying some of these questions than their more common counterparts due to the frequently controlled nature of their proliferation in the archaeological past.

TECHNOLOGICAL CONSIDERATIONS

Often the method which uses copper based powder metallurgy in joining granules to a substrate (copper salt reaction soldering) is understood to be the only true method of granulation, especially in more art historically oriented literature. When the intent of the craftsman, based on visual effect, is taken into account, several techniques must be considered in addition to what is commonly considered 'true-granulation'. Even 'true-granulation' has several variations often indistinguishable by the naked eye. These techniques can be broken down into several groups.

Proto-granulation

This category is limited to the very earliest period of the development or introduction of granulation to a region. It is indicative of an atmosphere of experimentation and the development of light reflective gold-smithing techniques. Included within this class is an eclectic group of instances that are often unique to the sites that produce them, and are often unique to the objects themselves as well. The most famous examples of this class are the daggers from the Royal Cemetery of Ur which use nail studs that have melted heads in order to achieve the effect (Weber & Zettler, 1998, 164, cat. 146; Zettler, 1998, 17, Fig. 15). Another example is a bracelet from Early Bronze Age Alacahöyük in Anatolia. It has been shaped around a mold while the gold has been worked up into spikes and grooves have been cut into the surface to further catch the light (Kosay, 1944, 115, pl. XCII; Bingöl, 1999, 73, cat. 195). This technique uses similar methods to those employed in the manufacture of some earrings from Ur during the Isin-Larsa Period. In addition to the spiked effect, the Ur earrings also exhibit true-granulation surrounding the suspension wire where it attaches to the main body (Maxwell-Hyslop, 1971, 83, pl. 57a–c). Yet another early group of examples from the Anatolian Early Bronze Age indicate an environment of experimentation with a mixture of technical variations aiming for the same stylistic forms. Basket earrings from the site of Troy on the North Aegean coast exhibit rosette patterning using two technical methods. The first type is created by using shots or large granules in the center framed by a wire reflectively enhanced through deckling (Tolstikov & Treister, 1996, 48, cat. 13, 50, cat. 14). This method of creating rosettes is also evident in a Trojan hair-ring (Tolstikov & Treister, 1996, 138, cat. 157) and a bracelet (Tolstikov & Treister, 1996, 114–115, cat. 123). The second technical variation in basket earring decoration creates the rosettes completely out of true granulation (Tolstikov & Treister, 1996, 136–7, cat. 155), while other basket earrings from the site use linear granulation patterning (Tolstikov & Treister, 1996, 51–53, cat. 15–16, 116–117, cats. 125–126). The contemporary Anatolian site of Eskiyapar, located much further inland produces yet another technological variation on the basket earring. Here domed sheet gold, resembling granules and equally light reflective, is attached in a pattern paralleling the Trojan versions (Bingöl, 1999, 86, cat. 72–73; Özgüç & Temizer, 1993, 615, pl. 106).

Pseudo-granulation

Many of the techniques in the category of 'pseudo-granulation' are also included in the previous grouping of 'proto-granulation' but post date the initial, formative period of experimentation and innovation that characterizes the very earliest inception of the technique. The techniques encompassed by the term 'pseudo-granulation' are all technologically easier to accomplish and are readily and visibly distinguishable from the group of techniques referred to as 'true granulation'. Yet all of the these techniques are aiming to create the same effect which maximizes the light reflective and visibility qualities of the raw material. Frequently, as was illustrated with the basket earrings from Troy and Eskiyapar, the same forms, iconographies and styles are captured with these recognizably simpler technologies. Among these is the much familiar dot punchwork frequently seen, for example, in Levantine medallion pendants (e.g. Maxwell-Hyslop, 1971, pls. 100, 101, 108, 109, 115). Another technique involves the carving or incising of solid metal to create a raised surface pattern resembling granulation. Many times this technique is evident on the shanks of toggle pins (e.g. Maxwell-Hyslop 1971, pl. 73, pl. 130). A third technique involves casting an object in a form that looks like granulation.

True granulation

The difference between this class and the others is that it uses heat to fuse small metallic balls, or granules to a metallic substrate. It can be executed in gold, silver or copper, but gold is the usual medium. Silver examples do exist as early as the beginning of the second millenium B.C. in Elam (Amiet, 1966, 262, Fig. 193a) and silver becomes the favored medium for granulation in Early Iron Age Palestine (Golani, 1996; Platt, 1972). 'True granulation' is the most difficult group of techniques to perfect and should be classified as a prestige technology requiring significant knowledge and skill to achieve. As discussed above copper salt reaction soldering is commonly thought to be the only 'true granulation' technique. However, especially during the early development of the effect, a range of other techniques are employed. While copper-salt reaction soldering does predominate in the Classical and later periods which are better

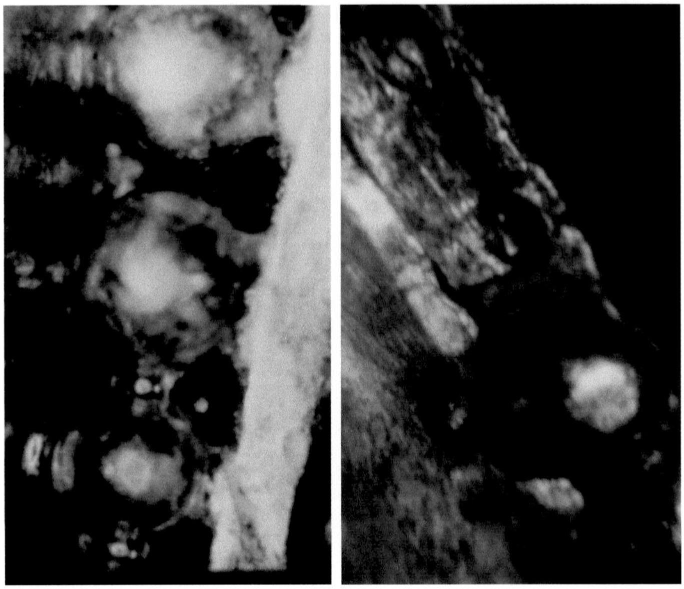

Figure 9.1: Sintering and soldering; A (left): Sinter join. Tell el 'Ajjul, Ashmolean 1949.308; B (right): Pitting from missing granule attached with copper salt reaction solder. Mycenae. National Museum, Athens

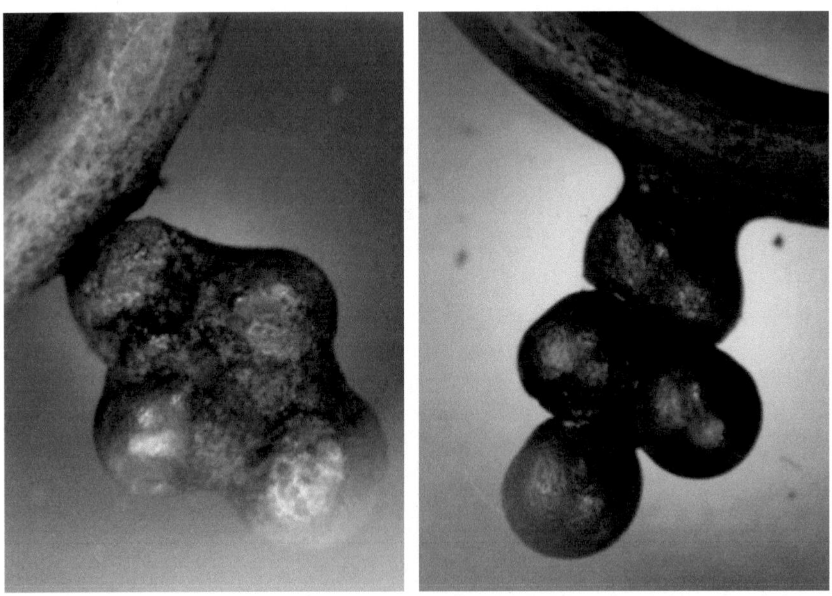

Figure 9.2: Details of mulberry earrings from Tell el 'Ajjul; A (left): Soldered cluster with copper salt reaction soldered join between cluster and hoop. British Museum WAA 130778; B (right): Copper salt reaction soldered cluster with solder join between cluster and hoop. British Museum L.1012

studied than the third and second millennium B.C. examples, Robert Baines notes that even in this later period a mixture of techniques may be used on the same piece due to the differing temperature needed for each of the techniques (Baines, 1998). Three techniques fall into the category of 'true-granulation.'

SINTERING (Fig. 9.1A)

This method can also be referred to in the literature as sweating, or fusing. Unlike the next two joining methods, sintering does not require an agent. In this approach a join is made by heating the metal surfaces to melting point — without reaching the point at which the bulk of the metal becomes liquid. This method works best with 95–100% purity of gold which would require reaching temperatures of approximately 1000° to 1063° C. The degree of skill, experience and temperature control necessary to perfect such a technique is extremely high, and this is probably the most difficult method to master. In this process the granules have a tendency to heat up before the substrate, and consequently the joins between granules are apt to be more flooded than those between the granules and the substrate.

COPPER-SALT REACTION SOLDERING (Fig. 9.1B–9.5)

This process is also referred to in the literature as diffusion bonding, brazing or colloidal hard soldering. This method achieves bonding through surface alloying. Finely powdered copper minerals are laid on the surface to be joined along with some carbon or an organic flux that transforms into carbon when heated–as the reducing agent. Organic glues that could also have contained a flux necessary in promotion of the flow of the solder may have been used to affix the granules prior to heating, but any of these are nearly impossible to detect. Under heat the copper mineral is reduced to metallic copper and at approximately 890° C, alloys with the gold in the surface zone and lowers the melting temperature of the gold in the join area. The surrounding gold remains untouched. Powder metallurgy requires a greater level of technical innovation to initially invent than sintering, but once introduced it requires less skill on the part of the artisan to perfect. Copper-salt reaction solders may be detected through several tell tale signs. Fig. 9.2A illustrates the small area of contact in the fine join created by this method in the area between the cluster and the hoop. Fine dendritic crystalline structures may still be evident, along with charcoal dust if this was used as the carbon reducing agent. Other signs may be a very distinctive haloing effect around an individual granule (Fig. 9.3) that results from overheating or the presence of too much copper in the agent (Untracht, 1985, 362–363), or a faint shadowing of the copper if it bleeds out of the join area. If a granule is missing, pitting may also be evident (Fig.9.1B).

SILVER-GOLD OR COPPER-GOLD SOLDER

In this method small pieces or chips of the chosen soldering alloy are laid into position, around the joint to be made. Next, a flux is applied to the surfaces to be joined. The piece is then placed into a charcoal fire, or held by tongs in the fire until the soldering alloy flows. The best choice of solder material melts at approximately 100° C below the fusion temperature of the alloy being soldered and matches closely in color to the gold

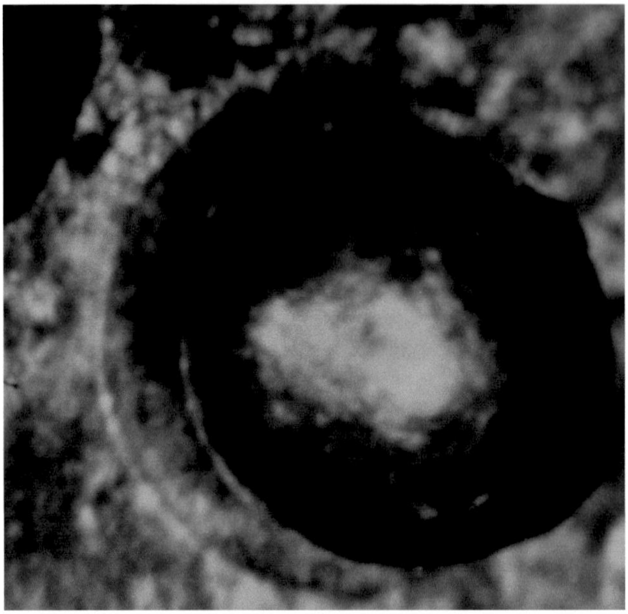

Figure 9.3: Halo effect from copper salt reaction solder. Tell el 'Ajjul, Ashmolean 1949.307.

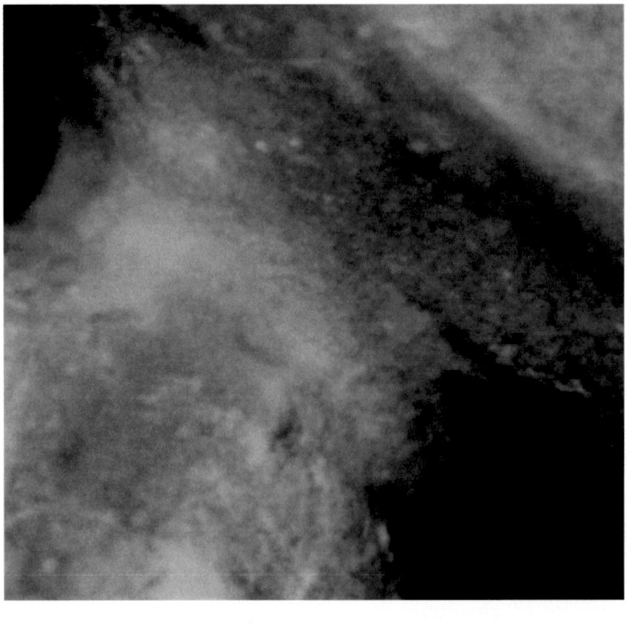

Figure 9.4: Silver-gold solder join. Tell el 'Ajjul. Ashmolean 1949.320.

elements being joined. The entire piece is preheated to solder flow temperature and then the flame is concentrated on the joint causing the soldering alloy to flow evenly (Untracht, 1985, 415). This method is often identifiable in simpler pieces by a large thickened join area with a slight color variation (Fig. 9.4). With a charcoal fire and blow-pipe or bellows, the amount and direction of heat would have been very hard to control successfully. This aspect causes another visual cue for the presence of the technique, especially in more complex pieces. The entire joining system, can become flooded (Fig. 9.2A). Fig. 9.5A shows evidence of a soldering process gone very wrong. A large chunk of solder chip that hasn't melted is visible between the granule and the substrate.

More than one type of joining process may be used in production of the same object. A simple example of this concept is often evident in the construction of mulberry earrings. Figure 9.2 illustrates two variations of dual join technique construction from Tell el 'Ajjul. Figure. 9.2A shows a granule cluster with flooded joins indicative of a simple solder join, while the join between the cluster and the hoop is very fine and indicative of a copper salt reaction solder. The methods are reversed in Fig. 9.2B.

Furthermore, there is a fourth option not discussed above. Instead of heat processes, a cold process that attaches granules through organic glues is possible. Usually this is done in preparation for one of the other methods in order to hold granules in place during heating, however it is possible that either certain components in more complicated pieces or all the granules may be attached this way.

Solders

Several of the techniques discussed above use different solders. There are choices available to the craftsman both between techniques and within the techniques themselves. Copper salt reaction solders require a copper mineral in ground powder or paste form as an agent. Several options are possible. The first, malachite, contains iron traces and zinc sometimes can make up as much as a quarter of the chemical composition (de Callataÿ, 1983, 189). Azurite, a blue rather than green copper mineral, is another option, and like malachite was often used as a pigment in antiquity. Another possibility is vendigris which is produced according to Pliny (*Naturalis Historia,* Book 33:29, Book 24: 26, 28) through a reaction between vinegar and copper the chemical composition of which is Cu $(OH_3CO_2 \cdot 5H_2O)$ (de Callataÿ, 1983, 190).

Silver-gold solders can be made from natural alloys of electrum or auriferous silver. Man-made solder alloys are made by adding the components one at a time in a crucible, in order of melting temperature from highest to lowest. Overheating or prolonging the melting time unnecessarily changes the nature of the alloy. During the entire process a reducing flame is necessary. Once cooled in ingot form, the solder is then made into sheet and either rolled into wire or cut into sheet snippets (Untracht, 1985, 395), a process not unlike some of the initial stages of some methods of granule production.

Granule Production

In addition to the range of choices available to an artisan in the attachment of granules, there are a numbers of possible methods for the production of the granules themselves. Carroll (Carroll, 1974, 33–34), a practicing goldsmith notes that granules are easily

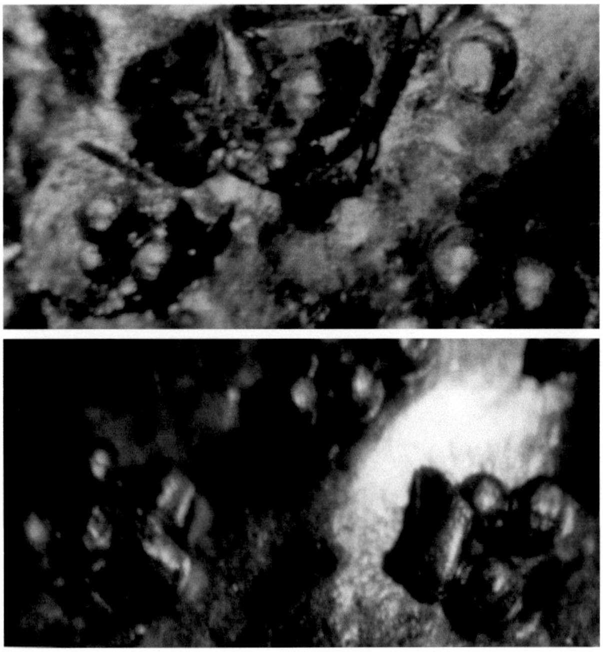

Figure 9.5: Undermelting; A (top): Under-melted solder chip, Tell el 'Ajjul, Rockefeller 35.3848; B (bottom): Under-melted strip metal produced granulation, Tell el 'Ajjul, Rockefeller 35.3848.

formed by dropping small quantities of molten metal on a flat surface. This method produces a range of grain sizes that must then be sorted. A variation of this method involves pouring molten metal which is passed through a sieve into a jar filled with water. The height from which the metal is poured is an important factor (de Callataÿ, 1983, 185). A combination of the first two processes is described in 1556 by Agricola in *De Re Metallica* (Book 10). He prescribes pouring molten metal on a slab with running water. Carroll (1974, 34) claims that the only reliable method of producing granules of equal sizes is by placing wire clippings of equal size in a crucible between layers of charcoal and heating to 1093° C, thirty degrees above the melting temperature of gold. The higher temperatures required to create perfect spheres with this method, along with the necessity to produce wire in advance makes this process particularly difficult and time consuming. This is especially true during the period under discussion since wire was not yet produced with a drawplate (Oddy, 1987). A variation on this method involves thin metal flakes or cut sheet metal strips, rather than cut wire (de Callataÿ, 1983, 186; Untracht, 1985, 359–360). Figure 9.5B illustrates that metal strip rather than wire may have been used by some craftsmen in antiquity. The coating of charcoal around the granules during this process creates a reducing atmosphere that prevents oxidation and maintains the bright reflective surface, a condition not present in the other methods so far discussed. Without a reducing atmosphere firescale may blacken the granules requiring a pickle – or weak acid solution such as acetic or citric acid – to remove it. A final technique, often referred to as shotting, produces metal balls one at a time. A charcoal block must be prepared in advance with a depression the size of the sphere being made. A length of wire or scrap metal is dipped in flux and then heated – preferably with a reducing flame. Spheres become flattened if they are too large, or if the metal is dropped on a flat surface rather than in a round depression (Untracht, 1985, 341). In the latter case, if a perfect sphere is sought the granules must be re-melted a second time (de Callataÿ, 1983, 185–186). Clearly this is a cumbersome and time consuming technique producing only one granule at a time and requiring several steps. Furthermore it works best when large granules are desired.

Analytical Methods

Being well versed in the range of possible variations available to the ancient artisan is only the first step. The next challenge is to find methods with which to accurately identify when each of the various processes are in operation and which choices the artisan has made. Since, care must be taken to develop non-destructive laboratory techniques, Oddy has recently emphasized the benefits of such conservation tools as optical and Scanning Electron Microscopes (SEM) to the museum curator and the archaeologist (Oddy, 1996). Ogden's books discuss many of the available techniques (Ogden, 1982; Ogden, 1992). Many details such as grain shape and size, joint characteristics and tool marks are visible with a simple stereo microscope with the magnification of x5 to x50 power. This type of examination relies quite heavily on observations of how the metals are behaving, and signs of any technological problems such as firestaining, insufficient spheroidization in the granules, or over-melting. Sensitivity to colors can be quite useful as different alloys create different color effects, in addition to differing melting temperatures. Because it is possible to obtain portable stereo microscopes, this is often the best option available when working on projects in the field or in museum collections which restrict the transport of objects to laboratory facilities.

An SEM is preferable when it is available for two reasons. First the depth of field is greater, especially for photography purposes, than the stereo microscope capabilities. The other advantage to the SEM is that it can have an attached analyzer for elemental analysis. Better still in the elemental analysis of fine metallic joins is a technique known as Proton Induced X-ray Emission (PIXE). Demortier and his colleagues have conducted considerable work on granulation with this method (Demortier, 1984; Demortier, 1988; Piette, Demortier & Bodart, 1986; Salamanca, *et al.*, 1998).

While PIXE analysis does yield very good results of the chemical composition of metals it should *not* be considered a tool for the tracing of the origins of the gold. Gold is notoriously difficult to source, although in recent years a new technique has been developed by South African mining companies to recover stolen bullion (Grigorova, *et al.*, 1998). Two factors prevent the linking of the gold in the object to the original sources when dealing with archaeological materials in the Mediterranean and Near Eastern Worlds. First, a databank does not exist of either all the mines and placer deposits used in antiquity, or of the chemical composition of the gold that comes from those deposits that are known. Secondly, and much more problematically, gold was recycled. It is highly unlikely that any single gold object from ancient contexts was manufactured completely of gold from a single source, precluding the ability to identify a reliable fingerprint. Ingots are rare, and even they may be suspect.

What PIXE analysis is able to accomplish is the identification of which specific elements are present in each component: granule, substrate, and, if present, solder. This helps give excellent information on such issues as the choices that the artisan was making and the pyro-technology involved in production. Different alloys heat to different temperatures; different joining techniques also require different temperatures; and, there may be different solders or no solder at all present. The exception is that any organic elements such as glues and fluxes burn off in the heating process and leave little or, more frequently, no trace. Furthermore, it is often difficult to distinguish between copper salt reaction solders, sinters and sometimes other solders through solely visual methods. PIXE analysis is a much more certain method for these identifications.

MOVING BEYOND THE TECHNOLOGICAL TO THE SOCIAL

People vs. Process: Choice, Analytical Hands, and the Artisan

Finding the individual and peopling the past is a growing concern within the archaeological community. The theoretical concepts of agency (Bell, 1992; Callinicos, 1988; Dobres, 2000; Dobres & Hoffman, 1999; Dobres & Robb, 2000; Saitta, 1994) and choice (Hill & Gunn, 1977; Lemonnier, 1993) are intimately linked to this problem. Art history and the art historical approach in Classical Archaeology have a long tradition of classification of objects according to individual artisans or workshops, and a solid methodology for this type of work has been developed within these fields (Cherry, 1992; Elsner, 1990; Maginnis, 1990; Morris, 1993). Yet until recently emphasis has been more stylistic than technological and predominantly has been on the classification of *the objects* rather than the social implications of the patterning revealed in the material record.

Much of what is outlined above deals with the range of choices available to the artisan in the creation of the granulation effect, and the means by which it is possible to distinguish the techniques. Technological choice is but one of a complete package of choices used in the execution of a piece that may lead to the discernment of individual analytical hands on a site. Furthermore, neither style nor technology should be considered in isolation from one another. The goldwork from the site of Tell el 'Ajjul in Southern Palestine illustrates this concept. It is the first site in the region that exhibits true granulation techniques. Punchwork creating similar light effects exists prior to the period of true granulation introduction, but the much more difficult prestige technique appears here for the first time sometime in the Middle Bronze Age. A minimum of three analytical hands, and possibly a fourth are visible in the granulated goldwork from the site. These distinctions are drawn by looking at several elements. Three joining methods are evident. PIXE analysis indicates both copper-salt reaction solders and sintering in operation on different pieces of the same morphological type, and visual microscopy indicates the presence of the third technique of silver-gold solder used at the site. (Fig. 9.4) There are differences in wire choices for the same type elements.

Figure 9.6: String holes, Tell el 'Ajjul; A (left): Clean cut, British Museum L. 1007; B (right): Ragged, Ashmolean 1949.318.

Some are square; some are round; some are twisted; and, some are plain. Some hair-rings with floating disk beads along the wire demonstrate a neatly cut string hole (Fig. 9.6A), while others are simply punched through and left ragged (Fig. 9.6B). A similar lack of finishing detail is witnessed in other pieces such as the evidence of stray tool marks that are not polished out of the backing of a piece (Fig. 9.7). Four types of suspension loops are used for granulated earrings and falcon pendants (Fig. 9.8). Two employ strip metal, a third round wire, and a fourth is a unique and complicated slide bolting system. The difference between the two which use strip metal is that one solders the broad flat surface of the strip to the object, while the other solders the narrow end of the strip to the object. The latter method tends to create a neater and more finished look, although the former sometimes is blended into the substrate after attachment. The round wire suspension loops are carefully inserted between sheet gold layers of the main body of the pendant in a very different process. Other differences include the hot-working of the sheet gold substrate in order to obtain an undulating surface, while similar style pendants obtain three dimensional effects through the use of doming blocks and swaging.

The date of the Tell el 'Ajjul jewelry has been placed as early as the early Middle Bronze Age (Ziffer, 1990, 54) and as late as the destruction layer (Town II) in which it was found – attributed either to the campaigns against the Hyksos led by Ahmose (Kempinski, 1983, 63; Mazar, 1990, 196, 221), or to the Asiatic campaigns of the Egyptian pharaoh Tuthmose III. Relationships have been postulated between both Ajjul and Ebla and Ajjul and Megiddo. Ebla dates to the earlier part of the Middle Bronze Age, while the Megiddo material is clearly Late Bronze I-II (See Lilyquist 1993: 47–50 for a more extensive review.). If there are links between the granulation from Ajjul and Ebla and/or Ajjul and Megiddo the plausible possibilities are:

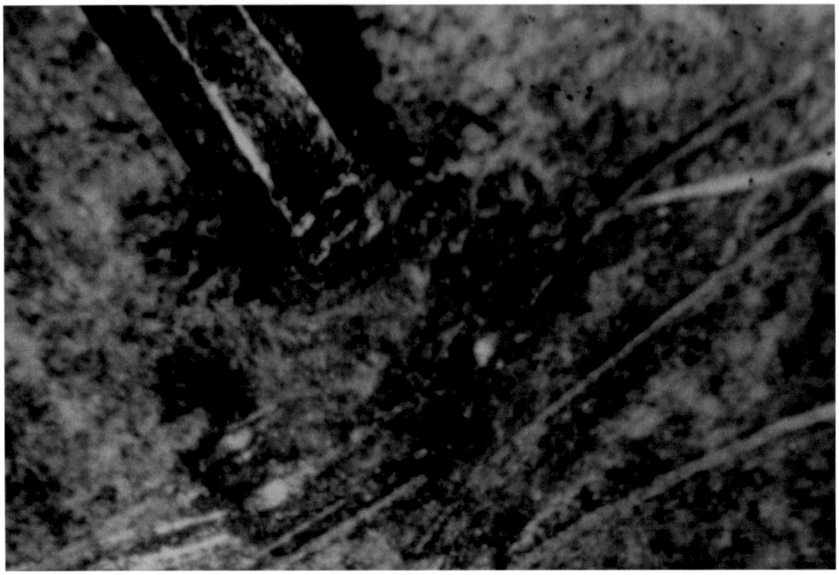

Figure 9.7: Tool marks under suspension loop solder, Tell el 'Ajjul, Ashmolean 1949.314.

Figure 9.8: Suspension loops from Tell el 'Ajjul pendants.

1. Travelling artisans coming from Ebla to Ajjul and/or from Ajjul to Megiddo (Zaccagnini, 1983);
2. Granulated objects travelling from Ebla to Ajjul and/or from Ajjul to Megiddo as gifts, commerce, or scrap metal;
3. Interaction between contemporary elite patrons who encourage the acquisition and interpretation of styles from other sites by their own goldsmiths.

Based purely on stylistic and iconographic grounds the possibility of traveling artisans seems appealing (Collon, 1985, 57). Yet there are stylistic details which are problematic. Many of the Ajjul pendants have a granulated edging (Fig. 9.9B) completely lacking in the Ebla repertoire, as are both the falcon/wryneck iconography of some of the pendants and the winged glass inlay earrings (Negbi, 1970). If there are links at all based on purely stylistic or iconographic grounds it would have to be through the elaborate crescent earrings that lack an edging. An argument might be made for an Eblaite artisan catering to local tastes, but it would be necessary to support such a hypothesis on technical grounds which is even more problematic. Although I have not examined the Ebla jewelry in person, based on the excellent color photographs in the 1995 exhibition catalog from Rome (Matthiae, Pinnock & Matthiae, 1995), it is quite clear that the Eblaite smiths were technically much more competent than those who produced the Ajjul material. Even the very best of the Ajjul pieces exhibit technical

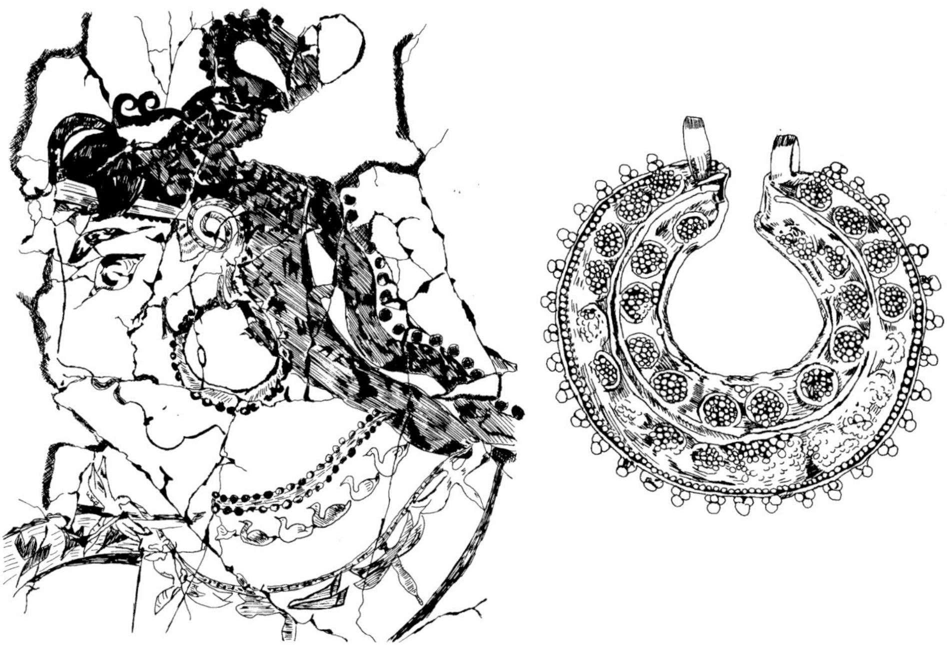

Figure 9.9: A (left) Detail of Goddess fresco, Xeste 3, Akrotiri, Thera. (After Doumas 1992, 163); B (right) Granulated earring from Tell el 'Ajjul.

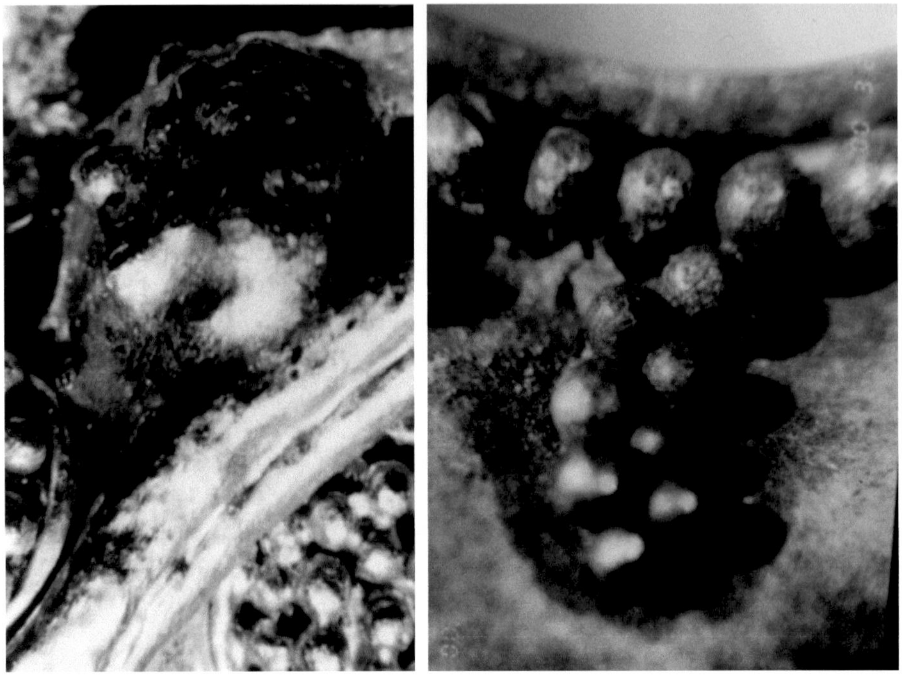

Figure 9.10: Technical mistakes; A (left): Over-melted granule cluster, Tell el 'Ajjul, Ashmolean 1949.308; B (right): Firestaining around granules, Tell el 'Ajjul, Rockefeller 38.493.

mistakes such as overmelting (Fig. 9.10A), the underfiring of granules (Fig. 9.5B), or firescale (Fig. 9.10B) which is completely lacking in the very fine Ebla examples. Instead, the Ajjul corpus illustrates an environment of artistic and technological experimentation and skill acquisition indicative of first, and perhaps second generation, smithing with the technologies necessary to create the prestigious effect of true granulation. Furthermore, suspension loop systems are constructed in a vastly different manner from the Ebla examples. The pendant from Ajjul with the sliding bolt mechanism is the most technologically sophisticated from that site, but again it is very different from any Eblaite pieces.

The situation between Ajjul and Megiddo is slightly better. At first glance, stylistically the Megiddo and Ajjul winged glass inlay earrings with granulated settings, look so similar that a shared workshop origin seems plausible. (Maxwell-Hyslop, 1971) However, the construction of the cloisonnes in which the glass is inlaid is conceptually very different between the Ajjul and Megiddo pieces. The Ajjul construction is based upon a sheet metal substrate (Fig. 9.11A) and soldered strip metal forming the inlay areas. In contrast, the Megiddo construction of the glass inlay area is with individual sheet metal cups of the appropriate shapes. These are then arranged and held together through an intricate system of strip metal straps (Fig. 9.11B). The workmanship is more finished and sophisticated.

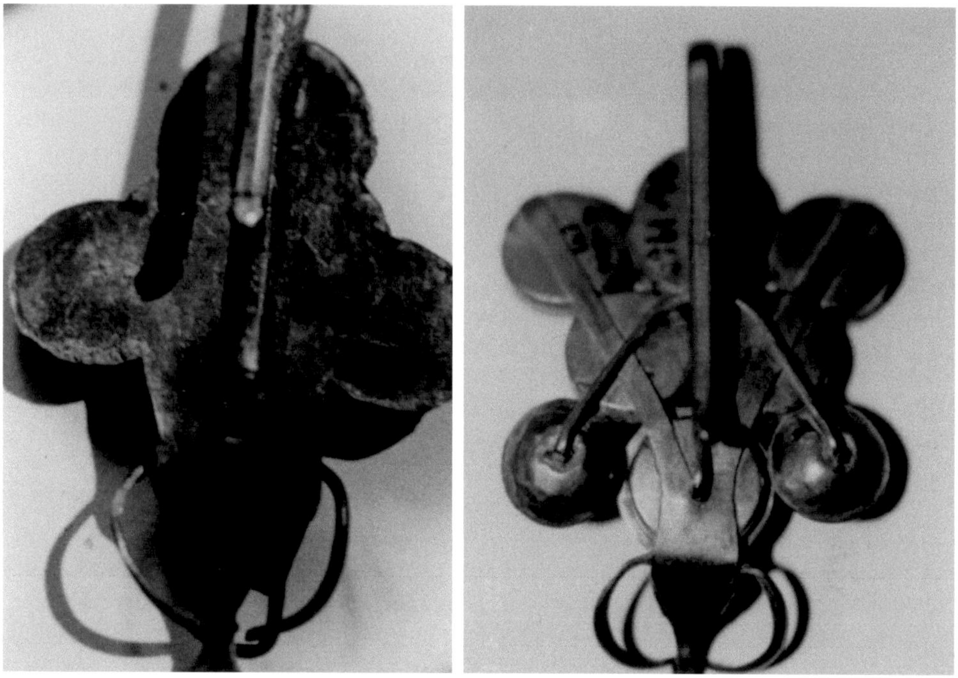

Figure 9.11: Backs; A (left): Glass inlay with granulation pendant, sheet metal backing, Tell el 'Ajjul, Rockefeller 38.495; B (right): Glass inlay with granulation pendant, cup settings with strip metal binding, Megiddo, Rockefeller 36.1968.

Based on the evidence of the glass inlays, the Ajjul earrings have been dated earlier than the Megiddo examples (Lilyquist, 1993, 53–55). It is tempting to agree, but as has been demonstrated between the Ebla and Ajjul goldwork technical virtuosity does not aide in chronological concerns between sites. At best, the Ajjul granulation would be contemporary with the Ebla gold which is far superior in craftsmanship, and in all likelihood the Ajjul granulation is later in date. Yet, as has been pointed out (Lilyquist, 1993) the granulation from tombs at Ajjul is of better quality that that found in the hoards in the destruction layer of Town II. For example, the granules are all perfectly formed, sizing is consistent, and attachment to the substrate bears little evidence of mistakes. All of the glass inlay earrings from Ajjul are from tombs, while most of the crescent earrings and all of the falcon/wryneck pendants are from hoards. If skill and quality are to be used in chronological considerations, it would be more reliably used as an internal indicator within the corpus of a single site. This difference between tomb and hoard groups may suggest that the date of production of the granulation in the Ajjul tombs is later that that found in the hoards. It is also possible that the Ajjul and Megiddo glass is roughly contemporary but clearly stemming from independent local workshops.

Therefore, in both the Ajjul-Ebla and Ajjul-Megiddo relationships the actions of traveling artisans and with them the transference of technology are not evident.

Consequently the patronage of such people by the elite of the site is not a possible explanation. Other agents for the appearance of granulation at Ajjul and its connections with other sites must be sought. The second possibility posed above is that the granulated objects found at Ajjul were not, in fact, manufactured there but were introduced through either the agency of gift exchange between elite individuals or the activities of merchants catering to the needs of customers at Ajjul and trading in either finished objects or scrap metal as raw material. Petrie identified many of the hoards as belonging to a peddlar, although recently the predominant opinion has been that they were hidden by owners or looters in a time of stress (Negbi, 1970; Tufnell, 1983, 60). In either instance individual pieces may have been manufactured elsewhere. The same holds true of any objects found in the tomb groups. However, although there is great variety in the technical execution of the Ajjul granulated goldwork, large enough groups of forms and styles seem consistent enough within the corpus that a majority of the large repertoire must have been manufactured in the local region, if not at Ajjul itself.

A pair of crescent earrings stand out as being anomalous in terms of form and technical execution. They exhibit the unique slide bolting suspension mechanism mentioned above (Fig. 9.8). Another feature that is unusual is the firestaining around the granules (Fig. 9.10B). This redish surface is caused by an oxidizing rather than reducing atmosphere during the heating of a join area that has copper. The resulting black scale of copper monoxide can be relatively easily removed in a weak acid pickle. The inner reddish layer, copper oxide, is less oxygen rich but is extremely difficult to remove. In addition, excessive amounts can form if oil is on the metal while heating (Untracht, 1985, 416). Only one other piece from Ajjul that I have examined leaves evidence for this practice in the artistic process. It is a granulated button with additional glass granule inlays (Petrie, 1934, pl. 18:89). Glass spheres or granules are used as inlays for the eyes of some of the elaborate glass inlay pendants with animal head terminals from Megiddo, but not in those from Ajjul. The loop on the backing of this button has broken off and reddish staining surrounds the attachment area. It might be worth investigating in the future if this is a technological characteristic observable in significant quantities within the Megiddo corpus. In any event, it is a strong possibility that these earrings were manufactured somewhere other than Ajjul. Therefore, at least on occasion, granulated jewels from other sites or regions made their way into the Ajjul cultural sphere. How they came is open to debate. Clearly this pair was acquired and accepted by the owner of the tomb as appropriate for the social activities in which s/he was engaged. The technique with which they were manufactured however does not seem to have been transferred with them. There are relationships in form and probably function, but not technological practice. This then leaves open the possibility that granulation was acquired through external stimulation and the patrons' demand for the technique combined with the innovation of local artisans.

The analysis of the goldwork from the Ajjul seems to point to local manufacture and local active experimentation to achieve the light reflective granulation effect, rather than the presence of any travelling artisans. The artisans of the site may have been inspired by granulated objects obtained through exchange with other regions, but fundamentally they developed the skills and knowledge necessary to produce

Figure 9.12: A (left): Detail of Saffron Gatherer fresco, Xeste 3, Akrotiri, Thera. (After Doumas 1992, 155); B (right) Granulated earring from Shaft Grave III, Mycenae.

granulation independently. The situation in the Mycenaean Shaft Graves of Greece in the mid-second millenium B.C., on the other hand, seems different.

Evidence for the existence of granulation on a site is not always direct. The famous Theran frescoes depict an enormous quantity of jewelry – much more than the Minoan counterparts on Crete. Parallels for certain types, although generally assumed, are completely lacking in the Cretan repertoire. An earring in the Theran frescoes has long been recognized as having its only parallel in the contemporary Mycenaean Shaft Graves (Televantou, 1984) (Fig. 9.12). Although the Theran version does not indicate granulation, the extant Shaft Grave version is the *only* example of granulation in all of the rich treasure of the Shaft Graves. Significantly, this earring consciously uses different grain sizes, organizing the largest at the bottom of the earring and the smallest at the top. Granulated crescent earrings in the Theran frescoes also consciously organize granules from largest at the bottom to smallest at the top (Fig. 9.9A). Furthermore granulated earrings of this type are completely lacking in Crete where the practice of wearing earrings is a rarity.

In contrast, the social practice of wearing earrings appears to be important within the local Theran context. A double hoop earring from the Theran frescoes is also granulated (Fig. 9.13A). Double hoop earrings play a role in local Theran symbolism as is demonstrated by their occurrence, not only in the frescoes, but also on nipple ewers. For example Fig. 9.13B illustrates a ewer from Thera that is painted with another ewer of the same type wearing double hoop earrings. Clearly earrings are locally more important on Thera than on Crete where earring types are limited to the rare simple

Figure 9.13: A (left): Detail of Red Haired Girl fresco, Xeste 3, Akrotiri, Thera. (After Doumas 1992, 167); B (right) Ewer depicting nippled ewer wearing double-link earring, Akrotiri, Thera.

hoop and occasional mulberry earring (Sakellarakis & Sakellarakis, 1997, 641–642). Furthermore, the frescoes illustrate elaborate granulated types within the Theran repertoire. This coupled with the fact that the only niello type dagger found outside of the Mainland (again these are thought to be Minoan imports) was found during the old excavations on Thera, and the best parallel for some of the 'Nilotic' scenes on them come from the famous West House Ship Fresco of Akrotiri (Doumas, 1992, 64–68, 71), also suggests a metallurgical link between Thera and the Greek Mainland.

Theran art and architecture is sufficiently Minoanized in spite of many local, indigenous characteristics, to lead some scholars to support the idea of Minoan political domination of the island. A recently analyzed crucible from Thera demonstrates both a local working of copper and a possibly worked copper source on Thera (Bassiakos & Michailidou, 1998). Some of the pigments used in wall painting are based on copper elements (eg. azurite) in precisely the proper form needed for copper-salt reaction solders, while a shift in pottery production in the LB Ia period indicates a consistent firing temperature of 850–1050° C (Papagiannopoulou, 1995) – the controlled temperature range appropriate for either copper salt reaction soldering or sintering. Therefore, the presence of the raw materials and the pyro-technology necessary for granulation are in evidence at the site.

Another example of the conscious use of grain size in granulation decoration comes from the Mainland of Greece. Gold bezelled finger rings are a hallmark of Minoan-Mycenaean Late Bronze elite symbolism which usually depict religious or military

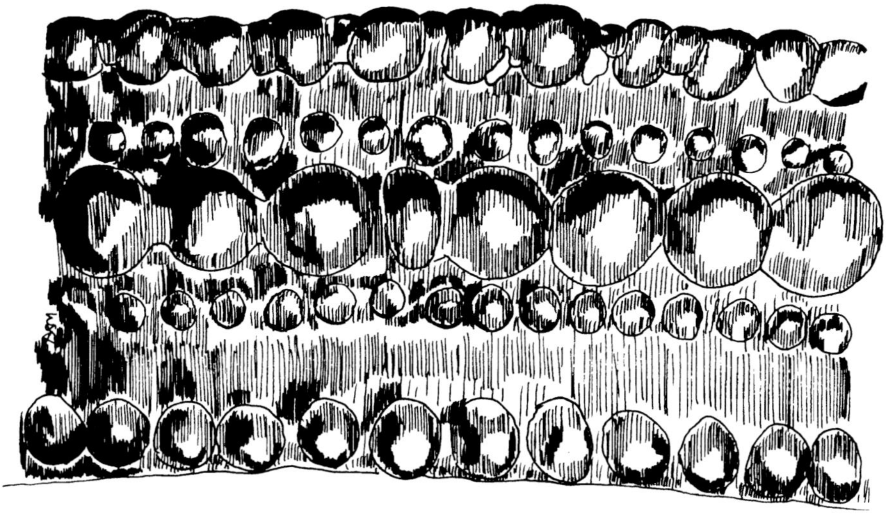

Figure 9.14: Granulated gold ring hoop, Mycenae.

iconography. Again, the assumption is that the early examples found at the beginning of the Mycenaean period are either imports from Crete or evidence of Minoan craftsmen working for the emerging Mycenaean elite. In the period just succeeding the Theran eruption, a group of these rings that have granulated hoops appear in Mycenaean contexts. Figure 9.14 illustrates one that, while not graduated, as in the Theran wall painting crescent earrings or the examples in the Shaft Graves, the craftsman has chosen to organize the largest granules in the center, the next largest along the edge, and the smallest flanking the large central granules. None of the granulation from Crete is organized according to grain size and this is either a purely Mycenaean design innovation or it is introduced from somewhere else. In light of the Shaft Grave earring as the first instance of granulation in a Mycenaean context, the parallel of its form in the Theran frescoes, the iconographic evidence for granulation in the Theran frescoes – including the organization of grains according to size, two conclusions may be drawn. First, a goldsmithing workshop independent from Crete operated on Thera and catered to local demands and tastes. And, second Theran artisans played an active role in the introduction of prestige technologies and iconographies of the Aegean *koine* to the Mycenaean Mainland specifically in the medium of metalworking. Might this be evidence of Theran craftsmen displaced from their homes by the explosive volcanic eruption of the Late Minoan Ia period, working for a new elite on the Mycenaean Mainland? If so, serious reconsideration of some Aegean relationships and interactions which, to date, have often understood dynamics as wholly Minoan or Mycenaean should be revised. More thought ought to be given to the smaller, but clearly thriving communities scattered throughout the Aegean islands.

Politics, technology and artistic environments

Both Tell el 'Ajjul and Mycenaean Greece are relative late-comers within the Mediterranean and Near Eastern Worlds to the use of granulation. Crete (Seager, 1912; Marinatos & Hirmer, 1960, Fig. 13), Troy (Tolstikov & Treister, 1996), and Poliochni (Bernabò-Brea, 1976) all incorporate true granulation techniques into their smithing repertoires by the Early Bronze Age in the Aegean. Mesopotamia, Anatolia and parts of the Northern Levant (Matthiae, *et al.*, 1995; Maxwell-Hyslop, 1971) use granulation long before the appearance of the technique at Tell el 'Ajjul in Southern Palestine. This situation raises the questions: Why was the acquisition of the technology so late? And, what spurred the acquisition of the technique and technology at the time that it was introduced?

Tell el 'Ajjul was part of an urban phenomenon following a three hundred year settlement pattern of region-wide rural and nomadic existence. Initially fortified cities were built along the coast of Palestine, and then spread inland and to the more arid southern portions of the region. Tell el 'Ajjul – especially if it is to be equated with Sharuhen, capital of southern Palestine – rose to prominence as a trading center linking Lebanon and Syria with the Asiatic cities of the Egyptian Nile Delta (Ilan, 2000, 34; Mazar, 1990).

The actors in this scenario are an elite who are building power structures through dynamic relationships with the external world. Merchants and foreign goods are crucial components. Elite families in other regions and cities are vital partners. The ability to project wealth, power and access to sought after commodities is of crucial concern. The capacity to attract and maintain the external relationships that drive and shape the internal life of the city is a matter of political and economic survival. Competition with neighboring cities for commercial ties must have been fierce. Public display, spectacle and the ritualization of power and wealth is a means through which elite classes and polities commonly build, solidify and maintain power bases (Cannadine & Price, 1987). The acutrements of such activities are sumptuous costumes, symbolic objects, and dramatic settings. True granulation is both easy to recognize and extremely difficult to achieve. This feature makes the technology attractive to patrons seeking ways in which to convey prestige. It is worked out of raw materials that are in themselves valuable both economically and symbolically. The technique adds to the brilliance, showiness and already highly visible light reflective qualities of gold. And finally, the technique is particularly appropriate for portable items of personal adornment.

The gaudiness of much of the Tell el 'Ajjul granulated goldwork, particularly that from the hoards, has been commented upon in the past, and it has been suggested that the concentration by the smith on the showiness of style rather than in precision of technique may indicate manufacture for a lower, more bourgeois class of patron (Lilyquist, 1993, 49). While much of the goldwork from the hoards does technically focus on achieving brilliant surface effects and technical mistakes abound, its usefulness in public spectacle would have been unmatched within Palestinian spheres. Rather than being produced for the middle class or the masses, it is more likely to have been produced for elite display in a period of political and economic solidification (Helms, 1993). Similarly showy pieces can be witnessed in use in contexts of ritual public display within the jewelry-rich Theran frescoes (Doumas, 1992; Marinatos, 1984) (Fig. 9.9A, 9.12A and 9.13A). The Theran community, like the one at Tell el 'Ajjul appears to have also gained its wealth as a intermediary for goods traded throughout the regions. Roughly contemporary with each other, it may be no accident that some of the Theran forms bear a marked resemblance to the Ajjul examples. Comparing Fig. 9.9A and 9.9B it is possible to see a striking similarity between the crescent earrings with granulated edging from both sites. The Theran examples are graduated, while the Ajjul examples use granule clusters, indicating separate workshops. Yet, conceptually the forms are very similar, and more notably so for the lack of any other parallels in the entire Mediterranean region during this time. Granulated, edged crescent shaped earrings become widespread in later periods (Goldman 1996) but Thera and Tell el 'Ajjul are the only sites that produce the form in the Middle to early Late Bronze Ages.

The period in which granulation first makes an appearance on the Mainland of Greece is also during the development of urbanism and a rising elite in the region. Here elite competition manifests in sumptuous burial practices and probably the display of overt militarism. The objects of the Shaft Graves are drawn directly from or inspired by the artistic environments of several better established neighbors with differing traditions. Minoan objects and inspiration have been long recognized and seen to dominate the repertoire (Matthäus, 1980). Other connections with more northerly European neighbors have been identified (Coles & Harding, 1979, 151, Fig. 52; Davis, 1983; Hiller, 1991); Kassite connections in the Near East have been proposed (Erlenmeyer, 1964);and even Egyptian relations (Hiller, 1994–1995) have been speculated upon. The first granulated pieces to appear in the region are the pair of earrings discussed above (Fig. 9.12B) of probable Theran origin. It is likely that these earrings were acquired as an import and were not manufactured by a Theran artisan working at Mycenae at this time, based on current evidence. The period just succeeding the Shaft Graves and after the Theran eruption which destroyed the island community, however, witnesses an increase in granulation in Mainland Greek contexts, including the granulated ring hoop in Fig. 9.14, which is being produced specifically to local tastes.

The craftsmen working in such politically charged artistic environments have a power of their own as they interact with both the merchants providing the raw materials for production, and the patrons for whom their work is produced while making choices in technology, style and iconography. The choices that they make and the technologies they acquire either through direct transfer or individual invention can directly effect the trajectories of the modes of artistic expression available within the community. Clearly during times when individualism and new innovations are actively encouraged the process is more dynamic, and *kudos* might be gained through the dramatic introduction of new styles, technologies or iconographies. Craftsmen may be able to gain more public recognition, and therefore garner more power and/or social standing than colleagues, for the introduction of new and novel items. Furthermore, successful innovators may set trends that are emulated by peers and become integral elements of the larger artistic environment. Yet, those innovations that are successful respond to specific needs and conditions already present in the community and are dependent to a certain extent on economic factors like the availability of the necessary raw materials. In the case of prestige materials that very often come from long distances, other groups such as traders may act as catalysts. Therefore, while a single individual may be able to significantly effect modes of expression it is, more often than not, the direct result of a synergistic relationship among several social groups.

In the case of the early rise of Mycenaean Greece, the Shaft Grave period exhibits a mélange of artistic styles, iconographies and objects. The social factors involved appear to be early competition based on display of wealth collected from elsewhere which gradually shifts to a practice of the transformation of exotic raw materials into locally manufactured prestige goods. In the process a manifestation of the Aegean *koine* becomes the dominant mode of artistic expression. The fact that the Theran eruption occurs approximately at the time that the Mycenaean elite are beginning to make the shift and are seeking their own artisans may be significant. In response to this need, the Mycenaeans probably acquired craftsmen from several areas in their quest to both control production and compete with each other. The eruption, if significant numbers of people were able to escape, would have made available a skilled labor force trained in the Aegean artistic *koine* who had knowledge of the crafting technologies that impacted the Mycenaean artistic environment during this time. Among those technologies was granulation.

Tell el 'Ajjul, in contrast, acquires the new technology of granulation through what appears to be a period of experimentation by local artisans stimulated by external relations. Whether the stimulation occurred at the patron or artisan level is difficult to discern. The craftsmen are clearly responding to local and/or regional conditions and needs with unique iconographic elements such as the falcon/wry-neck pendants and the winged glass inlay pendants with animal head terminals. Furthermore experimentation with a technology which by nature enhances light reflectivity, and the showy style of the elements and surface effects suggests a deliberate response to the intended function of the objects. The interpretation and specific execution of those community demands vary slightly from artisan to artisan.

Winter (1994) has effectively demonstrated a link between Mesopotamian literary language which used a rich imagery of light and shining in connection with divinity, kingship, and weaponry and the physical manifestations of the artistic environment. Granulation is an excellent technique for physical expression of these concepts, and may have been developed for royal and divine regalia in response to these beliefs. It is particularly striking that one of the earliest forms of proto-granulation is on Mesopotamian weaponry from Ur (Weber & Zettler, 1998; Zettler, 1998, 17, fig 15). Furthermore, ceremonial stone axes from third millennium B.C. Troy are highly polished and carry designs emulating granulated fittings (Tolstikov & Treister, 1996, 148–152). The granulation from Middle Bronze Age Byblos is predominantly on ceremonial axe fittings (Maxwell-Hyslop, 1971, 103, Fig. 75, color pl. D). In addition, very convincing proto- or pseudo-granulation occurs on a sword hilt from early Mycenaean Greece (Papadopoulos 1998, pl. A:32) and true-granulation appears on hilts of late Second Intermediate Period and Eighteenth Dynasty Egypt, as well as on the fittings of scepters from Ebla and Egypt. This hints that not only was granulation, as a highly light reflective technological effect, developed as a manifestation of a particular Mesopotamian ideology, but that also the concept may have spread with the technology.

The question at Tell el 'Ajjul is: Who sparked the technological acquisition of granulation, and were they consciously aware of the symbolism? Much about the Tell el 'Ajjul granulation speaks of a response to purely local agendas. The iconography is almost exclusively local or regional. Granulation is not found on any ceremonial weaponry or staffs from the site but that would not be expected unless weaponry and military might were relevant features of the local power structures. Were the Ajjul pieces created in response to a need for showy, highly visible regalia with local artisans seeking new means and innovations to fulfill this need? And, if so, were the patrons who required the regalia consciously adopting and adapting wider Near Easter concepts of the divine and rulership into local contexts? Or was the concept already present but not expressed in granulation until the skill, innovation, and political needs of a rising elite were present? Either is possible, since it is a clear that the elite at Ajjul need to both maintain internal power while negotiating externally with older, more established political identities. The presence of a technologically diverse body of granulation in large quantities would indicate that the presence of the technology, itself, may have been symbolically meaningful, independent of form, style or iconography. This technology may have been meaningful to the local population (or become meaningful), but it was already meaningful – unlike much of the iconographic elements present – to parts of the wider, well-established elite of communities in the Northern Levant, the Near East and the Aegean with whom the Tell el 'Ajjul elite would have been interested in establishing ties.

Technology and social boundaries

Style has been employed in archaeology to identify social relationships and boundaries, primarily through the analysis of ceramics and lithics (Conkey & Hastorf, 1990; Hegmon, 1992). Spatial analysis involving long distance trade (Zeitlin, 1987; Zeitlin, 1994), political legitimization, and other aspects of social ideology (Chang, 1983; Winter, 1982) have been another use of style as a means to identify and define social ordering. Technology and technological choices have been used as a similar marker, a method that has already been successfully applied to metalwork (Lechtman, 1977; Lechtman & Merrill, 1977; Stark, 1998).

If, as is stated in the introduction, true granulation is seen as a sub-class within a group of techniques which strive to achieve the same visual effect, then to use it as a type of material culture through which social boundaries are identified, it is necessary to take into account all of the techniques. The appearance of granulation, together with less difficult imitation techniques, delineates the boundaries of a specific elite negotiation or 'playing' field in the Bronze Age. The evolution of the shape and size of the playing field can be traced over time as new players join and/or change status. In the Early Bronze Age granulation occurs in Mesopotamia, moves across Anatolia toward the Northwest Coast and the site of Troy, out into the Aegean to the island of Lemnos and across to Crete. The site of Byblos on the Levantine coast also indicates early participation. In the Middle Bronze Age a shift seems to occur to the south incorporating North Syria and the Levantine coast while Crete is the only Aegean area demonstrating actual examples of granulation. Egypt has a few examples dating solidly to the Middle Kingdom, but the technique is rare and usually seen as 'foreign' in the Egyptian repertoire (Trokay, 1968). Although Egyptian smiths clearly practiced all of the techniques necessary to produce granulation since the Old Kingdom, joining granules to a substrate in decorative patterns is a choice and style that does not appear until quite late. Middle Kingdom examples, if made in Egypt either by Egyptian or foreign artisans, or acquired as gifts, war booty or trade, are a clear break with Egyptian traditions. An interesting and perhaps significant corollary is that like granulation, earrings are not popular in elite personal adornment and symbolic systems in Egypt until the Second Intermediate Period and New Kingdom when they become *de rigueur*. This shift in material culture and personal adornment practices, especially in Egypt where tradition is slow to change and symbolism is rich, is important.

By the time of the New Kingdom, Egypt has been fully incorporated into the system, as has Cyprus. The more northerly boundaries reach as far as the Trans-Caucasian region. The magnificent barrows at Trialeti are extremely difficult to date accurately, and it is unclear exactly how early this region should be considered the upper boundary, but it is definitely part of the system by the mid-second millennium B.C. The incidence rate of granulation explodes within the entire system during this period. It is also in the mid-to-late second millennium B.C. that the International Style takes a firm hold, texts speak of diplomatic marriages and trading agreements, and battles are fought between empires. In short, interaction between elite social groups of the far-flung regions around the Mediterranean and Near East have become frequent and intense. Presumably, this is not a situation that has happened overnight. For this reason, it is important to look at early material patterning in the periods leading up to this more complicated picture in order to understand how it evolved and what factors played important roles. Relationships and motivations are more easily understood when interaction is less frequent and behavior patterns are first emerging. Furthermore, the variability offered in the initial stages in the development of granulation effects and techniques is not only greater and easier to trace, but also presents an opportunity for greater refinement of relationships within those boundaries.

The broad boundaries defined by the granulation class of techniques delineates an area tied loosely together by a shared technologically produced visual effect. This does not imply that the people from all the regions encompassed by the boundaries at any given time were in direct contact with one another, or even aware of the outer limits of the system. The actions and choices of one of those groups, however could impact other members of the system either directly or indirectly. Instead it is a material culture marker that signals to the archaeologist the limits of shared past social practices, economic systems, political playing fields, and/or belief systems. By singling out one carefully chosen type of material culture and tracing its limits an artificially induced boundary is created, but one that cuts across traditionally examined boundaries such as culture, polity or state. This angle begins to allow the investigation of systems formed by those more traditional groups and the manner in which they interact with each other to produce larger entities such as economic networks or political events. The actions and choices of each player can effect both larger and smaller entities within the structure, the system as a whole, and themselves equally. It is through these dynamics that the character, size and shape of the system change over time.

Although all sub-classes of granulation are included when determining the boundaries, a hierarchy among the techniques exists after the earliest period of experimentation. True granulation techniques require the most skill and specialization to produce and should be considered a prestige technology. Sites within regions belonging within the defined boundary which produce this particular version of the visual effect are considered primary sites within the network. Sites which lack evidence of a production workshop of true granulation, but which have acquired granulated objects from other areas and perhaps produce items with pseudo-granulation in similar styles are considered second order sites within the system. Sites that produce examples of pseudo-granulation only form the class of tertiary sites within the system. It is through looking at the objects themselves, the contexts in which they were found, and many of the issues discussed in the previous sections that such determinations may be made.

Since expression of the technique is adapted and tailored to the local agendas, closer relationships within the system are discernable through the choices made on each site. If technological variation, style, form and iconography are shared between sites, the relationship would be considered quite close. The addition of shared practices in the use of the granulated object would prove a very intimate link between the sites. The fewer similarities between these variables that are present, the less close the link between the sites within the system. Upon close examination of the components sometimes expected relationships are not as near as had been thought based on form and style. The situation between the Tell el 'Ajjul and Megiddo granulated glass inlay pendants with animal head terminals are a good example of this. Stylistically and iconographically the two are very close, leading some to suggest the same workshop. Upon closer examination, technologically the same workshop is not possible. Instead the two sites are in close interaction, but not to the extent of sharing artisans.

Unexpected results revealing closer ties are equally possible. For example, according to geography, granulation from Egypt should be more closely related to Western Asiatic granulation. Much of it is, based on stylistic grounds, yet Aegean elements are visible in conjunction with the Western Asiatic and Egyptian stylistic elements on some of Tutankhamun's jewelry in the New Kingdom. Two bracelets from the tomb exhibit both the granulated triangle forms popular in Western Asiatic granulation, in addition to filigree running spirals of Aegean inspiration. The heavy stone inlay is purely Egyptian. (Edwards, 1976, Fig. 13) Considering the fact that the International Style is in full bloom in this period, and both granulated triangles and running spirals have become common in many regions, this is not surprising. Rather than a direct relationship between regions, this only definitively speaks of participation in the same system. However, another piece from the repertoire – an openwork buckle – (Fig. 9.15) is more unusual. Again the workmanship is probably Egyptian, but certain design elements are not commonly found either in Egyptian work or throughout the International Style system. The granules are organized according to size. Large granules stud the horse's blanket and form elements of the royal collar. Medium granules outline

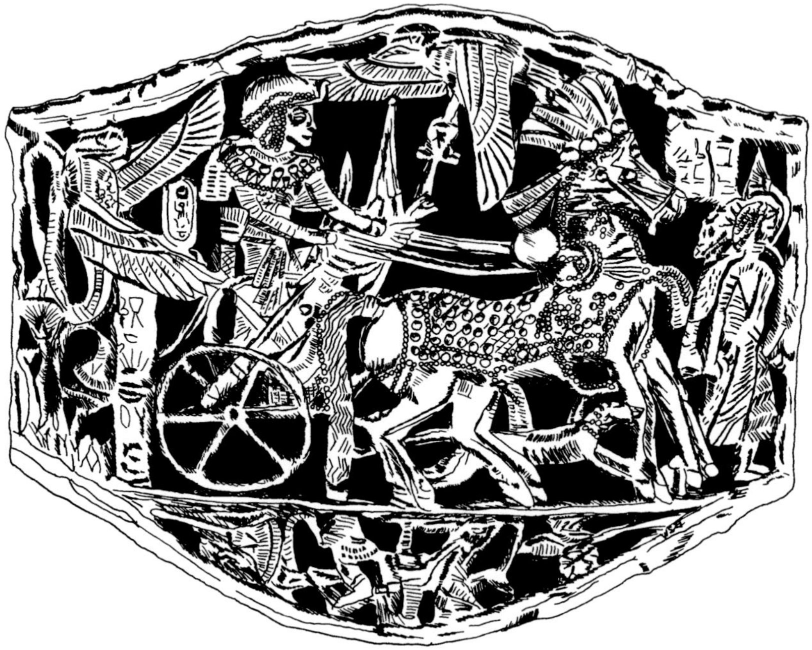

Figure 9.15: Granulated open-work buckle, Tutankhamun's tomb, Egypt (After Edwards 1976, fig. 4).

details. As discussed in relationship to the interaction between Thera and the early Mycenaean Mainland, granulation is not handled in most workshops according to size. This is an extremely unusual feature during the periods under discussion. The use of granulation in this manner pointed to very close connections between Thera and the Greek Mainland and suggested the migration of artisans. Another distinctive feature is the use of granulation to highlight and define details on an animal figure. Again this is an Aegean practice, not a Western Asiatic feature which tends to limit granulation to geometric forms. The other site in the Eastern Mediterranean with evidence of this design choice is Tell el 'Ajjul and the distinctive falcon/wryneck pendants. These also show iconographic, but certainly not stylistic or technological, similarities to a pair of earrings from the Tutankhamun group (Edwards, 1976, Fig. 7). The Ajjul examples considerably pre-date the Egyptian, but it would appear that the iconographic features are longer lasting. Another example (Bass, 1987, 693), more closely related to the Ajjul examples in execution, comes from the Ulu Burun shipwreck off the coast of Turkey which is closer in date to the Tutankhamun material. Certain technological details indicate that it was not manufactured in an Ajjul workshop contemporary with the Ajjul hoards, and therefore production of these pendants must have lasted longer and/or were manufactured at other sites than Tell el 'Ajjul. The unique stylistic similarities between the Tell el 'Ajjul and Theran granulated edged crescent earrings (Fig. 9.9A and

9B) hinted at a closer relationship between the two mercantile polities than has been previously suggested. Does the organized design principles witnessed in the Egyptian granulation indicate more direct relationships with the Aegean than is commonly accepted? Or, does the Southern Levant serve as an intermediary , as part of the same sub- system with trading vessels such as the Ulu Burun wreck the mechanism for interaction?

Once boundaries have been defined, the challenge is to determine what exactly they mean. While it is simple enough to observe that the use of granulation – which may hold a symbolic meaning connected with the divine and elite social standing – is practiced in certain regions during particular periods, the meaning behind the practice that links these disparate regions together is more enigmatic. It is clear that granulation crosses many, more traditional boundaries. More than one discreet cultural group, language group, and certainly many independent polities use the technique. Iconography and forms are not consistent enough throughout the system to suggest a connection with a single religion.

Economic considerations in the period might be an explanation. Metals, especially, are known to have been traded over very long distances, particularly tin. Mesopotamia was already negotiating access to these raw materials in the third millennium B.C., as the region is completely lacking in metals. Tin bronzes, rather than arsenical, are already in use at other third millennium B.C. sites that exhibit granulation such as Troy and Poliochni in the Northern Aegean (Pernicka, Begemann, Schmitt-Strecker & Grimanis, 1990) Muhly has pointed out that tin is often found in the same geologic environment as alluvial gold (Muhly, 1980, 30). Texts from Mari and Ugarit – the western ends of long distance networks – speak of trading in tin with men of Kaphtor (Crete), in addition to Near Eastern partners (Heltzer, 1989; Malamat, 1998). The Ebla archives not only discuss alloy recipes for tin bronzes and qualitative gradations of gold, they also indicate far-flung long distance relations. These include all of Syria and Palestine, as far south as the Sinai, parts of metal-rich Anatolia, and even Cyprus. Some of the most intensive relations are with Byblos which Pettinato suggests is an indirect source of Egyptian gold (Pettinato, 1981).

It is not until the New Kingdom in Egypt that texts referring to extensive trading networks exist, a number of which address Cyprus and the trade in copper. (Moran, 1992) The Kamose Stela from Karnak (Pritchard, 1975, 91) that celebrates the destruction of the Hyksos hints at extensive external trade relations under the auspices of the Hyksos in the Second Intermediate Period, but prior to that time external trade relations appear to be focused almost exclusively on Byblos as a gateway community. The opening of Egypt and her vast gold resources to the Near Eastern network may account for a shift from a silver to a gold standard in Mesopotamia during the Middle Babylonian Period (Muhly, 1980). How Egypt remained aloof to the Near Eastern metals trade until the Second Intermediate Period and New Kingdom is not problematic. Extensive copper reserves existed in the Egyptian Eastern Desert and Sinai regions that were exploited in pharaonic times. Recent evidence suggests that copper mining at Timna in the Sinai resumed, after a long hiatus, during the Middle Bronze Age (Mellado, 1995) at about the time polities such as Tell el 'Ajjul were making an appearance in Palestine. Mining operations there intensified in the New Kingdom, and, at the same time, copper began to be heavily imported from Cyprus. This expansion of Egyptian copper interests beyond the resources available within Egypt suggests both an inability to meet supply demands and a socio-economic impetus for external relations.

Tin, the scarcest metal distributed in the Near Eastern trade network, is also available locally in the Eastern Desert. Furthermore, tin is not commonly alloyed with copper to form tin-bronzes until quite late (Middle Kingdom) in Egypt compared to many Near Eastern and Aegean counterparts. Cassiterite, a tin bearing ore, is commercially mined in the Eastern Desert today, while there is some indication that placer tin deposits from the region were exploited as early as the Middle Kingdom (Sabet, Tsogoev, Shibanin, el-Kadi & Awad, 1976; Rothe, Rapp jr. & Miller, 1996). Heltzer notes a dramatic drop in the price of tin at Ugarit in the second millennium B.C.E (Heltzer, 1978), which would indicate new, plentiful and more immediate tin resources becoming available at the time Egypt first begins to exploit her resources.

The similarities between the dynamic shape of the long distance metals trade networks and the boundaries mapped through granulation are striking. Fundamentally, both are characterized by predominantly Western Asiatic players. The earliest participants in both are Mesopotamia and parts of Anatolia and the Aegean, with Syria and the Levantine Coast following immediately behind. Southern Palestine, Egypt, Cyprus and Mycenaean Greece are the last to join the system. The early 'foreign intrusions' of granulation in Middle Kingdom Egypt might be linked with the Asiatic presence in the Nile Delta region. These communities may have become established there as members of a trading diaspora (Sherratt & Sherratt, 1991) interested in gold and perhaps even copper and tin. The rise of Tell el `Ajjul in Southern Palestine may have been connected with this phenomenon, as a coastal trading depôt.

CONCLUSIONS

The social study of the prestige technology of granulation requires multiple and sliding scales of analysis. Large patterns and systems can be refined and humanized as the direct result of individual actions through the detailed study of the objects and contexts. Furthermore, it is possible to glimpse how people and polities either create known events in the archaeological record through a series of choices and actions, or react and adapt to known events.

The specific implications of the larger picture are particularly relevant to current debates and research on both the rise and role of the Hyksos relations (Davies & Schofield, 1995; Oren, 1998), and the rise and meaning of the International Style of the Late Bronze Age (Cline & Harris-Cline, 1998; Kantor, 1947). By focusing on the patterns, events and social practices leading up to these complicated and problematic periods, insights into the developmental process and setting of the trajectory is illuminated. This, in turn, aides in understanding the mature phases. By refining local and specific relationships within the wider system, the behavior of the system becomes better resolved. Clearly when dealing with complicated cross-cultural behaviors, issues of chronology are of utmost concern in bringing final resolution to the relationships. This is a dynamic and hotly contested field within Mediterranean Bronze Age studies, and great strides have been made in recent years (See: Åström, 1987; Manning, 1999; Warren & Hankey, 1989. Also see Manfred Bietak's ambitious chronology project: http://www.nhm-wien.ac.at/SCIEM2000).

Long distance networks that cross cultural and language boundaries have a particular need to develop a shared form of visual and symbolic means of communication. Granulation is an excellent candidate for several reasons. First, it is a highly visible and easily identifiable technique. Second, it is an appropriate addition to portable objects of ceremonial and symbolic significance such as personal adornment, ritual regalia, and ceremonial weaponry that would be natural items of display during negotiations and meetings of elite peers. Third, it is a transformation of metal – the material under negotiation – from a raw to a highly worked and controlled state. And finally, as true granulation is the most difficult smithing technique to master, access and proliferation was easily controlled. The technology of granulation may embody the intersection of elite local and international negotiation and power maintenance symbolism within this system. The ideological symbolism of divinity, kingship or elite status originating in Mesopotamia may have taken a systemic hold, translating the symbolism into locally meaningful forms and iconography. The very presence and display of such a prestige technology may have been the international signal, regardless of form.

ACKNOWLEDGEMENTS

I would like to thank the Samuel H. Kress Foundation, the American School of Classical Studies, Athens, and the W.F. Albright Institute of Archaeological Research for funding much of the research presented in this paper. Thanks is also due to the Ashmolean Museum, the British Museum, the National Museum, Athens and, the Rockefeller Museum, Jerusalem for access to material in their collections. I would like to especially thank the Oxford University Department of Material Science for conducting the PIXE analysis, the detailed results of which will be made available elsewhere. Wendy Goodman executed all the drawings for the figures. All photography was done by the author.

REFERENCES

Amiet, A., 1966, *Elam*, Archée Editeur, Paris.

Åström, (ed.), 1987, *High, Middle or Low? Acts of an international colloquium on absolute chronology held at the University of Gothenburg 20th–22nd August 1987*, Aström Gorlag, Gothenburg.

Baines, R., 1998, Technical antecedents of early Hellenistic disc and pendant ear ornaments, in D. Williams (ed.), *The Art of the Greek goldsmith*, British Museum Press, London,122–126.

Bass, G. F., 1987, Oldest known shipwreck reveals splendors of the Bronze Age, *National Geographic*, 172 (December), 693–733.

Bassiakos, Y. & Michailidou, A., 1998, Metallurgical evidence for local copper exploitation at Akrotiri, Thera: first results, *Abstracts of the 31st International Symposium on Archaeometry (Budapest, Hungary) 27 April-1 May 1998.*

Bell, J., 1992, On capturing agency in theories about prehistory, in: J.-C. Gardin & C. S. Peebles (eds.) *Representations in archaeology*, University of Indiana Press, Bloomington, 30–55.

Bernabò-Brea, L., 1976, *Poliochni II; Città preisotirica nell'isola dii Lemnos*, Monografie della scuola archeologica di Atene e delle Missioni italiane in oriente 2, 'L'Erma' Bretschneider, Rome.

Bingöl, I., 1999, *Ancient jewellery. Museum of Anatolian Civilizations*, Ministry of Culture, Republic of Turkey, Ankara.

Callinicos, A., 1988, *Making history: agency, structure and change in social theory*, Cornell University Press, Ithaca and New York.

Cannadine, D. & Price, S. (eds.), 1987, *Rituals of royalty: power and ceremonial in traditional societies*, Cambridge University Press, Cambridge.

Carroll, D., 1974, A classification for granulation in ancient metalwork, *AJA*, 78, 33–9.

Chang, K., 1983, *Art, myth and ritual. The path to political authority in ancient China*, Harvard University Press, Cambridge, Mass.

Cherry, J. F., 1992, Beazley in the Bronze Age? Reflections on attribution studies in Aegean Prehistory, in R. Laffineur & J. L. Crowley (eds.), *Eikon. Aegean Bronze Age iconography: shaping a methodology. Proceedings of the 4th International Aegean Conference, University of Tasmania, Hobart, Australia 6–9 April 1992*, Aegaeum, vol. 8, Liège, 123–144.

Clark, G., 1986, *Symbols of excellence: precious metals as expression of status*, Cambridge University Press, London.

Cline, E. & Harris-Cline, D. (eds.), 1998, *The Aegean and the Orient in the Second Millennium. Proceedings of the 50th Anniversary Symposium, Cincinnati, 18–20 April 1997*. Aegaeum, vol. 18, Université de Liège, Liège.

Coles, J. & Harding, A., 1979, *The Bronze Age in Europe*, Methuen, London.

Collon, D., 1985, A North Syrian cylinder seal style; evidence of North-South links with Ajjul, in J. Tubb (ed.) *Palestine in the Bronze and Iron Ages: papers in honor of Olga Tufnell*, Institute of Archaeology, London, 57–68.

Conkey, M. & Hastorf, C. (eds.), 1990, *The uses of style in archaeology*, Cambridge University Press, Cambridge.

Davies, W. V. & Schofield, L. (eds.), 1995, *Egypt, the Aegean and the Levant*, British Museum, London.

Davis, E., 1983, The gold in the shaft graves: the Transylvanian connection, *Temple University Aegean Symposium*, 8, 32–38.

de Callataÿ, F., 1983, Granulation, in T. Hackens & R. Winkes (eds.), *Gold jewelry. craft, style & meaning from Mycenae to Constantinople*, Brown University & Université Catholique de Louvain, Providence, Rhode Island & Louvain-la-Neuve, 185–191.

Demortier, G., 1984, Analysis of gold jewellery artifacts. characterization of ancient gold solders by PIXE, *Gold Bulletin*, 17(1), 27–38.

Demortier, G., 1988, The soldering of gold in antiquity, *Material Research Society Symposium Proceedings*, 123, 193–198.

Demortier, G., 1989, Ancient gold solders: what was chrysocolla? *Archaeological Chemistry*, IV, 249–263.

Dobres, M.-A., 2000, *Technology and social agency*, Blackwell Publishers, Oxford.

Dobres, M.-A. & Hoffman, C. R. (eds.), 1999, *The social dynamics of technology. practice, politics, and world views*, Smithsonian Institution Press, Washington, D.C.

Dobres, M.-A. & Robb, J. (eds.), 2000, *Agency in archaeology*, Routledge, London.

Doumas, C., 1992, *The Wall-paintings of Thera*, The Thera Foundation, Athens.

Echt, R. & Thiele, W. R. , 1995, Sintering, welding, brazing and soldering as bonding techniques in Etruscan and Celtic goldsmithing, in J. Northover & G. Morteani (eds.) *Prehistoric gold in Europe*, Kluwer Academic Publishers, The Netherlands, 435–451.

Edwards, I., 1976, *Tutankhamun's jewelry*, The Metropolitan Museum of Art, New York.

Elsner, J., 1990, Significant details: systems, certainties and the art-historian as detective, *Antiquity*, 64, 950–52.

Eluère, C., 1989, A 'gold connection' between the Etruscans and Early Celts? *Gold Bulletin*, 22(2), 48–55.

Erlenmeyer, H., 1964, Kassitische Goldarbeiten aus dem Schachtgrab III in Mykene? *Kadmos*, III, 177.

Golani, A., 1996, The jewelry and the jeweler's craft at Tel Miqne-Ekron, unpublished Masters Thesis , Hebrew University.

Goldman, B., 1996, Nabataean/Syro-Roman Lunate Earrings, *Israel Exploration Journal*, 46(1–2), 77–99.

Grigorova, B., Anderson, S., de Bruyn, J., Smith, W., Stülpner, K. & Barzev, A., 1998, The AARL gold fingerprinting technology, *Gold Bulletin*, 31(1), 26–29.

Hegmon, M., 1992, Archaeological research on style, *Annual review of anthropology*, 21, 517–36.

Helms, M., 1993, *Craft and the kingly ideal: art, trade and power*, University of Texas Press, Austin.

Heltzer, M., 1978, *Goods, prices and the organization of trade in Ugarit*, Dr. Ludwig Reichert Verlag, Wiesbaden.

Heltzer, M., 1989, The trade of Crete and Cyprus with Syria and Mesopotamia and their Eastern tin-sources in the XVIII–XVII century B.C., *Minos*, 24, 17–23.

Hill, J. & Gunn, J. (eds.), 1977, *The individual in prehistory: studies of variability in style in prehistoric technologies*, Academic Press, London.

Hiller, S., 1991, The Mycenaeans and the Black Sea, *in Thalassa. l'Egée préhistorique et la mer. Actes de la 3e Rencontre Égéenne Internationale de l'Université de Liège, Calvi, Corse, 23–25 Avril 1990*, Aegaeum, Liege, vol. 7, 207–16.

Hiller, S. , 1994–1995, Die Schachtgräber von Mykene: In ägyptischen Kontext, *Minos*, 29–30, 7–31.

Hoffman, G. L. , 1997, *Imports and immigrants. Near Eastern contacts with Iron Age Crete*, the University of Michigan Press, Ann Arbor.

Hosler, D., 1994, *The sounds of colors of power: the sacred metallurgy of ancient West Mexico*, MIT Press, Cambridge.

Ilan, O., 2000, *Image and artifact. treasures of the Rockefeller Museum*, The Israel Museum, Jerusalem.

Kantor, H., 1947, The Aegean and the Orient in the second millennium B.C., *AJA*, 51, 1–103.

Kempinski, A., 1983, *Syrien und Palästina (Kanaan) in der letzten Phase der Mittelbronze IIB-Zeit (1650–1570 v. Chr.)*, Ägypten und altes Testament, vol. 4, Harrassowitz, Wiesbaden.

Kosay, H., 1944, *Ausgrabungen von Alaca Höyük*, Ankara.

Lechtman, H., 1977, Style in technology: some early thoughts, *in* H. Lechtman & R. Merrill (eds.) *Material culture: styles, organization, and dynamics of technology*, St. Paul, MN, American Ethnological Society, 3–20.

Lechtman, H., 1984, Andean value systems and the development of prehistoric metallurgy, *Technology and Culture*, 25, 1–36.

Lechtman, H. & Merrill, R. (eds.) , 1977, *Material culture: Styles, organization and dynamics of technology*, Proceedings of the American Ethnological Association, St. Paul, 21–52.

Lemonnier, , 1993, Introduction, *in* Lemonnier (ed.) *Technical choices. Transformation in material cultures since the Neolithic*, Routledge,London/New York, 1–35.

Lilyquist, C., 1993, Granulation and glass: chronological and stylistic investigations at selected sites, ca. 2500–1400 B.C.E, *BASOR*, 290–291, 29–94.

Maginnis, H., 1990, The role of perceptual learning in connoisseurship: Morelli, Berenson and beyond, *Art History*, 13, 104–17.

Malamat, A., 1998, Mari and its relations with the Eastern Mediterranean, *in* M. Lubetski, C. Gottlieb & S. Keller (eds.) *Boundaries of the Ancient Near Eastern World. A tribute to Cyrus H.*

Gordon., Journal for the Study of the Old Testament Supplement Series, vol. 273, Sheffield Academic Press, Sheffield, 411–18.

Manning, S. W., 1999, *A test of time: the volcano of Thera and the chronology and history of the Aegean and East Mediterranean in the mid second millennium B.C.*, Oxbow Books, Oxford.

Marinatos, N., 1984, *Art and religion in Thera*, Athens.

Marinatos, S. & Hirmer, M., 1960, *Crete and Mycenae*, London, Abrams.

Matthäus, H., 1980, Minoan influence on the Greek Mainland during the sixteenth century B.C. and the origins of Mycenaean Civilization, in Betancourt, P. (ed.), *Minoan foreign relations at the beginning of the Late Bronze Age*, Temple University Symposium 5, Temple University, Philadelphia, 37–44.

Matthiae, P., Pinnock, F. & Matthiae, G. S., 1995, *Ebla. alle origini della civilità urbana*, Electa, Rome.

Maxwell-Hyslop, K., 1971, *Western Asiatic jewellery c. 3000–612 B.C.*, Methuen, London.

Mazar, A., 1990, *Archaeology of the land of the Bible 10,000–586 BCE*, Doubleday, New York, The Anchor Bible Reference Library.

Mellado, E., 1995, Egyptian mining in Sinai during the Middle Kingdom. *Abstracts of papers presented at the Seventh International Congress of Egyptologists 3–9 September 1995*, Cambridge.

Moran, W. L., 1992, *The Amarna letters*, John Hopkins University Press, Baltimore.

Morris, C., 1993, Hands up for the individual! The role of attribution studies in Aegean prehistory, *Cambridge Archaeological Journal*, 3, 41–66.

Muhly, J. , 1980, Metals and metallurgy in Crete and the Aegean at the beginning of the Late Bronze Age, in Betancourt, P. (ed.), *Minoan foreign relations at the beginning of the Late Bronze Age*, Temple University, Philadelphia, 25–36.

Negbi, O., 1970, *The hoards of goldwork from Tell el- 'Ajjul*. SIMA 25, Göteborg.

Oddy, A., 1996, Jewelry under the microscope. A conservators' guide to cataloguing, *in* A. Calinescu (ed.) *Ancient jewelry and archaeology*, University of Indiana Press, Bloomington, 185–197.

Oddy, W., 1987, Does 'strip-drawn' wire exist from antiquity, *MASCA Journal*, 4(4), 175–77.

Ogden, J, 1982, *Jewellery of the ancient world*, Trefoil, London.

Ogden, J., 1992, *Ancient jewellery*. Interpreting the Past, University of California Press/British Museum, Berkeley.

Oren, E. (ed.), 1998, *The Hyksos: new historical and archaeological perspectives*, Philadelphia University Museum, Philadelphia.

Özgüç, T. & Temizer, R. , 1993, The Eskiyapar treasure, *in* M. Mellink, E. Porada & T. Özgüç (eds.) *Aspects of art and iconography: Anatolia and its neighbors. Studies in honor of Nimet–Özgüç*, Ankara, 613–28.

Papadopoulos, T. J., 1998, The Late Bronze Age daggers of the Aegean I, The Greek Mainland. Prähistorische Bronzegunde, Abteilung VI, 11.Band, Franz Steiner Verlag, Stuttgart.

Papagiannopoulou, A. , 1995, The change in LC pottery production and its trade implications, *in* C. Gillis, C. Risberg & B. Sjöberg (eds.) *Trade and production in premonetary Greece: aspects of trade,*. Proceedings of the Third International Workshop, Athens 1993, Paul Aströms Forlag, Jonsered, 55–60.

Pernicka, E., Begemann, F., Schmitt-Strecker, S. & Grimanis, A., 1990, On the composition and provenance of metal artefacts from Poliochni on Lemnos, *OJA*, 9(3), 263–298.

Petrie, W. M. F., 1934, *Ancient Gaza IV*, British School of Archaeology in Egypt, London.

Pettinato, G., 1981, *The archives of Ebla. an empire inscribed in clay*, Garden City, Doubleday & Company, New York.

Piette, M., Demortier, G. & Bodart, F., 1986, PIXE analysis of solders on ancient gold artifacts with a nuclear microprobe, *in Proceedings microbeam analysis conference*, San Francisco Press, 333–36.

Platt, E., 1972, Palestinian Iron Age jewelry from fourteen excavations, unpublished Ph.D. dissertation, Harvard University.

Pritchard, J. B. (ed.) , 1975, *The Ancient Near East. a new anthology of texts and pictures, volume II*, Princeton University Press, Princeton.

Renfrew, C., 1986, Varna and the emergence of wealth in prehistoric Europe, *in* A. Appadurai (ed.) *The social life of things. Commodities in cultural perspectives*, Cambridge University Press, Cambridge, 141–68.

Rothe, R. D., Rapp jr., G. & Miller, W. K., 1996, New hieroglyphic evidence for pharonic activity in the Eastern Desert of Egypt, *JARCE*, 33, 77–104.

Sabet, A., Tsogoev, V., Shibanin, S., el-Kadi, M. & Awad, S., 1976, The placer tin-deposits of Igla, Aby Dabbab and Nuweib, *Annals of the Geological Survey of Egypt*, 6, 213–221.

Saitta, D., 1994, Agency, class and archaeological interpretation, *Journal of Anthropological Research*, 13, 201–27.

Sakellarakis, Y. & Sakellarakis, E., 1997, *Archanes. Minoan Crete in a new light*, Ammos Editions, Eleni Nakou Foundation, Athens.

Salamanca, M. O., Demortier, G., Gomez, F. F., Coquay, , Ruvaldaba-Sil, J.-L. & Respaldiza, M., 1998, PIXE and SEM studies of Tartesic gold artefacts, *Nuclear Instruments and Methods in Physics Research*, B 136–138, 851–857.

Seager, R. B., 1912, *Explorations in the island of Mochlos*, American School of Classical Studies at Athens, Boston.

Sherratt, A. & Sherratt, S., 1991, From luxuries to commodities: the nature of Mediterranean Bronze Age trading systems, *in* N. Gale (ed.) *Bronze Age trade in the Mediterranean. papers presented at the conference held at Rewley House, Oxford in December 1989*, SIMA 90, Jonsered, Paul Åströms Forlag, 351–86.

Stark, M. T. (ed.) , 1998, *The archaeology of social boundaries*, Smithsonian Institution Press, Washington DC.

Televantou, C., 1984, Jewellery from prehistoric Thera (in Greek, English summary), *ArchEph*, 14–54.

Tolstikov, V. & Treister, M., 1996, *The gold of Troy*, C. Sever & M. Bonnichsen (trans), Harry N. Abrams, Inc. Publishers, New York.

Trokay, M., 1968, Les origines du décor à granulations dans l'orfèvrerie égyptienne, *Cd'E*, 43, 271–280.

Tufnell, O., 1983, Some Gold Bird Ornaments. Falcon or Wryneck? *Anatolian Studies*, 33, 57–66.

Untracht, O., 1985, *Jewelry concepts and technology*, Robert Hale, London.

Warren, & Hankey, V., 1989, *Aegean Bronze Age chronology*, Bristol Classical Press, Bristol.

Weber, J. A. & Zettler, R. L., 1998, Tools and weapons, in R. L. Zettler & L. Horne (eds.) *Treasures from the royal tombs of Ur*, University of Pennsylvania Museum of Archaeology and Anthropology, Philadelphia, 163–173.

Williams, B., 1975, Archaeology and historical problems of the Second Intermediate Period, unpublished Ph.D dissertation, University of Chicago.

Winter, I., 1982, Art as evidence for interaction: relations between the Assyrian Empire and North Syria, *in* H. Nissen & J. Renger (eds.) *Mesopotamia und seine Nachbarn*, Berlin, D. Reimer, 355–381.

Winter, I., 1994, Radiance as an aesthetic value in the art of Mesopotamia (with some Indian parallels), *in* B. Saraswati, S. Malik & K. Madhu (eds.) *Art, the integral vision: a volume of essays in felicitation of Kapila Vatsyayan*, D.K. Printoworld, New Delhi, 123–32.

Zaccagnini, C., 1983, Patterns of mobility among ancient Near Eastern craftsmen, *Journal of Near Eastern Studies*, 42, 250–254.

Zeitlin, R., 1987, The sociocultural perspective on the spatial analysis of commodity and stylistic distributions, *in* G. d. G. Sieveking & M. Newcomer (eds.) *The human uses of flint and chert*, Cambridge University Press, Cambridge, 173–81.

Zeitlin, R. N., 1994, Accounting for the prehistoric long-distance movement of goods with a measure of style, *World Archaeology*, 26(2), 208–33.

Zettler, R. L. , 1998, Ur of the Chaldees, *in* R. L. Zettler & L. Horne (eds.) *Treasures from the royal tombs of Ur*, University of Pennsylvania Museum of Archaeology and Anthropology, Philadelphia, 9–19.

Ziffer, I., 1990, *At that time Canaanites were in the land*. The Israel Museum, Jerusalem.

Chapter 10

Minoan Foreign Relations and Copper Metallurgy in Protopalatial and Neopalatial Crete

Zofia Stos-Gale

Abstract

The 18th to 17th centuries B.C. on Crete was the period of great changes and social upheavals. Minoan Crete was a leading military power in the Aegean in the MBA and the wide mercantile activity depended on a political system co-ordinating private and public interests and providing security. The rise in wealth and social stratification provided markets for exotic materials, including a wide spread introduction of tin bronze. The island of Crete has no sources of copper and tin, or other metals. Lead isotope analyses of 118 Prepalatial and Old Palace periods weapons and tools indicate that in this period the rapid development of metallurgy was based to a large degree on the exploitation of ore deposits and the extraction of copper metal on the Cycladic islands, amounting to 52% of the artefacts sampled. However, another 18% of the copper came from Cyprus, and 17% from Laurion in Attica. The remaining 13% originated from other, possibly quite distant sources.

Around 1700 B.C., after the widespread destruction, the palaces are rebuilt and improved. This change is followed by a dramatic change in the sources of copper. Lead isotope analyses of 300 Minoan metal artefacts dated to 17th–12th century show that the major source of copper in this period (around 50%) are ores from Laurion, around 20% from Cyprus, and 30% from non-Aegean sources.

INTRODUCTION

The question of Minoan foreign relations and trade should be reflected in the origin of copper metal used on Crete in the Bronze Age, because all geological and archaeometallurgical evidence indicates that copper extraction from local ores was not undertaken, which is not surprising since copper ores on Crete are of poor quality and quite scarce (Stos-Gale 1993). The only copper extraction workshop on the island on the site of Chrysokamino was dated to FN–EMIII (Betancourt *et al.* 1999) and represents small scale copper production from imported ore (Stos-Gale 1998a). The sources of metals used on Crete from the 17th century B.C. are of a particular interest, because this is the time when, after the destruction of the first palaces, the island enters the period of unparalleled prosperity and related rapid development of technology, arts and crafts.

The main difference between this study and many others published in the recent decades is its reliance on the extensive lead isotope data base accumulated during the last 20 years, allowing direct comparison of the isotopic composition of the regional copper and lead ores with the Minoan copper based alloys, lead metal and silver. The following discussion is based exclusively on the above mentioned database. The reason for exclusion of other published analytical work (for example Mangou and Ioannou 1998) is lack of parallel lead isotope data. The chief aim of this paper is a comprehensive discussion of metal composition from the technological and provenance perspective. It is hoped that this approach will contribute to the understanding of the relation between the changes in social systems in the Minoan Crete and the significant increase in metal consumption and introduction of tin bronze in the second millenium B.C.

METHODOLOGY

Lead isotope compositions of ore deposits can give a very specific 'fingerprint' of ores from a given geographical location. Samples of ores collected from the mines are analysed for their lead isotope ratios with very high accuracy. Each of the lead isotope datapoints has its unique position in the three-dimensional space (usually: $^{208}Pb/^{206}Pb$–Y, $^{207}Pb/^{206}Pb$–X and $^{206}Pb/^{204}Pb$–Z). Groups of points representing ores from different mines form patterns restricted to a certain range of lead isotope ratios characteristic for each geological ore formation. For identification of ore sources used for production of metals in antiquity, the lead isotope ratios measured for ancient artefacts are compared point-by-point with the ore data (for more detailed description of the methodology see: Gale and Stos-Gale 2000).

The lead isotope data base of copper, lead and silver ore sources relevant to the Bonze Age Aegean is particularly well developed. The number of ore samples and artefacts that were used for comparisons with 411 copper-based artefacts from 46 Minoan sites are summarised in Tables 10.1 and 10.2.

The interpretation of lead isotope compositions in provenance studies has been discussed in depth during the last decade (see, for example, *Archaeometry* 34 (2), 1992 and *Archaeometry* 35 (2), 1993). It is now widely accepted that the most appropriate interpretation of lead isotope ratios measured for a sample of an ancient artefact is their complete identity, within the error of measurement, with data for samples of ores from a deposit that might have been the source of this particular metal in the antiquity. A very helpful addition to the information about lead isotope data of ore deposits are analyses of slags from ancient smelting sites (for example: Hauptmann *et al.* 1993, Stos-Gale 1998a). The large number of lead isotope data published in the last five years allows now to identify, with a reasonable accuracy, copper from Cyprus (Gale *et al.* 1997 and Stos-Gale *et al.* 1998a), the lead-silver and copper deposits on the Greek mainland and the Cyclades (Stos-Gale *et al.* 1996, Stos-Gale 1998a and Wagner *et al.* 1986) some of the Turkish deposits (Pernicka *et al.* 1984, Yener *et al.* 1991, Hirao *et al.*

Table 10.1: Summary of the lead isotope data for ore samples available comparison with the Minoan metals

Country	No. of occurrences	No. of ore samples
Greece	33	867
Cyprus	46	700
Turkey	130	290
Near and Middle East	36	226
Italy and Spain	73	610
The Balkans	20	220
Total	338	2,930

Table 10.2: Summary of the lead isotope data for Bronze age metal artefacts used for comparison with the Minoan metals

Country	Sites	Cu or Cu/Sn	Pb	Ag
Greece and Aegean	61	780	690	70
Cyprus	20	420	70	6
Egypt and the Near East	12	200	90	30
Turkey	10	700	125	0
The Balkans	45	500	2	0
Total	148	2,600	977	106

1995, Wagner *et al.* 1989), the ancient mines of Timna and Feinan in Wadi Arabah (Gale *et al* 1988 and Hauptmann *et al.* 1994 respectively), copper and lead from Sardinia, some of Spanish deposits (Stos-Gale *et al.* 1995a) and the Balkan ores (Stos-Gale *et al.* 1998b and Pernicka *et al.* 1993). A group of samples of ores and slags from an archaeological expedition to the region of Kerman in Iran given to Noel Gale by Pleiner (1967) was analysed at Oxford giving a much needed preliminary 'fingerprint' for copper and silver ores from this region (Stos-Gale 2001).

Since each set of lead isotope data for one sample consists of three numbers, three lead isotope ratios, the most convenient way of comparing them is by using two 2-dimensional plots. Such lead isotope diagrams represent mirror images of the datapoints on two planes positioned at a 90° angle. Therefore each analysed artefact, or ore, is represented by two symmetrical points; one on the upper, the other on the lower diagram. An example of such diagram is presented on Fig. 10.1. In this configuration it is easy to see if two given points are identical in all three ratios.

The direct comparison of 'ore-artefact' lead isotope ratios gives the correct answer only if the metal artefact contains lead from only one source. If the copper metal circulating on Crete in the Bronze Age originated from many geographically, and isotopically, different sources, and when re-used, the metals from different sources were melted together, then the lead isotope 'fingerprint' of the new artefact will consist

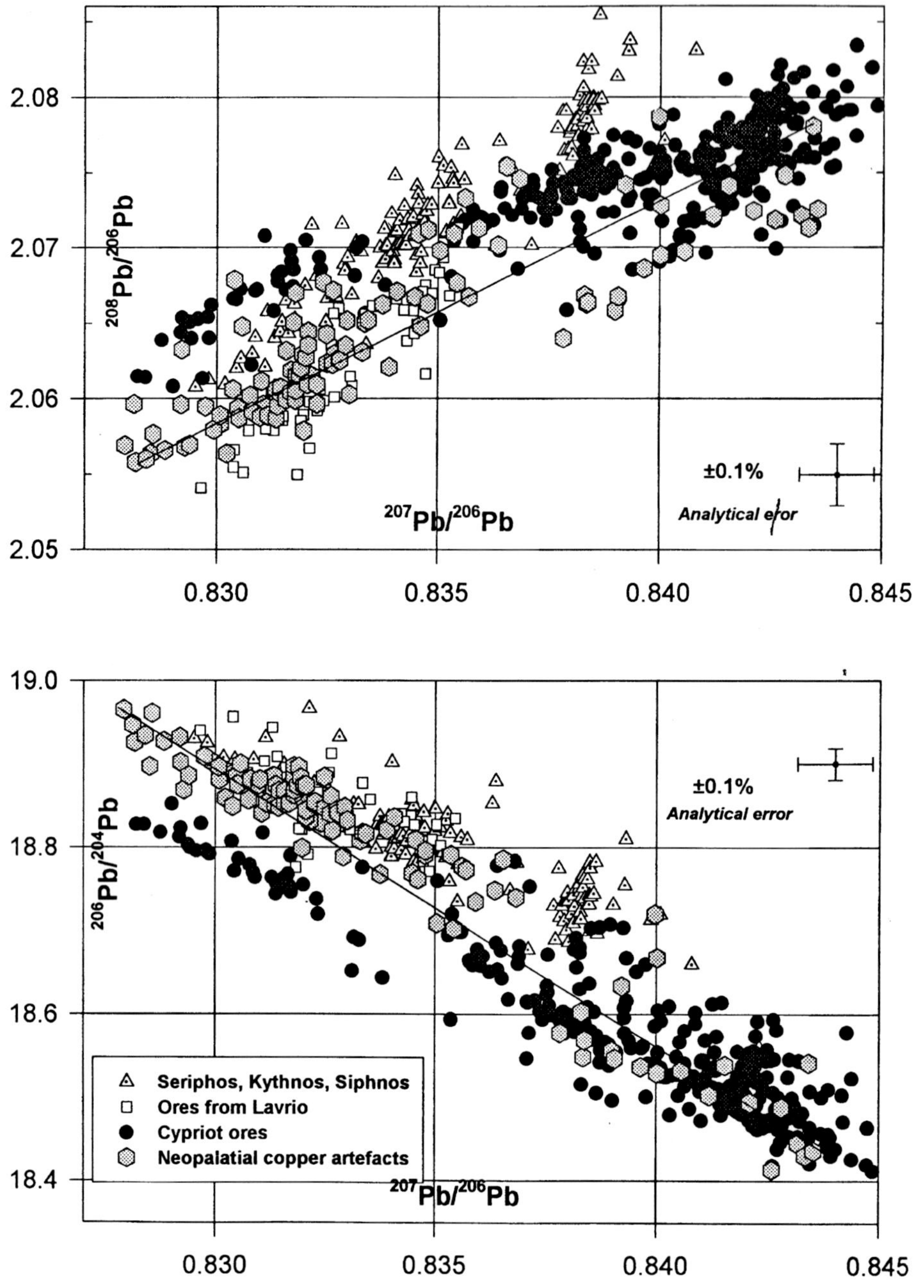

Figure 10.1: Lead isotope ratios for ore samples from Cyprus, South Turkey and the Aegean compared with the data for Neopalatial artefacts. The line running across the data-points is the theoretical 'mixing line' for the extreme lead isotope ratios of the main group of artefacts.

of three lead isotope ratios of values somewhere between each of the two original ratios. On the lead isotope diagrams such point will be positioned on a line joining the two original points. Figure 10.1 presents data for ores from Cyprus, Southern Turkey (Bolkardag range in the Taurus Montains) and the Aegean compared with the lead isotope ratios measured for a number of Neopalatial tin bronzes and copper artefacts. The pattern of these data-points shows that two larger groups of artefacts are identical point-by-point with the ores from Cyprus and Laurion. A small group in the left-hand corner is largely consistent with ores from the Bolkardag ($^{206}Pb/^{207}Pb$ ratios between 0.825–0.83). While some of the objects falling into the central group (identical with Laurion ores) might have been made from re-melted Laurion and Bolkardag scrap metal, the pattern of lead isotope ratios of all Neopalatial artefacts does not represent what would be expected from a group of metals originating from 'mixed' sources. If copper originating from all these deposits available in the Bronze Age Aegean was frequently mixed, the resulting lead isotope patterns will be looking far more as a continuous scatter of points, rather than separate groups. The same conclusion applies to the alloy compositions of the Minoan copper and bronzes – they are far from uniform.

Two of the deposits indicated by lead isotope ratios, Laurion and Bolkardag, have raised some discussions about their suitability as copper sources. Both of these regions are accepted as lead-silver deposits, but there is no direct evidence of copper smelting in the Bronze Age. The Bolkardag area is a wild mountainous region which seems very important for archaeometallurgy and definitely should be further explored (Yener *et al.* 1989a and b, 1991). Perhaps, it should be mentioned here that a few lead isotope analyses of ores from the Ergani Maden mine, which is often quoted in the literature as important source of copper for the ancient Near and Middle East (Muhly 1977), are superficially similar to the 'Bolkardag' LI compositions, but close comparisons show them to be quite different. So far no copper metal BA artefacts consistent with the ratios measured in Oxford and Mainz for the ores from Ergani Maden have been noted. However, both Bolkardag and Ergani Maden, have not been yet sufficiently well characterised isotopically and it is quite possible that further lead isotope analyses of copper ores from these mines will match some Bronze Age artefacts.

The data points for a large group of Neopalatial artefacts on Fig. 10.1. are identical on both diagrams with ores from Laurion. The presence of rich copper ores in the Laurion mines is well know to geologists and only a few years ago the local miners were selling copper minerals from Kamareza mine to the tourists. It is true that copper ores were never exploited in modern times though, because in parallel with many other comparatively small copper deposits (for example Kirki in northern Greece) they are not large enough for modern mining. On the other hand, the long history of lead and silver mining in Laurion is so well documented, that it overshadows the possible copper exploitation. There is no reason, however, why in the Bronze Age copper might not have been mined there also. Ulrich Zwicker, one of the most prominent archaeometallurgists, mentioned that he found copper slag on the hill of Velatouri, near the archaeological site of Thorikos (private communication 1986). It is interesting that on the opposite side of the road, quite near to the coast and behind the building of the Belgian Archaeological Mission, there is a large accumulation of lead slag, a heap reaching over 20m in height. This slag originates largely from modern smelting and is fenced off within the perimeter of the Laurion Mines Ltd., and therefore it was never examined by archaeometallurgists, but it is quite possible that there is ancient slag (may be also copper slag) at the bottom.

THE SOURCES OF COPPER IN PREPALATIAL AND OLD PALACE PERIODS

The great majority of copper-based artefacts from Prepalatial Crete consists of weapons from the tombs in Ayia Triada, Koumasa, Platanos, and others. In itself the quantity of these weapons deposited with the dead is quite substantial for the 3rd millenium B.C. Aegean. It indicates that copper became widely available then and became part of everyday life. The earliest metals from this group are associated with the EMII pottery (before 2200 B.C.), but it is not really possible to say which of them are the earliest, because the tombs usually contain pottery dated to quite a wide chronological band. For example, the Ayia Triada communal tomb falls in the range EMIIa–MMIa, Koumasa and Marathokephalo EMI–MMII, etc. The stylistic features are not much help in this respect either, because most of these weapons are of a simple 'triangular' shape that persisted throughout the early part of this period.

The chronology of Bronze Age Crete is perhaps best represented by the series of events rather than the pottery styles named from the Early to Late Minoan. The earliest Bronze Age, until 2000 B.C. is called the Prepalatial period. The period between c. 2000 B.C. and 1700 was the first period on this island when the palaces were build and played the central role in regulation of food supplies in their region. This period is usually referred to as the Old Palace, but it is also called Protopalatial or First Palace period. The first palaces were large and designed to house not only the residential quarters, but also, sanctuaries, storerooms and workshops. The palatial pottery workshops, particularly those working for the palaces of Knossos and Phaistos, were very advanced and produced exquisite pottery. Vessels made in this workshops and the agricultural produce of the island were the main items of the trade, which was controlled by the palaces. The finds of Minoan pottery of the Old Palace period suggest that the Minoans were trading with the Cyclades, the Greek Mainland and every area of the Eastern Mediterranean including Egypt, the coast of Syria and Palestine, and Asia Minor.

When metals are concerned, there is a small difference in shapes and types of copper-based weapons and tools between the ones from the Old Palace period and the earlier, Prepalatial metals. The chronological summary of dagger types (Branigan 1968) indicates that both, long and triangular daggers appear commonly in the Prepalatial EMII context (around 2500 B.C.), but by the end of the Prepalatial period, MMIa, the triangular daggers disappear making way to more developed weapons. The lead isotope data indicates that the main source of copper used for all these objects were the Cycladic islands (Stos-Gale 1993 and 1998). Out of 118 copper-based metal artefacts from Prepalatial and Old Palace periods analysed to date, 60 are consistent with the copper ores from Kythnos, Siphnos and Seriphos. The next large copper source for these weapons is Cyprus (25 artefacts). Copper from Lavrion comes third (17 artefacts) and 5 daggers (4 Ayia Triadha and 1 from Platanos) have lead isotope characteristics of ores from the Kerman region of Central Iran (Caldwell ed. 1967). Three of these daggers belong to a group of the earliest Minoan tin bronzes. Other imports from distant sources include copper from the Arabah valley in Sinai (either Timna or Feinan mines) and in the Old Palace period there are 2 artefacts consistent with the ores from the Taurus mountains in Southern Turkey (a pin and a long dagger Type V). The pie-charts on Fig. 10.2 show that there is no great difference in the distribution of copper sources used for the tools and weapons in the Prepalatial and Old Palace periods. The major sources: the Cyclades, Cyprus and Lavrion remain the same from around 2500 B.C. to 1700B.C., the end of Old Palaces.

Already in the EMI period (Prepalatial) some of the copper artefacts contain tin, leading way to the most prominent Bronze Age metal – the tin bronze. The increase in the use of tin between the Prepalatial and Old Palace is insignificant amongst the artefacts analysed in Oxford, remaining at about 20–30%. There is no marked predominant source of copper used for the objects containing tin: amongst 29 artefacts with tin content above 1% there are 11 isotopically consistent with Cypriot copper ores, 6 with the ores from Kerman in Iran, and 10 with ores from Lavrion in Attica. One bronze long dagger from the Old Palace tombs in Kalathiana is isotopically identical with the ores from the Arabah valley in Sinai, and a P fishhook from Ayia Photia is consistent with the Cycladic copper. It is quite surprising here though, that in view of an overall predominance of Cycladic copper amongst all analysed Prepalatial and Old Palace artefacts, tin bronzes appear mostly amongst the metals imported from the Near East.

THE NEW PALACE PERIOD

Around 1700 B.C. a wave of destruction swept across Crete, the Old Palaces were destroyed, but the order returned without delay. Quite soon New Palaces emerged, even bigger and better. The archaeological evidence is not sufficient to be sure of the nature of this destruction. It is not certain at present if it was caused by natural disasters, or as a result of internal unrest. Foreign invasion can be rather ruled out because after the destruction there is a complete cultural continuity into LMIb. However, the cult habits have changed gradually in LMII/III (Driessen and Macdonald 1997, 37 and 61). The New Palace period is regarded as the most brilliant age of Minoan Crete. The palaces were re-built in a much grander style, with majestic colonnades and staircases connecting several storeys. The arts and crafts developed rapidly. This is the period when the earliest figurative frescoes were used for decoration of the palaces, the luxury objects include beautifully carved stone vessels, sophisticated faience figurines and bas-relief plaques, exquisite gold and ivory objects. Seal carving flourished greatly in the Neopalatial Crete.

In parallel with other arts and crafts metallurgy reaches a high level of professionalism in the Neopalatial times. The new metal forms include large vessels made of sections of sheet metal riveted together and repoussé decoration is used frequently. The tools and weapons become much larger and include new forms like adzes, hammers, hooks and saws. This period also marks the first appearance of a sword.

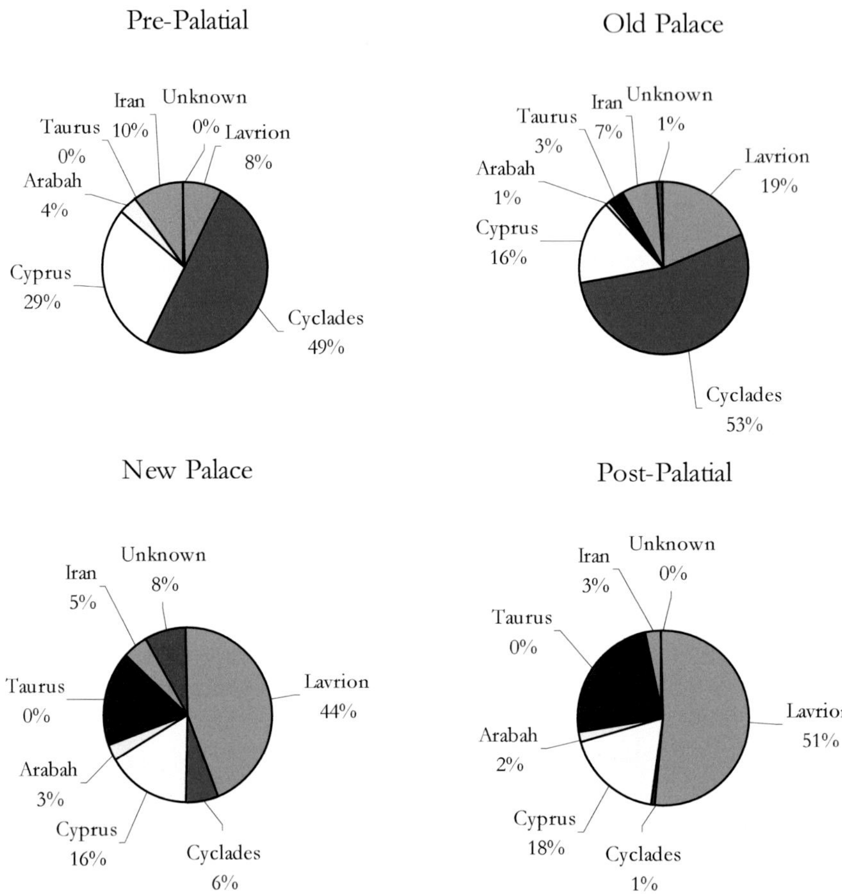

Figure 10.2. Distribution of copper ore sources for metals used in Minoan Crete.

The amount of copper metal used was quite considerable. Sometimes around 1550 B.C. (LMIb) copper arrives to Crete in a quite new form; numerous large (around 25 kg in weight) 'oxhide' shaped ingots were found in the LMI storerooms of the palace of Ayia Triadha and Zakro. In the coastal site of Mochlos in Mirabello Bay the metallurgical workshop contained many fragments of copper ingots – some of them undoubtedly originating from the 'ox-hide' shaped, some from a round 'bun' ingots (Soles and Davaras 1996).

The tin bronze is the predominant type of metal used during the Neopalatial period. Excluding the pure copper ingots, nearly 60% of 132 copper based artefacts are of tin bronze. No pure tin ingots, or ingots of tin bronze were identified amongst the Minoan metal finds. The pie-chart on Fig. 10.2 shows the proportion of copper sources identified amongst the Neopalatial metals. The difference in the sources of copper used after 1700 B.C. on Crete in comparison with the earlier periods is staggering. Copper from

Lavrion expands to nearly 50% of the total, Cyprus copper remains at the same level as in the Old Palace period (16%), a few percent of the artefacts are consistent with Iranian and Arabah copper, but a new, it seems quite important, source appears: copper from the Bolkardag valley in the Taurus Mountains of Southern Turkey (18%). In the later Bronze Age, in the 13th–12th century B.C. all oxhide ingots analysed for their lead isotope composition have isotope ratios identical with the copper ores from the Solea region in northwest Cyprus (Stos-Gale *et al.* 1997). All but one (Gale 1991, p. 202 and Pl.1), of the ingots from Ayia Triadha have lead isotopic composition incompatible with all the ores from Greece and Near and Middle East analysed so far. The same lead isotope ratios were found amongst ingots from Kato Zakro and two 'bun' ingots from Arkalohori and Nirou Khani. This 'unknown' source of metal has been also found amongst other Bronze Age Aegean artefacts, notably early tin bronzes from Troy and Kastri on Syros (Stos-Gale *et al.* 1984 and Gale *et al.* 1985). It would be very tempting to relate this 'unknown' source of copper to the earliest trade in tin, however, all Minoan objects with this lead isotope fingerprint analysed so far, are of pure copper. On the other hand, the majority of copper artefacts with the lead isotope characteristics of copper from Lavrion contain high percentage of tin (the lead content in nearly all these artefacts is below 1%, and in the great majority well below 0.1%).

The results of lead isotope analyses of the Minoan copper based artefacts indicate that in the first half of the 17th century B.C. nearly half of copper used on Crete was coming from the east. Cypriot copper is quite prominent, but not exclusive. Copper from another two districts: the Arabah, some 600 km south from Ugarit, and from the southern Anatolian mines, also by that time was available on Crete.

THE COPPER SOURCES OF THE NEIGHBOURS – AEGEAN AND THE NEAR EAST

Lead isotope analyses of nearly 800 Bronze Age copper-based metals from Mainland Greece (Stos-Gale *et al.* 1999a, Kayafa *et al.* 2000) and the islands (Stos-Gale 1993, Stos-Gale *et al.* 1999b) show that in the other parts of the Aegean the pattern of copper supplies is quite similar to the one observed on Crete. Cycladic copper dominates in the Early Bronze Age and copper from Lavrion takes over in the Late Bronze Age with about 50% of metal coming from Cyprus, Iran and Arabah throughout the Bronze Age. Metal (copper, lead and silver) consistent with ores from the Bolkardag mines appears only in the LBA, but in quite considerable quantities. On the Mainland copper from Lavrion, as well as Cypriot copper, appears amongst the earliest artefacts.

It is quite interesting, if not unexpected, to see that copper artefacts from Cypriot BA sites are not all made of Cypriot copper. In the Early Cypriot period, around 30% of analysed artefacts are consistent with the Iran-Kerman ores. In the MC, the copper from Taurus forms a considerable group and the Kerman ores are also still present. In the LC period to these foreign sources copper from the Arabah can be added, and a few objects are consistent with Sardinian copper. Throughout the BA copper metal artefacts from the Aegean are very scarce on Cyprus. Only a few (less than 10) objects can be regarded as consistent with the Cyclades and Lavrion.

Unfortunately, lead isotope analyses of Bronze Age copper-based artefacts from

Egypt and Syro-Palestine are still very few – 177 altogether. For Egypt, there are only analyses of LBA artefacts from Qantir and el Amarna. These show predominantly lead isotope ratios characteristic for the Arabah, Cyprus and, what now is identified as Taurus copper, but it is still open to discussion if some other deposit in the Sinai peninsula could not have similar lead isotope compositions (Stos-Gale *et al.* 1995b).

The analyses of copper metal finds from Ugarit and another few sites along the coast show again a pattern of sources including the Arabah (since the Chalcolithic), Taurus, Cyprus and Iran. Copper from the Cyclades has not been found, and only a few pieces from the LBA are consistent with Lavrion (unpublished data, Isotrace Laboratory, Oxford).

In summary, the picture of copper sources in the Bronze Age Eastern Mediterranean obtained from lead isotope analysis indicates that Cyprus played an important role in supplying copper both to the Aegean and the Near East throughout the BA. However, people in the Aegean relied predominantly on its own sources, the Cyclades in the EBA and Lavrion in the LBA. This 'Aegean copper' was not widely exported to the east, where apart from Cypriot copper, much of this metal was coming from the Arabah valley (Timna or Feinan, or both). Imports from southern Anatolia and Iran played also an important role.

TIN BRONZE: THE POSSIBLE SOURCES OF TIN

Another important question is the association of the earliest tin bronzes in Crete with the sources of copper, and other early tin bronzes from the surrounding area. A summary of the EBA copper artefacts analysed in Oxford and the number of tin bronzes amongst them is presented in Table 10.3, together with their copper sources, based on their lead isotope data.

The earliest tin bronzes appear in Crete in EMI–II period. Out of 118 Prepalatial and Old Palace artefacts analysed in Oxford 29 from various sites across Crete contain tin (25% of the total analysed). The average tin content throughout the Prepalatial and Old Palace period remains quite low at about 1.5%. The earliest tin bronzes found on the EBA sites located on the Aegean island (Kastri on Syros, Ayia Irini on Kea, Emporio on Chios, Manika on Euboea, Trianda on Rhodes, Thermi on Lesbos) show a somewhat different pattern of lead isotope ratios. Out of 162 artefacts 25 contain tin. The highest proportion of these tin bronzes – and the highest tin contents – have very high $^{208}Pb/^{206}Pb$ lead isotope ratios (above the value of 2.11). Such isotope ratios are so far unknown amongst all analysed ores from the Aegean, Near and Middle East. These lead isotope ratios appear also amongst the EBII bronzes from Troy. Tin bronzes amongst the analysed Early Helladic artefacts are very few, and the two artefacts with high tin content (4.5% and 10%, a needle and some decorated fragments from Lithares) have lead isotope ratios consistent with Kerman region in Iran. Early tin bronzes on Cyprus (EC-MC) also contain comparatively large proportion of artefacts consistent with ores from Kerman.

The evidence presented above is based on a comparatively small number of EBA artefacts analysed for their lead isotope ratios, but if one can draw any conclusions from these comparisons, it seems that in the EBA tin bronze might have been coming

Table 10.3: The lead isotope pattern of the EBA tin bronzes in the Aegean and Cyprus

Totals	Crete	Islands	The Mainland	Cyprus
Total of EBA-MBA copper artefacts analysed	118	162	90	94
Total of analysed EBA artefacts containing tin	29	25	3	14
Sources for artefacts containing tin	**Crete**	**Islands**	**The Mainland**	**Cyprus**
Lavrion	10	4	1	0
The Cyclades	1	5	0	3
Cyprus	11	3	0	5
Iran	6	3	2	6
Arabah	1	2	0	0
Unkown 'high' 208Pb/206Pb	0	8	0	0

to the Aegean from two directions: from the north, with copper carrying the 'Unknown' high lead isotope ratios found amongst the Trojan and Kastri metals, or from the south, with Cypriot and Iranian copper. There are also EBA artefacts from Kastri, Manika and Thermi consistent with the Cycladic copper containing high tin (Stos-Gale *et al.* 1984, Stos-Gale *et al.* 1999b and Stos-Gale 1992). This evidence suggests that pure tin was available in the Aegean and the alloy was procured on the islands. The only evidence of pure tin is the famous tin bracelet from Thermi. It is interesting to note that this bangle shows lead isotope compositions similar to the 'Unknown' high $^{208}Pb/^{206}Pb$ ratios mentioned above (Begemann *et al.* 1992).

In the Neopalatial period there is a considerable increase in the number of objects containing tin, and the average tin content in them. Figure 10.3 shows the tin content and the standard deviation in groups of the Minoan artefacts from different periods. The differences are quite considerable: from 1.5% in the Prepalatial and Old Palace, to 3.7% in the Neopalatial, and raising further in the later periods. Tin bronzes from all eastern Mediterranean sites have lead isotope ratios reflecting all used copper sources. However, there is one interesting development: from the MBA on the islands and the Mainland, and from Neopalatial period on Crete, a new source of copper emerges: the Bolkardag-Taurus copper. It seems rather more than a fortuitous coincidence that this is the highly disputed source of tin (Yener *et al.* 1989a). It is somewhat disconcerting though, that Yener and her colleagues reported the Kestel mine as the Early Bronze Age tin source only. From the current lead isotope records it emerges clearly that copper from this region enters the Aegean and Cypriot trade in metals only in the LBA. Is there a reconciliation of this problem possible? Perhaps the miners from Kestel exhausted their tin supplies by the end of the MBA, but started to act as middlemen for

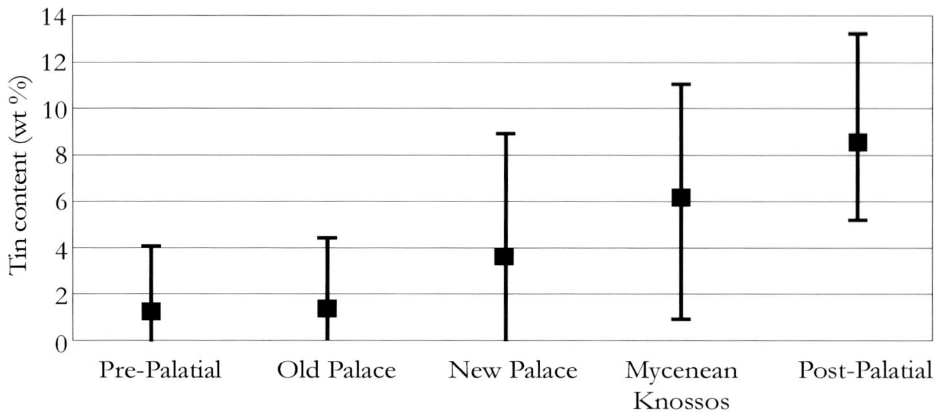

Figure 10.3: Mean tin content in copper artefacts from five Minoan periods.

tin from another location, adding their copper to the metal shipments. Or is there another copper mine, so far not sampled for lead isotope measurement, covering the same range of lead isotope ratios as the Bolkardag minerals? On geological grounds, it is possible that the region of Ergani Maden might contain ores isotopically identical with Bolkardag. There are also mines and ancient copper slags in the Egyptian part of Sinai which have not been characterised isotopically. Unfortunately there are no lead isotope analyses available of copper artefacts from Egypt dated before the 18th Dynasty. Much more archaeometallurgical research, supported by an extensive program of lead isotope analyses is needed to solve this question.

The other question, which is even more interesting, concerns the social mechanism of this very important change in Minoan metal industry: it seems that with the emergence of a new ruling elite in the New Palace period, the majority of copper started to be imported from Attica and simultaneously a considerable amount of tin became available. Were these three points: New Palaces, copper from Lavrion and the availability of tin, at all connected? There is no noticeable change in sources of metals at that time on Cyprus, or Syro-Palestine. Nothing can be said about Egypt, because there are no lead isotope analyses of EBA Egyptian copper artefacts.

CONCLUSIONS

It is interesting to compare the results presented above with a statement by Jim Muhly published 23 years ago in *The Search for Ancient Tin* (Muhly 1977, 44):

> 'For the Sumerians, copper came from Magan and Meluhha, that is from the south (I think from Kerman in Iran and the Bihar district in India) … What the Egyptians called Asiatic copper may have come from Cyprus or perhaps from Wadi Arabah in the Negev…. References to copper from Alashia (are) putting the use of Cypriot copper back into the third millenium B.C. … the Hittite texts speak of copper and bronze coming from Mount Taggata in Alashia. … Egyptian texts also refer to tin coming from Alashia, but

> there is no tin on Cyprus. Yet Cyprus could have been a centre for the transhipment of tin as well as copper.'

The lead isotope and alloy analyses of Bronze Age copper artefacts give evidence in support of these ancient texts. Perhaps a large number of scientific analyses combined with archaeological and textual evidence, can lead the way towards a better understanding of the question of Bronze Age economy, social interactions and trade.

And finally, one more quotation from Muhly's seminal article (1977, 47):

> 'I feel that, in general, the importance of the metals trade in the bronze Age has been greatly exaggerated. Even Stuart Piggott ... continually uses the search for or trade in metals as explanation of diffusionist theories and hypothetical migrations.'

My feelings are very much in support of this statement. From the perspective of 20 years of lead isotope analyses of the BA Mediterranean metals it seems to me that the Eastern Mediterranean, Anatolia and the Middle East of the second millenium B.C. was very much the 'global village' of its time. Goods including metals and pottery were circulating in all the neighbouring countries and the influential classes of merchants and professional craftsmen must have been quite prominent. The smelting of copper was a highly specialised craft and, most likely, quite separate from the palace copper-smith, and the merchant supplying him with metals. Copper was a necessity and it seems that each region had access to its own supplies. In the Aegean the mines in Laurion might have been contested between the Mycenans and Minoans, who had no ore sources worth exploiting on Crete itself. 'Minoan Thalassocracy' might have been induced by 'the search for copper' and possibly the final LBA Mycenean domination in the region had as its backing the Laurion mines. However, copper from other rich and well developed extraction (smelting) centres like Cyprus, Kerman, or Wadi Arabah was 'diffusing' into all parts of the region. The source of tin in the LBA is still a mystery. A project in Oxford based on isotopic analyses of over 100 ingots of tin from the Uluburun shipwreck did not provide an answer as to the geographical location of the tin ore source. The lead isotope compositions of these ingots do not match convincingly cassiterites from Spain, Bohemia, Cornwall, or the Far East (N.H. Gale and R. Clayton, private communication 1999). They are quite closely related to lead and copper ores from the Taurus mountains in Southern Turkey, but until cassiterites of the same compositions are found it is impossible to make this connection (Stos-Gale *et al.* 1998c).

ACKNOWLEDGEMENTS

The project on BA Mediterranean metal sources and trade was masterminded by Noel Gale and he was instrumental for many years in raising funds for the analytical work in Oxford and geological and archaeological sample collections. We are very grateful for funding to the University of Oxford, the British Academy, the Leverhulme Trust, SERC and mostly to INSTAP, whose generous support allows us to continue this research just when the results are starting to emerge. This short paper is a preliminary study of data that will be fully discussed in a monograph currently being prepared by the author with Noel Gale and Don Evely. I would like to thank Jeff Soles and Nicoletta Momigliano for most helpful comments and corrections.

REFERENCES

Begemann, F., Schmitt-Strecker, S. and Pernicka, E., 1992, The metal finds from Thermi III–V: a chemical and lead-isotope study, *Studia Troica* 2, 219-239.

Betancourt, P. Muhly, J., Farrand, W., Stearns, C., Onyshkevych, L., Hafford, W. and Evely, D., 1999, Research and excavation at Chrysokamino, Crete 1995–1998, *Hesperia* 68/3, 343–370.

Caldwell, J. R. (ed.), 1967, *Investigations at Tal-I-Iblis*, Illinois State Museum Preliminary Reports No.9, Springfield, Illinois.

Driessen, J. and Macdonald, C. F., 1997, *The troubled island – Minoan Crete before and after the Santorini eruption*, Universite de Liege and University of Texas at Austin.

Gale, N. H., 1991, Copper oxhide ingots:their origin and and their place in LBA metals trade in the Mediterranean, in Gale, N.H. (ed.), *Bronze Age trade in the Mediterranean*, SIMA XCII, Paul Astroms Forlag, 197–239.

Gale, N. H., Stos-Gale, Z. A. and Gilmore, G. R., 1985, Alloy types and copper sources of Anatolian copper alloy artefacts, *Anatolian Studies*, Vol. XXXV, 143–173.

Gale, N. H., Bachmann, H. G., Rothenberg, B. and Stos-Gale, Z. A., 1990, The adventitious production of iron in the smelting of copper in Timna, W. Arabah, in Rothenberg, B. (ed.), *The ancient metallurgy of copper*, IAMS, UCL London, 182–191.

Gale, N. H., Stos-Gale, Z. A., Maliotis, G. and Annetts, N., 1997, Lead isotope data from the Isotrace Laboratory, Oxford: *Archaeometry* data base 4, ores from Cyprus, *Archaeometry*, 39 (1), 237–246.

Gale, N. H. and Stos-Gale, Z., 2000, Lead isotope analyses applied to provenance studies, in Ciliberto, E. and Spoto, G. (eds.), *Modern Analytical methods in Art and Archaeology*, Chemical Analyses Series, Vol. 155. John Wiley and Sons, 503–584.

Hauptmann, A., Begemann, F., Heitkemper, R., Pernicka, E. and Schmitt-Strecker, S., 1992, Early copper produced in Feinan, Wadi Araba, Jordan: the composition of ores and copper, *Archeomaterials*, 6.1, 1–33.

Hirao, Y. Enomoto, J. and Tachikawa, H., 1995, Lead isotope ratios of copepr, zinc and lead minerals in Turkey in relation to the provenance study of artefacts, in H.I.H. Prince Takahito Mikasa (ed.), *Essays on Ancient Anatolia and its surrounding civilisations*, Harrassovitz Verlag, Wiesbaden, 89–114.

Kayafa, M., Stos-Gale, Z. A. and Gale, N. H., 2000, Chemical and lead isotope analyses from early Helladic Tsoungiza, in Pare, C. (ed.), *Metals make the world go round: supply and circulation of metals in Bronze Age Europe*, Oxbow Books, Oxford.

Mangou, H. and P. V. Ioannou, 1998, On the chemical composition of prehistoric Greek copper-based artefacts from Crete, *BSA* 93, 91–102.

Muhly, J. D., 1977, New evidence for sources and trade in Bronze Age tin, in Franklin, A. D., Olin, J. S. and Wertime, T.A. (eds.), *The search for ancient tin*, Smithsonian Institution, Washington, 43–48.

Pernicka, E., Seeliger, T. C., Wagner, G. A., Begemann, F., Schmitt-Strecker, S. Eibner, C., Oztunali, O., Baranyi, I., 1984, Archeometallurgische untersuchungen in Nordwestanatolien, *Jahrbuch des Romisch-Germanischen Zentralmuseums*, 31, 533–599.

Pernicka, E., Begemann, F., Schmitt-Strecker, S., Wagner, G. A., 1993, Eneolithic and Early

Bronze Age copper artefacts from the Balkans and their relation to Serbian copper ores, *Praehistorisches Zeitschrift*, 68(1), 1–54.

Pleiner, R., 1967, Preliminary evaluation of the 1966 metallurgical investigations in Iran, in Caldwell, J. R. (ed.), *Investigations at Tal-I-Iblis*, Springfield Illinois, 340–405.

Soles, J. S. and Davaras, C., 1996, Excavations at Mochlos 1992–1993, *Hesperia*, 65, 175–230.

Stos-Gale, Z. A., 1992, The origin of metal objects from the Early Bronze Age site of Thermi on Lesbos, *Oxford Journal of Archaeology*, 11/2, 155–178.

Stos-Gale, Z. A., 1993, Lead isotope provenance studies of the early Minoan weapons, in Scarre, C. and Healy, F. (eds.), *Trade and Exchange in Prehistoric Europe*, Oxbow Books, Oxford, 115–129.

Stos-Gale, Z. A., 1998a, The role of Kythnos and other Cycladic islands in the origins of Early Minoan metallurgy, in Mendoni, L. and Mazarakis, A. (eds.), *Meletimata 27, Kea-Kythnos: History and Archaeology: Proceeding of the Kea-Kythnos Conference, Kea June 1994*, Diffusion de Broccard, Paris&Athens, 717–736.

Stos-Gale, Z. A., 2001, The impact of the natural sciences on studies of Hacksilber and early silver coinage, in Balmuth, M. (ed.), *Hacksilber to coinage: new insights into the monetary history of the New East and Greece*, 51–71 in press.

Stos-Gale, Z. A., Gale, N. H. and Gilmore, G. R., 1984, Early Bronze Age Trojan metal sources and Anatolians in the Cyclades, *Oxford Journal of Archaeology*, 3.3, 23–44.

Stos-Gale, Z. A., Gale, N. H., Houghton, J. and Speakman, R., 1995a, Lead isotope analyses of ores from the Western Mediterranean, *Archaeometry*, 37.2, 407–415.

Stos-Gale, Z.A., Gale, N.H. and Houghton, J., 1995b, The origin of copper metal excavated in El Amarna, in Davies, W. V. and Schofield, L. (eds.), *Egypt, the Aegean and the Levant: interconnections in the 2nd millenium B.C.*, British Museum Press, London. 127–135.

Stos-Gale, Z. A., Gale, N. H. and Annetts, N., 1996, Lead isotope analyses of ores from the Aegean. *Archaeometry*, 38.2, 381–390.

Stos-Gale, Z. A., Maliotis, G., Gale, N. H. and Annetts, N., 1997, Lead isotope characteristics of the Cyprus copper ore deposits applied to provenance studies of copper oxhide ingots, *Archaeometry*, 39.1, 83–124.

Stos-Gale, S., Maliotis, G. and Gale, N. H., 1998a, A preliminary survey of the Cypriot slag heaps and their contribution to the reconstruction of copper production on Cyprus, in Rehren, T., Hauptmann, A. and Muhly, J. (eds.), *Metallurgica Antiqua, in Honour of Hans-Gert Bachmann and Robert Maddin*, Deutches Bergbau Museum, Bochum, 235–262.

Stos-Gale, Z. A., Gale, N. H., Annetts, N., Todorov, T., Lilov, P, Raduncheva, A. and Panayotov, I., 1998b, Lead isotope data from the Isotrace Laboratory, Oxford: Archaeometry Database 5, Ores from Bulgaria, *Archaeometry*, 40.1, 217–226.

Stos-Gale, Z. A., Gale, N. H., Bass, G., Pulak, C, Galili, E. and Sharvit, J., 1998c, The copper and tin ingots of the Late Bronze Age Mediterranean: New scientific evidence, in *Proceedings of The Fourth International Conference on the Beginning of the Use on Metals and Alloys (BUMA–IV)*, The Japan Institute of Metals. Aoba, Japan. 115–126.

Stos-Gale, Z. A., M. Kayafa and N. H. Gale, 1999a, The origin of metals from the Bronze Age site of Nichoria, *Opuscula Athenesia*, 24, Lund, 99–120.

Stos-Gale, Z. A., Sampson, A. and Mangou, E., 1999b, Analyses of metal artefacts from the Early Helladic cemetery of Manika on Euboea, *Aegean Archaeology*, 3, Art and Archaeology, Warsaw.

Wagner, G. A., Pernicka, E., Vavelidis, M., Baranyi, I. and Bassiakos, I., 1986, Archäometallurgische Untersuchungen auf Chalkidiki, *Der Anschnitt*, 38, 166-240.

Wagner, G. A., Begemann, F., Eibner, C., Lutz, J., Oztunali, O., Pernicka, and Schmitt-Strecker, S. F., 1989, Archäometallurgische Untersuchungen an Rohstoffquellen des frühen Kupfers

Ostanatoliens, *Jahrbuch des Römisch-Germanischen Zentralmuseums Mainz,* 36, 637–686.

Wertime, T., 1967, A metallurgical expedition through the Persian desert, in Caldwell, J. R. (ed.), *Investigations at Tal-I-Iblis,* Springfield Illinois, 327–339.

Yener, K. A., Ozbal, H., Kaptan, E. Pehlivan, A. N. and Goodway, M., 1989a, Kestel: an early Bronze Age source of tin ore in the Taurus Mountains, Turkey, *Science,* 244, 177–264.

Yener, K. A., Ozbal, H., Minzoni-Deroche, A. and Aksoy, B., 1989b, Bolkardag: archaeometallurgy surveys in the Taurus Mountains, Turkey, *National Geographic Research,* 5.3.

Yener, K. A, Sayre, E. V., Ozbal, H., Joel, E. C., Barnes, I. L. and Brill, R. H., 1991, Stable lead isotope studies of Central Taurus ore sources and related artefactsfrom Eastern Mediterranean Chalcolithic and bronze Age sites, *Journal of Archaeological Science,* 18.5, 541–577.

Chapter 11

Social Influences on the Development and Spread of Glass Technology

Andrew Shortland

Abstract

Glass was first used in Egypt on a regular basis during the reign of Tuthmosis III (1479–1425 B.C.) for the manufacture of small glass vessels. 13 glass vessels are more-or-less securely dated to his reign and show that first colours of glass to be employed were light-blue (imitating turquoise), dark-blue (imitating lapis lazuli), yellow (probably imitating gold) and, rarely, white. This paper looks at the innovation of glass making in Egypt, especially what examination of these vessels can say about the whether this technology was imported from Mesopotamia or was a local invention. It places some of the techniques used to work this early glass within established Egyptian industries for the processing of stone and semi-precious stone and highlights those which are novel.

INTRODUCTION

The first man-made glass vessels to be found in Egypt date from the reign of Tuthmosis III (1479–25 B.C.). Several beads, amulets and other small objects are securely dated before this time, including beads from Qau, inlays in the Ahhotep treasure (last half of the sixteenth century B.C., Lilyquist and Brill 1993, 23) and several colourless, blue or green beads from the time of Hatshepsut (1473–58 B.C.), but their occurrence is sporadic and does not seem to represent the regular sustained production of glass (Peltenburg 1992; Nicholson 1993; Lilyquist and Brill 1993; and others). It is possible that these early examples of glass may represent reworking of accidental over-firing of faience objects (Tite *et al* 1998). It is only during the reign of Tuthmosis III, coincident with the introduction of glass vessels, that a proper industry seems to develop and glass is available in larger quantities. From then on, the amount of glass found in the archaeological record increases with time, reaching a peak during the reigns of Amenophis III and Akhenaten, evidenced by the large quantities of glass uncovered by excavation of sites at Malkata and Amarna respectively. In addition to the finds of broken glass, both of these sites have glass workshop facilities preserved (Petrie 1894, 25; Nicholson 1995; Keller 1983, 20–3) which at Amarna at least was on a large scale (Shortland 2000).

This paper briefly considers the evidence that places the invention of glass in Mesopotamia in the sixteenth century B.C. It then goes on to look at the social aspects that control the spread of glass and the glassmaking technology. It considers what information can be derived concerning the first one hundred years of glass production from study of textual, stylistic and technological and compositional evidence. It is probable that the invention of glassmaking also quite quickly led on to the invention of glazing on clay bodies. The social aspects controlling the spread of this technology is also considered and compared to those for glass.

INVENTION OF GLASS

It has long been considered that the technology for the production of glass was first developed in Mesopotamia around the mid-sixteenth century B.C. (Lucas and Harris 1962; Oppenheim 1970; Moorey 1995 and many others). Moorey (1995, 192) states that 'there is nothing before the Kassite period [post-1600 B.C.] in Mesopotamia to indicate anything other than an infrequent and irregular production of glass, predominantly for personal ornaments, which showed little, if any, appreciation of the material's special properties'. The development of new colorants, special tools and new techniques recognising the unusual properties of glass occurred in Western Asia 'in the late sixteenth century B.C.' (Moorey 1995, 193). Glass was first prized for its resemblance to semi-precious stones, specifically lapis lazuli and turquoise, and was used in inlays and for small amulets and beads. Core forming, where glass rods are wound around a central core of sand, could be regarded as the key breakthrough in the production of glass vessels. The earliest evidence for core-formed vessels is from Tell Atchana (Alalakh) Levels VI–V (probably the late sixteenth century B.C.), but these levels have been demonstrated to contain pottery and faience from a demonstrably later date, so must be treated with caution. No other core-formed vessels dated earlier than the second half of the fifteenth century are known from Mesopotamia (Moorey 1995, 193). In Egypt, the evidence is remarkably similar. Two sherds of core-formed vessels were found associated with KV38, the tomb of Tuthmosis I (1504–1492 B.C.), but these may be from a reburial by of this king by Tuthmosis II, Hatshepsut or even Tuthmosis III, so again these fragments must be treated with caution. After this, the earliest core-formed vessels in Egypt are dated to the reign of Tuthmosis III, that is to say the mid-fifteenth century B.C. Where core-forming was invented is therefore uncertain, with either Egypt or Mesopotamia being possible.

THE ORIGIN OF EGYPTIAN GLASS

Textual evidence

A number of authors (Nolte 1968; Nicholson 1993; Lilyquist and Brill 1993) have pointed out that the first appearance of significant quantities of man-made glass in Egypt corresponds with a series of campaigns that Tuthmosis III undertook in the Levant. Tuthmosis III was the longest reigning king of the 18th dynasty, and one of the most influential. When he acceded to the throne when he was still very young and the regency fell to his aunt, Hatshepsut. He did not become sole king until 1458 B.C., following which he set out on a series of campaigns over successive years in the Levant and in Nubia. Accounts of these campaigns are recorded in reliefs in the temple of Amen-Re at Karnak. They record that Tuthmosis achieved notable victories against many states, culminating in his eighth campaign with the crossing of the Euphrates and the defeat of the King of Mitanni. As a result of these successes, many of the powers of the Levant considered it wise to send gifts to Tuthmosis, the Karnak texts listing gifts from the king of Ashshur (Assyria), the king of Sangar (Babylon) and the Hittites. Amongst the gifts given to Tuthmosis is glass. Reliefs in Tuthmosis III's Jubilee Hall in Karnak list 0.73 kg of 'true lapis lazuli' (*ḫsbḏ mꜣꜥ dbn 8*) and 2.2 kg of dark-blue glass (*ḫsbḏ iri dbn 24*, usually translated as 'man-made lapis lazuli') given to the King by sngr (Babylon) and other passages on the same relief suggest that other states may also have given glass (Nolte 1968, 8).

Another relief at Karnak records Tuthmosis dedicating the booty and tributes obtained from the Levant to Amen-Re in thanksgiving for his victories. The scene consists of ten registers of gifts, depicted as small pictures of the objects themselves along with an indication of the number of that particular object given and some explanatory text. The objects in the top five registers are marked as made of gold (*nbw*), followed by two registers of objects of silver (*ḥḏ*), one of semi-precious stones and lastly two registers with objects of copper (*biꜣ*). This suggests that the objects are probably placed in order of their value, from highest to lowest. Immediately after the silver objects and leading the list of semi-precious stones, are five baskets containing round or irregular lumps. In the first three baskets the lumps are shown as a blue colour, whereas in the last two baskets the lumps are blue/green. Each basket has an associated line of text as listed below:

1. *{ḫs}bḏ Men-ḫpr-Rꜥ ḥrtt 24* 24 lumps of dark-blue glass
2. *{ḫ}sbḏ mꜣꜥ dbn 569* 51.8 kg lapis lazuli
3. *{ḫs}bḏ Men-ḫpr-Rꜥ dbn 600 {...}* >54 kg, <90 kg dark-blue glass
4. *mfkt dbn 93 kt 2* 8.5 kg turquoise
5. *mfkt Men-ḫpr-Rꜥ dbn 10913.8* 993 kg light-blue glass

Various interpretations have been applied to this depiction of five baskets. Most of these agree that basket 2 contains true lapis lazuli (*ḫsbḏ mꜣꜥ*), and basket 4 contains true turquoise (*mfkt*). Where the phrase *ḫsbḏ mꜣꜥ* is used in a scene, the use of *ḫsbḏ* without the adjective *mꜣꜥ* is usually taken to mean dark-blue glass (Nolte 1968). Basket 3 is therefore thought to contain dark-blue glass. The text indicating the weight of glass in this basket is damaged, but enough remains to show that there must be at least 600 deben (six signs for 100 are preserved), but less than 1000 deben (since no 1000 sign is present), i.e. between 54 and 90 kg. Basket 5, like basket 4, is labelled *mfkt*, this time with the royal cartouche *Men-ḫpr-Rꜥ*. While this could indicate more turquoise, it seems more likely that this is light-blue glass. It should be noted that a substantial quantity of this glass is indicated, in all 993kg. Basket 1 is usually interpreted as representing lapis lazuli (ref, Nolte 1968). The reason for this interpretation is the fact that it is positioned first in the line and therefore probably of greatest value. However, it should be noted that this is an interpretation based on the modern value system, where glass is 'obviously' less valuable than lapis lazuli. There is only one other mention of both lapis lazuli and glass from the reign of Tuthmosis III, the gift of *sngr* described in Tuthmosis III's Jubilee Hall in Karnak (detailed above) and this lists lapis lazuli before glass, implying greater value. However, basket 1 is described simply as *ḫsbḏ* without the adjective *mꜣꜥ*, which as mentioned above, in this context usually implies glass. It also has the royal cartouche *Men-ḫpr-Rꜥ*, suggesting some special royal ownership. This is used twice more in the descriptions of the five baskets (baskets 3 and 5), in both cases within descriptions of baskets thought to contain glass. It is therefore certainly possible that the material in basket 1 is dark-blue glass rather than lapis lazuli as has been previously stated. Unusually, no weight is given for the material in basket 1, instead it is described as '24 lumps' (*ḥrtt*). They are depicted as quite large and very circular in comparison with the irregular shaped objects in baskets 2 and 3 and it is certainly true that one of the advantages of glass is that it could be made into larger pieces than the relatively small pieces of lapis that could be mined (Nolte 1968). Perhaps it is because they are large pieces that they are placed in the position of greatest value. It should also be noted that the glass ingots found in the later Ulu Burun shipwreck (dated to 1300 B.C.) were also round, suggesting that, even at the early stage in the history of glass, the objects depicted in the relief may be ingots.

To summarise the textual evidence, it is clear that during the reign of Tuthmosis III, at least some glass was imported into Egypt from the Near East (specifically Babylon, but probably also the Mitanni and others). This glass was of two colours, a light-blue similar to turquoise and a dark-blue similar to lapis lazuli. The same word was used for the glass and its mineral twin, showing that the two materials were closely linked in the Egyptian mind. If the weights of materials given in the dedication relief are real, then there was far more light-blue glass in circulation than dark-blue, perhaps of the order of 10 to 1.

Stylistic and Technological evidence

Nolte (1968, 46–50) lists 12 glass vessels or fragments dated to the reign of Tuthmosis III (see Table 11.1). Two fragments (Cairo 24960 and Brooklyn 53.176.4) are probably from the same vessel, making a total of 11 different vessels. Four of them are kohl pots or fragments of kohl pots, although BM 24391 is the only example that preserves both pot and lid. It is made of turquoise blue glass and decorated with a thin layer of gold leaf to the edge of the rim and foot of the vessel and the lid. The shape of the vessel is identical to that common in stone kohl pots of the Second Intermediate Period and early New Kingdom. This vessel is not directly dated, but is identical in shape to a stone kohl pot from the Wadi Qirud tomb of the foreign wives of Tuthmosis III, which is inscribed for Tuthmosis III. Nolte therefore dates this vessel to the same period. The vessel has very flat surfaces, is very regular in shape and shows the remains of scratch marks that show that it was made by drilling and grinding from a solid block of glass. Similar in many ways to BM 24391 is UC 19657. It is the same size, shape and colour, but has no gold leaf and the lid has been lost. It is mentioned in the excavation reports of Riqqeh and again dated by its similarity to the Wadi Qirud vessel. It was also made by drilling and grinding (Nolte 1968, 48). A lid from another kohl pot (Cairo 24959) was found in KV34, the Tomb of Tuthmosis III, from which it derives its date. It is once

Table 11.1: Glass vessels or fragments dated to the reign of Tuthmosis III

number	shape and type	body colour	decoration	production method
Cairo 24960 Brooklyn 53.176.4	rounded vessel	light blue	yellow leaves to upper part, black and yellow horizontal twist	core-formed
Cairo 24961	handled vessel	light blue	possible yellow leaves, vertical dark blue lines	core-formed, with handle support
Cairo 24959	kohl pot (lid only)	dark blue	none	cold worked
BM 24391	kohl pot with lid	light blue	gold leaf applied to rim, lid and foot	drilled and cold worked
UC 19657	kohl pot (no lid)	light blue	none	drilled and cold worked
MMA26.7.1179	kohl pot (no lid)	light blue	none	drilled and cold worked
MMA23.9	lotus chalice	light blue	gold rim, incised inscription	cast and cold worked
Munich AS630	chalice	light blue	yellow and dark blue trails drawn upwards into "U" shapes, dark blue cartouche at odd angle	core-formed
Ashmolean E2451	chalice	light blue	none	core-formed
MMA26.7.1175	krateriskos	marbleised	gold applied to rim and foot	probably core-formed
BM 47620	jug	light blue	yellow band on rim and foot, dots and feathery trees plus inscription, white and dark-blue bands on handle	core-formed, applied powdered glass decoration

again the same size and shape, but is a dark, lapis lazuli blue as opposed to turquoise. Again it was produced by grinding from a solid block. The fourth kohl pot, MMA 26.7.1179, is of a different shape. It is smaller and slimmer and of a shape popular in stone vessels of Tuthmosis III's time, on the basis of which it is dated to this reign. It too is drilled and ground from a solid block. Others are made by the core-formed technique, some decorated by wrapping further glass rods of contrasting colour around the core, techniques known as trailing and marvering.

It has long been realised that the shape of the glass vessels imitates vessels in other materials and evolves as they evolve. Therefore although several of the shapes of the Tuthmosis III vessels are very rare in glass and one (BM 47620) unique, these are skeuomorphs of earlier shapes of vessel. Five colours are used in the glass of the vessels from the reign of Tuthmosis III. These are, in order of abundance: light-blue, dark-blue, yellow, white, brown and black. These colours are bright and clean, with strong, consistent colours possible where the glass is well mixed. With the exception of Cairo 24961, which is dark-blue and MMA 26.7.1175, which is marbleised, all the vessels have a light-blue body colour. Yellow and white are used only for decoration, but dark-blue was used both for decoration and for one of the bodies. Several of the vessels are monochrome and undecorated, which is very unusual for Egyptian glass vessels. Many of the Tuthmosis III vessels show very unusual decoration, including a trailed cartouche, yellow leaf patterns, and feather or tree and dot designs. These are the sole examples of these types of decoration in glass.

Examination of the later glass from Malkata and Amarna (early to mid-fourteenth century B.C.) shows more consistency in the method of manufacture and the styles of decoration. All of the glass vessels from these sites are made by the core-formed technique and almost all are decorated. The most common form of decoration is trailing and marvering to give 'W' or 'U' shapes (Petrie 1894, 27; Nicholson 1993, 53 and many others). Handles and a foot were sometimes added to the vessels by adhering hot rods in a similar way. Around 90% of Amarna and Malkata glass vessels have dark blue bodies with trails of light blue, yellow and white. Rarer examples include light blue bodies trailed in dark blue and white, and yellow and white bodies decorated with blue 'eye' type patterns or plain with a lightly decorated rim. All of the vessel shapes from the long thin kohl tubes to rounder amphoriskoi and krateriskoi use similar techniques of decoration.

In summary, the vessels from the reign of Tuthmosis III are therefore predominantly light-blue in colour, and show a wide range of decorative styles. Some of them were produced by cold working, whereas others were produced by core-forming. Later at Malkata and Amarna, dark-blue glass is dominant and the range of decoration is more restricted, trailing and marvering to produce 'W' or 'U' shapes being overwhelmingly dominant. All the vessels from this later period were made by core-forming.

Analytical evidence

Lilyquist and Brill (1993) have analysed two of the vessels from the reign of Tuthmosis III, Brooklyn 53.176.4 and MMA 26.7.1175, and shown that they both contain around 0.7% CuO and 1.5% Sb_2O_5. This almost certainly indicates that the light-blue colour of these vessels was obtained by using a combination of a copper colourant and calcium antimonate opacifier. This pattern is seen in later vessels from Malkata and Amarna (Shortland 2000), where identical compositions can be found. Similarly, both early and later copper coloured glasses have high levels of potash and magnesia (>2.0% in both K_2O and MgO), which is consistent with the use of plant ash alkali in the manufacture of the glasses (Sayre 1967, Lilyquist and Brill 1993, Shortland and Tite 2000).

Lilyquist and Brill (1993) have analysed three examples of dark-blue glass from the reign of Tuthmosis III, and in these glasses cobalt is the dominant colorant. The presence of cobalt in these glasses is strongly associated with raised levels of other elements, principally alumina, manganese, nickel and zinc. Kaczmarczyk (1986) has shown that the source of cobalt of this composition is in the Oases of the Western Desert, specifically the Dakhla and Kharga Oases. These Oases were worked in antiquity for their deposits of alum (Lucas and Harris 1962), some of which is cobaltiferous. This alum yields a cobalt colorant which is accompanied by significant amounts of magnesia, aluminium, manganese, iron, nickel, zinc and sulphur, indicating that this is the source of the cobalt used in the glasses. Analysis (Lilyquist and Brill 1993, Shortland and Tite 2000) of dark-blue glass from Malkata and Amarna has shown that these are predominantly made using this same cobalt colorant, although some utilise high levels of a copper colorant instead, or some combination of the two. Where cobalt colourant is used, whether or not there is additionally a significant level of copper, the glasses have low levels of potash (<1.5% K_2O). This is consistent with the use of a natron based alkali in combination with the magnesium rich cobalt colourant in the manufacture of these glasses (Shortland and Tite 2000). The most obvious source of natron in Egypt is the Wadi Natrun and it is probable that this was the source of the natron in the cobalt coloured glass. Dark-blue glasses, coloured solely with copper have high levels of potash and magnesia suggesting a plant ash alkali and, other than a high copper content and the absence of calcium antimonate, are essentially identical to the light-blue glasses.

In summary, the copper blue glasses both from the reign of Tuthmosis III and later were made using a plant ash alkali. The cobalt coloured glasses from both periods have an Egyptian cobalt colorant and Egyptian natron based alkali. There is no variation in the composition of each of the colours throughout the 18th Dynasty.

Isotopic evidence

Lead isotopic analysis has the potential to trace the lead in the lead antimonate yellow colorant used in the glasses to the ore source where it was mined. Lilyquist and Brill (1993) have used lead isotopic analysis to analyse two glasses from reign of Tuthmosis III, one from the marbleised MMA 26.7.1175 and the other from the yellow glass decorating Brooklyn 53.176.4. They found that the lead isotopic ratios are similar to glasses, glazes and lead metals of Mesopotamian provenance and assumed by Lilyquist and Brill (1993) to be made from a lead metal extracted from a Mesopotamian ore source, as yet unidentified. All but one of the yellow glasses from Malkata and Amarna have lower Pb^{208}/Pb^{206} and Pb^{207}/Pb^{206} ratios, similar to galena from mines in the region of Gebel Zeit in Egypt which is known to have been worked for galena in the 18th Dynasty. The exception to this, a glass rod from Amarna, has Mesopotamian isotopic characteristics. Although it is only one sample and so interpretation should be treated with care, this suggests that 'Mesopotamian characteristics' glass was not restricted to the reign of Tuthmosis III, but that glass of Mesopotamian type was still available in the form of rods or ingots during the Amarna period and was still being incorporated into the batches. This may be because the glass was still being imported to Egypt during the Amarna period, or it may be that it represents 'old stock' that was moved around with the glass workers as they moved. Either way, this material would have contaminated any glass batch with Egyptian lead characteristics to which it was added.

DISCUSSION

There seems strong evidence from the reliefs of Tuthmosis III at Karnak that at least some of the first man-made glass in Egypt was brought in as tribute following his victories in the Levant. The texts give evidence of two colours of glass coming from Mesopotamia: light-blue intended to resemble turquoise and dark-blue resembling lapis lazuli. Lead isotopic analysis of the yellow glass in Brooklyn 53.176.4, dated to Tuthmosis III suggests that at least some of the yellow glass of this period was also possibly of a Mesopotamian source. There is also evidence that complete glass vessels were imported. The marbleised material of the krateriskos is unique in Egypt, but several are known from Nuzi. It has been convincingly argued on stylistic (Barag 1970), and scientific grounds (Lilyquist and Brill 1993 using compositional and lead isotopic analysis) that this is a Mesopotamian vessel imported into Egypt as a finished object. This opinion has been supported by a relief in the Tomb of Rekhmire, the vizier of Tuthmosis III, which probably shows two such marbleised glass vessels being given as tribute by the 'Chiefs of Retnu', identified as Syria (Shortland, in press). However, Nicholson (1993, 47) and other authors (e.g. Moorey 1995; Barag 1970) have suggested that not just glass, but Mesopotamian glass-workers or makers were brought to Egypt by Tuthmosis III. Nicholson has stated that the Egyptian glass industry 'arrives, apparently fully fledged', in Egypt in the New Kingdom 'without any period of experiment in glass' and other authors (Peltenburg 1992) have pointed out that the industry expands rapidly. However, several lines of evidence suggest that this may be, at best, only partly true.

Firstly, some of the early, Tuthmosis III, glass vessels contain cobalt coloured glass. Although four of these glass vessel are decorated with dark-blue and one has a body of dark-blue glass, it is not clear how many of these are cobalt coloured, since only one has been analysed: Brooklyn 53.176.4. The dark-blue/black decoration of this vessel was analysed by Lilyquist and Brill and does contain Egyptian Western Oases type cobalt. There are also two other examples of the use of Egyptian cobalt, an inlay and a plaque, both securely dated to this period (Lilyquist and Brill 1993). Egyptian cobalt blue glass is therefore definitely present among the first glass in Egypt. As pointed out in Shortland and Tite (2000), this glass has both a cobalt colorant and a natron based alkali which almost certainly originate in Egypt. While it is possible that these raw materials could have been exported to Mesopotamia and made into glass there, it is far more likely that they were used to produce glass locally in Egypt. Mesopotamian glass is made from copper colorant and plant ash alkali (Brill 1970; Moorey 1995; Shortland and Tite 2000) and seems likely that the hypothesised imported Mesopotamian glassworkers would have used these raw materials, which were freely available in Egypt. However, the use of Egyptian raw material, particularly Egyptian natron, suggests that there is a small, native Egyptian production of glass during the reign of Tuthmosis III. Therefore not all of the glass from the reign of Tuthmosis III was imported or even made by foreign craftsmen.

The second point concerns the methods of manufacture of the earliest glass vessels. Many of the Tuthmosis III vessels bear the scratches and drill marks indicative of being cold worked. Cold working was the technique used for the production of stone vases and kohl pots throughout Egyptian history and involved the rough shaping of a block of stone with chisels and drills and then grinding smooth the surfaces. Sharp edges, flat surfaces and very regular shape typically result from the use of this technique and very different to the rounded unevenness of core-forming. It is clear from the ancient

Egyptian words used to describe glass that they regarded glass as an artificial stone, e.g. xSbD iri, (literally 'lapis lazuli that is made'). These early vessels show that on initial contact with this material, they used existing, well established Egyptian stone working techniques to make vessels skeuomorphic on Egyptian stone types. There is therefore no reason to suspect foreign workmanship for these vessels, and they may represent the first glass to arrive in Egypt. This glass was used by well-established industries, as though it was any other hard stone. Other early vessels were made by other techniques. MMA26.7.1179 was cast in a mould (Lilyquist and Brill 1993, 34) and then decorated by inscribing a cartouche and fire polishing, techniques not seen in the Malkata and Amarna glasses. BM 47620 also shows a unique decoration technique. In this vessel, some of the decoration has been created by applying powdered glass and enamelling. It is difficult not to see this range of techniques as 'experimentation' in the new material. The remaining vessels were made using the core-forming technique which, as far as can be determined, is no different to the technique that became the norm for the rest of the New Kingdom.

Thirdly, the decoration of the vessels also shows interesting patterns. As discussed above, by the time of Malkata and Amarna, the decoration of glass vessels was fairly standardised with trailing and marvering of 'W' and 'U' shapes being dominant and only rare examples of other techniques. In contrast, the vessels from the reign of Tuthmosis III show a much wider range of decorations, some of which have been variously described as 'mannered', 'simple' and of a 'tentative nature' (Lilyquist and Brill 1993, 25-6). Barag (1970, 183) defined this period as 'Egyptian glass before the 'standardisation' of shapes and decorative designs' It is tempting to see this also as experimentation with a new technology (*contra* Nicholson), which then settles down by the reign of Tuthmosis IV to produce the small range of decorative designs seen through to the end of the 18th Dynasty. The styles themselves may indicate who was doing the experimentation, that is to say whether they were native Egyptians or Mesopotamian. As stated by Barag (1970), there is no evidence of foreign influence in most of the vessels of this period. However, there is one exception and one possible exception to this general pattern. As discussed above, the marbleised goblet (MMA 26.7.1175) is very Mesopotamian in design and similar to a number of marbleised vessels found in Nuzi Stratum II (probably dated to the first half of the fourteenth century B.C.). The evidence cited above suggests that it was perhaps brought into Egypt as a finished object. The possible exception is the British Museum jug with the feather/leaf designs (BM 476220). Lilyquist and Brill have argued that various aspects of the decoration of this vessel (the feathery trees, random dots, etc.) are suggestive of 'foreign style – if not a foreign craftsman' (Lilyquist and Brill 1993, 27). There are Egyptian examples of the use of such decorations, but they are much rarer. This seems to be the only possible evidence of foreign influence in the design of the vessels.

Summary

The first glass to be deliberately made on a regular basis was probably produced in Mesopotamia in the late sixteenth century B.C. It was used initially to produce beads, inlays and amulets. Evidence cited above suggests that some of the earliest glass to be found in Egypt was imported from Mesopotamia. This glass was mostly light-blue in colour and strongly resembled the mineral turquoise after which it was named (*mfkt*). It was made in Mesopotamia from a plant ash alkali and a copper colorant and given by Mesopotamian rulers to Tuthmosis III as tribute following his campaigns in the Levant. It is possible that it was transported in the form of round ingots, similar to those found in the later Ulu Burun shipwreck (1300 B.C.). A small amount of plant ash alkali glass with a higher copper content was probably also given. This was dark-blue in colour and resembled lapis lazuli. At the same time, some cobalt natron dark-blue glass was being made from Egyptian raw materials, almost certainly in Egypt, probably by local Egyptian glassmakers. This dark-blue glass represents only a small proportion of the glass during the reign of Tuthmosis III, but by the time of Amarna, became as abundant, if not more abundant, than the copper coloured glass.

The earliest glass vessels found in Egypt can be dated to the reign of Tuthmosis III. As can be seen from studying the method of production and decoration of the few vessels that remain from this reign, they are in all cases distinct from the glass vessels found so commonly at Malkata and Amarna. They also do not form a single group, since the only property they have in common is their age. Four of the vessels were manufactured using stone working techniques. In these vessels the unique properties of glass are not utilised and it is treated like any other stone, hence the Egyptian term for glass, *ḫsbḏ iri*, 'artificial lapis lazuli' or literally 'lapis lazuli that is made', seems particularly apt. There is every reason to suspect that the glass for these vessels was brought into Egypt as tribute to Tuthmosis III, and manipulated just like *ḫsbḏ m3ʿt*, 'true lapis lazuli' by Egyptian craftsmen skilled in these techniques. Perhaps at the same time as this, but more likely slightly later, the plastic nature of hot glass begins to be utilised to create the cast vessel MMA23.9. However, casting is an unnecessarily laborious process and core-forming is easier since it does not require the manipulation of a complex mould. The earliest reliably dated evidence for core-forming in Mesopotamia is contemporaneous with these Egyptian vessels. Where core-forming was first used is therefore uncertain. The simplest of the core-formed vessels is the monochrome blue chalice (Ashmolean E2451) which may represent one of the first attempts with the technique in Egypt. The first core-formed vessels appear in Mesopotamia at Tell Atchana, but their date is questionable. E2451 is practically identical in shape to ÄS630, with its trailed cartouche. While both are competent examples, they, or any of the other core-formed examples, do not show the level of technical skill of many of the later vessels. One might therefore see them as transitional, with the first cautious movements into the core-formed technology. Both the shape and decoration of Ashmolean E2451 and ÄS630 show no foreign influence and it is tempting to see them as products of Egyptian craftsmen. The marbleised goblet (MMA 26.7.1175) is stylistically and compositionally Mesopotamian, but a relief showing similar vessels in a tomb dated to the reign of Tuthmosis III suggests that it may have been imported as a complete vessel. Only one vessel, BM 47620 shows possible Mesopotamian stylistic influences.

CONCLUSIONS

Glass was first made on a regular basis in Mesopotamia in the late sixteenth century B.C. Glazes are applied to clay bodies in Mesopotamia on the same sites at the same time. There is good evidence that at least some of the earliest copper-coloured plant ash-alkali glass to be found in Egypt was imported from Mesopotamia, but, at the same time, small quantities of cobalt-coloured natron glasses is already being produced in Egypt by local workers exploiting local raw materials. These glasses could be produced in bright, clean colours and were used in the manufacture of beads, amulets and, slightly later, glass vessels. The glass vessels from the reign of Tuthmosis III show a wide range of designs. Some were made by established stone working techniques, others show the use of casting and the beginning of the core-forming technique. Similarly, the decoration is very varied and often applied with less sureness of touch than later vessels. The range and quality of both techniques and decorative styles suggest experimentation in this technology, which is then 'formalised' by the reign of Tuthmosis IV, and applied almost universally by the time of Malkata and Amarna where glass is relatively abundant. The evidence for foreign influence in this process in the form of foreign workers operating in Egypt is slim. Both texts and objects suggests that a better interpretation might be that the glass, at least in the beginning produced mostly abroad, was manipulated by Egyptian craftsmen, initially using techniques borrowed from other industries, but developing other methods as the unique properties of glass became evident.

REFERENCES

Barag, D., 1970, Mesopotamian core-formed glass vessels (1500-500 BC), in Oppenheim, A. L., Brill, R. H., Barag, D. and von Saldern (eds), *Glass and glass-making in ancient Mesopotamia*, Corning Museum of Glass, New York.

Keller, C.A., 1983, Problems of dating glass industries of the Egyptian New Kingdom: Examples from Malkata and Lisht, *Journal of Glass Studies*, 25, 19–28.

Lilyquist, C., and Brill, R.H., 1993, *Studies in Ancient Egyptian Glass*, New York, Metropolitan Museum of Art

Lucas, A. and Harris J.R., 1962, *Ancient Egyptian Materials and Industries*, London.

Moorey, P.R.S., 1995, *Ancient Mesopotamian Materials and Industries*, Oxford

Nicholson, P.T., 1993, *Ancient Egyptian Faience and Glass*, London, Shire publications

Nicholson, P.T., 1995, Glass-making and glass-working at Amarna: some new work, *Glass Technology*, 37, p11–19.

Nolte, B., 1968, *Die Glassgefässe im alten Ägypten*, Verlag Bruno Hessling, Berlin.

Oppenheim, A.L., Brill, R.H., Barag, D., von Saldern, A., 1970, *Glass and glass-making in ancient Mesopotamia*, New York, Corning Museum of Glass

Peltenberg, E. J., 1987, Early faience: recent studies, origins and relations with glass, in Bimson, M. and Freestone, I. C. (eds), *Early vitreous materials*, British Museum Occasional Papers, 56, London.

Petrie, W.M.F., 1894, *Tell el-Amarna*. London.

Shortland, A.J., 2000, *Vitreous materials at Amarna: the production of glass and faience in 18th Dynasty Egypt*, British Archaeological Reports International Series, S827

Shortland A.J., in press, Depictions of glass vessels in two Theban tombs and their role in the dating of early glass, *Journal of Egyptian Archaeology*, 86

Shortland, A.J. and Tite, M.S., 2000, Raw materials of glass from Amarna and implications for the origins of Egyptian glass, *Archaeometry*, 42.1, 141–53

Tite, M.S., Shortland, A.J., Nicholson, P.N. and Jackson, C., 1998, The use of copper colorants in vitreous material in ancient Egypt, in Colinart, S. and Menu, M. (ed.), *La couleur dans la peinture at l'émaillage de L'Égypte ancienne,* Centro Universitario Europeo per I beni culturali, Ravello, 111–121

Chapter 12

Qantir-Piramesses and the organisation of the Egyptian glass industry

Thilo Rehren, Edgar B Pusch and Anja Herold

Abstract

The most fundamental problem in the reconstruction of workshop activities is the rather small number of excavations which focus on workshops, one of the more mundane aspects of archaeology. Then, with complex workshops working a range of materials, one is typically confronted with an intensive mixing of raw, intermediate, waste materials, and remains of often non-diagnostic features. Interpreting this scrambled evidence can be guided by a functional approach, assuming a technologically based and (to a modern mind) sensible sequence, or co-existence, of processes. An 'iterative' approach, establishing a working hypothesis, and testing and refining it against more or different analyses, may result in a sensible, though not always correct, reconstruction.

Based on the Ramesside workshops in Qantir-Piramesses, a detailed model for the organisation of LBA glass production and working is proposed. The basic difference to be made is that of a producer workshop, producing monochrome glass ingots, and that of a consumer workshop, working such ingots into monochrome or polychrome objects. Coloured glass was produced at a number of sites, both in Mesopotamia and Egypt, in a number of common colours. More specific colours, though, were produced only at specialised sites, dependent on specific raw material access and know how. Long distance trade in glass of all colours then allowed artistic centres to work glass of all colours available.

INTRODUCTION

The most fundamental problem in the reconstruction of workshop activities is the rather small number of excavations which focus on workshops, one of the more mundane, and hence often neglected, aspects in Egyptian archaeology. Then, with workshops in complex societies typically working a range of materials and using a number of different operations, one is confronted with an intensive mixing of raw, intermediate, waste materials, and remains of often non-diagnostic features. Any attempt to read and eventually interpret this scrambled evidence has to be cautiously guided by a functional approach, taking into account environmental parameters as well as the wider economic, social and political setting. For this, we assume the undertaking in antiquity of a technically reasoned and functionally controlled sequence, or co-existence, of individual operations linking up to material-specific processes as part of the pattern of activities of any such workshop or industry. In the context of Qantir-Piramesses, we understand *processes* as a combination of individual *operations*, related to a specific transformation of material(s). Specific manifestations of such processes, or single operations, are referred to as *techniques*, while *technology* is used as a generic term to refer to any combination of operations or techniques related to a particular material type, such as bronze, glass, faience, *etc*. Parallel to this *chaine operatoire* approach to classifying operations, we use the term *group* or *type* of material to describe (typically fragmented) finds specific for a particular operation (e.g. bronze melting crucible), and *class* of finds to refer to similar, though not identical groups, such as furnaces or crucibles in general, including e.g. those used for bronze melting and glass colouring. An 'iterative' approach is followed, first establishing a working hypothesis by attributing groups of finds to operations, and linking operations to processes, and then testing and refining this hypothesis for each group and process against the constantly developing archaeological evidence both from Qantir and elsewhere. This approach often results in one or several possible reconstructions. Such reconstructions, however, will never be complete, let alone accurate or precise in detail. The main purpose of any such reconstruction can only be to facilitate further research and thought, and to stimulate other suggestions.

This paper addresses the problems and possibilities in workshop reconstructions using a given archaeological example as a case study, exploring the situation in a given chronological, regional and material setting. Accordingly, no overarching model will be developed for the organisation of LBA industries in general. Based on the Ramesside workshops in Qantir-Piramesses, a specific model for the overall organisation of LBA glass production and working in Egypt is proposed. The basic difference to be made as a starting hypothesis is that of primary workshops, producing glass ingots of a few different colours only, and that of secondary, artistic workshops, working the full range of colours of such ingots into often polychrome objects. The setting of both primary and secondary workshops has to be seen separately in the wider context of crafts and industries in general, and may vary for different situations. It appears that they need not necessarily be situated in proximity, but may operate independently from each other. It is argued that glass was produced at a number of sites, both in Mesopotamia and Egypt, in a number of common colours, primarily copper blue. More specific colours, though, were probably produced only at fewer, more specialised sites, depending on specific raw material access and know how. Long distance trade in glass ingots then allowed artistic centres to work glass of all colours available.

THE INDUSTRIAL COMPLEX AT QANTIR

The first mentioning of Qantir as a major production site for high temperature artificial materials was made by Mahmud Hamza in 1930, following the discovery by a local boy of large scale remains of faience objects, mostly tiles once lining entrances, and subsequent excavation of some areas around the find spot. The sheer quantity of both finished objects and remains of faience production, such as moulds, made it clear that Qantir was an important production site for architectural faience. Finds included tiles, inlays and statuettes, as well as beads and amulets of all kinds. In addition to these remains, M. Hamza had also excavated – though not published – a number of objects related to the production of glass, including a complete red glass ingot (see Rehren and Pusch 1999, 174), sherds from cylindrical crucibles with adhering glass, glassy slag, and various other fragments related to the melting of metals, and the production of Egyptian blue. Some of them are on exhibit in the Egyptian Museum in Cairo, but no information is available concerning the spatial distribution or stratigraphic relationship of these finds within the excavated area, or the relative quantities of this material excavated over 70 years ago.

Starting in 1980, the German team of the joint German-Austrian mission to Avaris-Piramesses in the area of the modern dwellings of Tell el-Dab^ca and Qantir respectively excavated a vast compound with bronze casting installations, unparalleled in size, level of organisation, and quality of preservation in the entire Old World (Pusch 1990; 1994; 2000). The paramount installations comprise 'melting batteries', channels of 15 meter length and a maximum width of 30 cm, and 'cross furnaces', complex features made up of one trench of approximately 9 metres in length, crossed at right angles by a set of three slightly shorter ones. Both installation types were built using mud bricks, while only the cross furnaces have a brick-laid floor and are internally lined with a lime-rich mud-based slurry or plaster. Among the numerous small finds made within and among these installations and relating to metal casting activity, such as fragments of hemispherical crucibles, scrap metal for recycling, and casting debris, were also a considerable number of finds related to other high-temperature processes. Important groups of finds included cylindrical crucibles, clearly different from the hemispherical casting crucibles in texture and heating mode (Rehren 1997), well over a thousand moulds for the production of faience objects (Engel forthcoming), and a group of obviously specialised, though difficult to interpret, highly porous and cream-coloured ceramic finds.

A major breakthrough in recognising the broad outline of this high-temperature jigsaw was the separation by G. Weisgerber of the German Mining Museum, Bochum, of the various crucible fragments into initially two types of crucibles, of cylindrical and hemispherical shape respectively, and the subsequent identification of the hemispherical crucibles as melting vessels for bronze (Pusch 1990), and the cylindrical ones as glass colouring vessels for the production of coloured glass ingots (Rehren 1997; Rehren and Pusch 1997). They do not only occur in site Q I (*"military workshops and exercise court for chariot troops built over a vast industrial bronze foundry"*), but also in the neighbouring site Q IV (*"horse stables built over a workshop area, which is most probably mainly connected with glass production"*) as well as in the latest excavation site Q V (*"administrative building with or erected over a multiple workshop area"*) opened in 1999. A specific feature of all but one or two of these crucibles was the striking red colour of the glass adhering to them, often weathered to a bright green. It became obvious that the crucibles all served to produce a rare and highly specialised colour in Egyptian glass technology: cuprite red. This extreme bias towards one colour only in the excavated material from the current work was further confirmed by the finds excavated half a century earlier by M. Hamza, including a complete glass ingot. With fragments of roughly 40 individual crucibles excavated so far, certainly representing only a fraction of the entire material once present, the production of red glass ingots proved to be a large scale operation.

More recently, it became possible to identify another type of crucible of porous, cream-coloured ceramic fabric, and to allocate it to the production of Egyptian blue (Pusch forthcoming; Rehren and Pusch forthcoming). The presence of Egyptian blue technology at Piramesses was already indicated by numerous finds of Egyptian blue cakes and fragments in various qualities and stages of production, excavated by M. Hamza as well as the current German mission from the Pelizaeus-Museum, Hildesheim. Now, we can prove that these semi-finished cakes were indeed produced on site, and not imported as such from elsewhere for further processing. Furthermore, having opened – or rather identified – another chapter of the Piramesses manual, we will hopefully soon be able to properly allocate some more pages to it.

READING THE ARCHAEOLOGICAL EVIDENCE

It soon became apparent that the high temperature material from the excavation at Qantir site Q I comprised remains of a number of different technologies. While the larger installations for bronze casting obviously were still *in situ*, the majority of the debris and small finds were not. The main task therefore of the scientific investigation of this material was to identify the range of materials produced and/or worked, and the nature of possible techniques involved, before allocating defined groups and classes of finds to these different materials and techniques. Due to the sheer quantities of material involved and the official regulations of the Egyptian Antiquities Department (the Supreme Council of Antiquities), this had to be done in Qantir and using a limited range of microscopic and analytical field techniques only. Reading material culture as text (Tilley 1991), and interpreting the different techniques identified at Piramesses as different chapters of a single book, each group of material and each operation represent an entry in this book. In the absence of pagination, we tried to read and arrange these pages in a sensible order. The ultimate aim is – metaphorically – to re-edit the workshop manual for the Ramesside workshops at Piramesses. This is an ongoing process, hampered by the loss of many pages, and the ambiguity of several of those which we are able to read. Furthermore, those pages we have apparently stem from at least two subsequent editions of the manual, following significant changes in the organisation and duty of the industrial estate. Its main function changed from a short-lived, large scale bronze foundry to a badly fragmented but rather complex maintenance area for chariot troops. Some details of this transition of stratum B/3 to stratum B/2 are given in Rehren *et al.* (1998: 229–232). Until recently stratum B/3 was dated to the late 18th or early 19th Dynasty, stratum B/2 to the reign of Ramesses II in the 19th Dynasty. However, this chronological setting has to be corrected. New investigations on the stratigraphical position of about 300 stone chariot fittings from site Q I, and a thorough study in regard to object stratigraphy within Qantir, point to a later date of the features than formerly believed. Thus, stratum B/3 can now be connected with the early reign of Ramesses II and his enlargement and construction activities of and within the city.

Stratum B/2 in its lower level may still be dated to the reign of Ramesses the Great, while the following substrata of B/2 date more likely to his successors, and therefore not exclusively to the reign of Ramesses II himself (Herold 1999, 172; Herold forthcoming). The most astonishing fact of the military workshops is the presence of foreign, explicitly Hittite craftsmen in stratum B/2. This is proven by the occurrence of several shield moulds, which were used for the local production of metal rim armaments for Anatolian 'Figure-Eight' shields. These are probably to be seen in the context of a diplomatic marriage of Ramesses II with a daughter of the Hittite king Hattusilis III around the midst of the pharaoh's reign. Beside the singularity of these finds in the whole eastern Mediterranean and the fact that they prove archaeologically the practical fulfilment of the peace treaty between the two former enemies, these Hittite tools are further evidence for the mobility of specialised craftspeople, as discussed in R. Moorey's contribution in more detail. At the same time, their presence adds further to the challenge to correctly allocate types of finds and possible operations to each other. Are we dealing not only with two subsequent editions of the workshop manual, but possibly also with two contemporary editions in different 'languages', Egyptian and Hittite?

The common denominator of all of these processes identified so far – bronze casting, red glass making, faience production, and Egyptian blue production – is the prominent role which is played in all of them by copper. It is the main constituent of bronze, forms the red precipitate colouring the glass, adds the blue or green colour to the faience glaze, and forms, together with lime and silica, the mineral phase cuprorivaite, the pigment in Egyptian blue. The scale of production and level of organisation obvious for at least three of these technologies at Qantir, namely bronze casting, glass colouring, and faience production, the intimate archaeological co-occurrence of the debris from the various operations within nearly all excavation sites at Qantir, in combination with the telling absence of evidence for the otherwise most common high temperature technology at all – the production of pottery – led us to postulate a copper-centred industrial complex at Qantir-Piramesses as the common originator of all these materials (Rehren *et al.* 1998). Copper, however, was by no means a scarce or specialised commodity in Ramesside Egypt. The specialisation therefore which is apparent at Piramesses seems to be governed not primarily by the availability of a specific raw material, but by the virtuosity with which this metal was mastered over a range of chemical compositions, redox conditions and temperatures. It appears reasonable to assume that the foundation of this industrial complex was a planned act, following the political decision to establish a major settlement in the eastern Delta. No technical or environmental reason for this location comes to mind, given the ubiquitous availability of copper within the New Kingdom, and since none of the raw materials employed originate from somewhere close to the site. One can only assume that skilled labourers were sent to Piramesses early on in the city's history, in a planned act to serve the needs of the rapidly growing capital in highly decorative architectural items: huge bronze doors to fit the temple entrance, vast numbers of faience tiles and inlay work for the facades of palaces and temples, and large amounts of red glass and Egyptian blue for smaller inlay work and jewellery (Pusch 1990).

Once these initial needs were satisfied, the foundry site at Q I was redeveloped as part of a major military compound, and the smaller attached workshops had to adapt to meet the new demand of production and maintenance of armour, weaponry and chariots (Herold 1999).

This new demand lasted for a considerable time. It started with the reign of Ramesses II, and may have ceased only after Ramesses III about a century later, but presumably even lasted as late as the end of the 20th Dynasty. Therefore Piramesses served as one of the main military centres in LBA Egypt well over a hundred years. Most of the campaigns against the Hittites and others in the eastern Mediterranean, including the so called 'sea people', started in the Ramesside capital on the Pelusiac Nile branch. During the late 19th and early 20th Dynasty the attacks of the 'sea people' on Egypt are to be seen in this context, which Ramesses III successfully refuted by battles at sea and at land in the eastern Nile Delta region (Drenkhahn 1984: 115). This military and political part in the history of the city is evident by ancient Egyptian texts and hymns, e.g. "The Praise on Piramesses". There the capital is described as "marshalling place of thy (the kings) chariotry. The mustering place of thy army. The mooring place of thy ships' troops" (Papyrus Anastasi III 7, 5–6). Tentatively this 'marshalling place' may be identified with the military installations at site Q I with its vast open exercise court for charioteers including the attached workshops. This site produced not only hundreds of stone chariot fittings, but also a pair of horse bits, a nave cap of a chariot wheel, and a linch pin, all made of bronze, and bronze knobs of the chariots' box and horse harnesses covered with gold leaf (Herold 1999) as well as the sand filled hoof prints of horses. In terms of related workshops, we should at least sum up the different installations in the various rooms with their fireplaces, crucible fragments, different tools for metalworking such as punches and stone hammers, debris like charcoal, unfinished objects, metal scrap, complete and crushed pebbles of nearly every flint stone variety, sawn animal bones as well as sharpening tools for bone arrow heads, and working tables from quartzite, and schist, granite and limestone for different purposes, etc.

The site of Q I corresponds in date with another excavation area near by. Site Q IV yielded parts of horse stables for approximately 460 animals. Again there appeared, although to a much smaller extent than in site Q I, stone chariot fittings. Since the two-wheeled Egyptian chariot was drawn by a team of two stallions, there might have been stationed at least 230 chariots in the military compound excavated in sites Q I and Q IV, a number vastly exceeding any other known New Kingdom settlement whether in Egypt or the Levant. One may safely assume that the workshops were able to provide the garrison with practically anything needed to keep the troops and their equipment in good order, and ready for combat whenever necessary. This chariot garrison was not the only one in Ramesside Egypt, even in Piramesses itself there is inscriptional evidence of another, second royal horse stud. Thus, the numbers of chariots mentioned in ancient Egyptian texts, about e.g. the battle of Qadesch, are emerging from the mere shadows of royal propaganda into the light of reality.

PRIMARY *VS.* SECONDARY WORKSHOPS

Any discussion of the organisation and position of workshops in a complex society has to distinguish between various possible major settings within the socio-economic pattern. Production and consumption of goods were embedded in a network of trade, and governed by a number of social, technological and economic constraints (e.g. Costin 1991). Following a process-orientated approach, the most basic distinction had to be made between workshops specialised in the primary production of artificial materials from raw materials, using chemical reactions to transform substances, and secondary workshops which re-shaped and worked a given material into a finished product, often physically combining various substances, albeit without much changing the composition of the materials involved.

In terms of the materials we are concerned with here, and in the framework of LBA Egypt, one can postulate the possible existence of several pairs of primary and secondary workshops, typically separated (or linked!) by elements of trade or exchange. For metals, there are the primary smelting sites where ore was transformed into metal, while the secondary workshops worked such metal into finished objects by melting, casting, hammering, cutting etc. Alloying is seen here as part of the secondary workshop practice, although it includes a certain element of chemical change in composition. The reason for this is that we have much evidence for trade in unalloyed metals in the Late Bronze Age, such as ingots of copper and tin (Bass 1997; Pulak 2000), but little if any for the trade of pre-alloyed metal other than as finished objects. Primary sites of relevance to Piramesses, producing copper from ores, thus would include Timna, Cyprus, the Sinai, and possibly Greece. Secondary workshops were probably present at almost all major settlements, and are evident from remains such as crucibles, casting debris and mould fragments (Davey 1985).

For glass, the separation is between primary sites where raw glass was made from silica and alkali fluxes, and secondary workshops where glass vessels and other objects were produced from existing glass. It has been argued elsewhere that the coloration of glass is more closely related to the primary production than to the artistic working (Rehren and Pusch 1999). This is further substantiated by the discovery of about 175 monochrome glass ingots as LBA trade items in the eastern Mediterranean (Bass 1997: 161). Some thirty years ago, Nolte (1968) already postulated the existence of different glass working schools or workshops in the New Kingdom, based on stylistic studies of finished objects. Here, we argue for the existence of several different primary glass production sites in Egypt and Mesopotamia as well. The relationship between these primary and secondary workshops are discussed below.

The separation is less clear cut for faience. This is due to the fact that there is little working to be done once a faience object is being made, except for the mechanical assemblage of individual pieces to more elaborate, or larger architectural, items. In terms of high temperature processes faience is a one-step product. The presence of large numbers of moulds is good evidence for the local production of faience, while the existence of faience objects as such only proves the local use of it. Shortland (this volume) argues for a decentralised nature of faience making within el-Amarna, possibly as a domestic activity, based on the distribution pattern of related debris.

Egyptian blue, finally, is produced by a delicate high temperature process, possibly involving repeated firings (Tite *et al.* 1987). Its working, however, is a cold process, involving either carving and cutting to make vessels and other small items, or grinding for pigment production and painting. Thus, one can distinguish between primary, high temperature, production sites making cakes and slabs, and secondary workshops where these intermediate products are processed rather than produced. Such cakes and slabs are long known from sites such as el-Amarna and Qantir, where they could have been either locally produced, or worked to shape, or both. Recent evidence from Zawiyet Umm el-Rakham, a Ramesside fort near the Libyan coast some 300 km west of Alexandria, indicates that such cakes were traded together with other pigment cakes for further working and consumption at some considerable distance from the primary production site (Thomas 2000).

Applying this somewhat formalistic distinction between primary and secondary, or producing and receiving, workshops to the industrial complex at Qantir reveals some interesting insights.

The casting installations are clearly part of a secondary workshop, despite their obviously massive scale. There is no primary production, no smelting of ore to metal taking place at Piramesses, but only casting and possibly alloying. Unfortunately, and despite several clear publications to this effect (e.g., Pusch 1990, 1994; Rehren *et al.* 1997), this has to be stressed here again, following the misleading figure captions in Ogden (2000: 156 and 167), referring twice to smelting of copper while showing drawings of melting installations and crucibles excavated at Qantir. There is no indication whatsoever for smelting at Qantir. The presence of an oxhide ingot fragment (Pusch 1994) which was most likely imported from Cyprus (Stos-Gale pers. comm.) to the contrary indicates that copper ingots were imported into Piramesses. The import of different types of metal as ingots is a well known part of the organisation of Egyptian metal working, as depicted e.g. in the often reproduced private Theban tomb paintings of New Kingdom officials like Rekhmire and others. Other sources of copper metal may have been the Sinai, where large scale Ramesside copper smelting activity is known (Rothenberg 1988; 1990), or possibly Laurion as indicated by lead isotope evidence from several metal objects from Qantir (Stos-Gale *et al.* 1995; Stos pers. comm.). At Piramesses, alloying copper and tin to produce bronze most certainly took place as well, since practically all of the objects found at Qantir and analysed so far are made of bronze rather than copper. In addition, however, recycling of existing bronze is evident from various finds of scrap metal, e.g. broken projectiles, a folded sword or a bent cheek piece of a horse bit, ready for re-melting, and this may have played a significant role in the metal procurement strategy followed at Piramesses.

A remarkable feature of the debris related to the glass industry at Piramesses is the almost entire absence of evidence for the working of glass into objects. None of the colourful glass rods, found in abundance at working sites such as el-Amarna, Lisht and Malqata, were discovered yet at Qantir. These rods are usually seen in the context of making core-formed vessels (Stern and Schlick-Nolte 1994); their absence thus indicates that these vessels were not produced within or near the areas excavated so far at Qantir. The only indication of working the glass found as yet are a number of drawn plates (Rehren and Pusch 1999), themselves being semi-finished products only, related to the manufacture of inlaid work. We have, on the other hand, indicators from site Q IV for the local production of glass from its raw materials, including slagged furnace pits which can not be linked to domestic use, pottery production or metal working, and a number of slags which seem to be closely related to glass production. This has been discussed elsewhere in more detail (e.g. Rehren and Pusch 1999). Hence, the glass workshops at Piramesses clearly represent primary production rather than secondary working.

The same is true for the production of faience. Thousands of moulds, excavated by M. Hamza and again more recently by the German team, clearly prove the vast extend of the local production of a wide range of faience objects. It has to be assumed that the objets were mostly used locally, and indeed significant numbers of faience objects are known from all over Qantir and its surroundings.

Finds of Egyptian blue were until recently restricted to cakes and fragments of slabs similar to those known from elsewhere, and could have been imported for secondary processing only. The identification of a type of highly specialised crucibles, however, with internally adhering remains of Egyptian blue, now undoubtedly proves the local primary production of this material as well. Details of this ceramic material, and its position within the multi-step process of Egyptian blue production, will be published elsewhere (Pusch forthcoming; Rehren and Pusch forthcoming).

Another aspect of workshop organisation shall be touched only briefly, that of size and level of organisation (Costin 1991). The distinction is to be made between a large scale, regularly working industrial workshop with a limited range of products, and a small scale, but more diverse craft or domestic workshop. The scale of labour and organisation evident from the casting installations (Pusch 1990; 1994), but also the relatively large output and colour specialisation of the red glass production all indicate an 'industrial' scale for both. Similarly, the sheer quantity of faience making remains indicate an 'industrial' level of production here as well, although a detailed appraisal of this aspect has still to be made. The production of Egyptian blue, on the other hand, appears at present to have occurred at a lesser scale, and may have played a minor role only within the complex.

THE LOCAL CONTEXT OF THE INDUSTRY AT QANTIR

Where were the products of these workshops used? Despite the known skills of the Egyptians in the long distance transport of large and heavy items, one may safely assume that any large copper objects such as temple doors that may have been cast at the foundry of stratum B/3 were destined for local use, and probably moved only locally over a relatively short distance. One may compare the situation at Qantir with the one at Thebes, with 'melting batteries' nearly identical in their management and appearance to the ones at Qantir within the enclosure of the mortuary temple of Seti I (Scheel 1989: 25–9; Pusch 1994). Although the metal working area in Thebes is much smaller, and lacks 'cross furnaces', the very short distance to the temple itself may be taken as indirect proof for our own assumption, that the final products of the Qantir bronze industry also where destined to be used not far away, wherever this may have been within the vast metropolis which Piramesses once was. The output of the faience factory similarly is likely to have served a local demand, assuming a large scale of production and the necessity of close communication between the factory and the receivers or consumers of tiles and inlays during the initial building programme of the enlargement of the city to a radiant royal capital starting with the reign of Ramesses II.

Both coloured glass ingots and cakes of Egyptian blue on the other hand may have been traded over any distance, from hundreds of meters to hundreds of kilometres, to other workshops specialised in the further processing of these substances. Indeed, the absence of such secondary working debris for glass, and the scarcity of it for Egyptian blue, in the archaeological record from Qantir indicate at least a partial, if not complete, export of these products for use elsewhere. This 'elsewhere', however, must not necessarily have been a long distance away from Piramesses. It may well lay just outside the excavated areas of the site, awaiting discovery, or destruction by the agricultural exploitation of the site. It has to be stressed here that less than 0.5 percent of the capital of Ramesside Egypt, with its centre alone covering about 15 square kilometres, is excavated so far. In addition, about one seventh of the whole area has been surveyed by a caesium magnetometer to develop the city's ground plan data. While this method helps to evaluate the overall appearance of ancient Piramesses, it fails in regard of developing a stratigraphy, the recovery of diagnostic finds and the identification of the function of features surveyed.

The later workshops of stratum B/2, to contrast, appear – in what is probably an oversimplistic and idealised model – as serving primarily the neighbouring chariot troops (Herold 1998; 1999). At present, and without further geophysical surveys, we can not localise these more closely. What is evident, is the presence of horse stables at a distance of only 250 meters to site Q I with its so-called equipment workshops. Within these workshops metal, stone, bone as well as leather and wood have been processed into aggressive and defensive armoury and weaponry, e.g. bronze and faience scales for body armour, arrow heads made from metal, flint, chalcedony, etc. as well as bone tips, daggers, swords and special bronze projectiles for fighting, hunting and fishing likewise. Additionally, the moulds for Hittite 'Figure-Eight' shields as well as for smaller Egyptian ones indicate the working of wood and leather. Unfortunately no traces have been found yet in the vicinity, indicating other features of the garrison such as troops shelters or store rooms for chariots or the like. The workshop rooms themselves are too small for storing larger items, even if they were kept dismantled. One can only guess whether some of the chariots were sheltered under the roofed colonnade within the exercise court.

Moorey (this volume) has already stressed the political importance of these troops, in addition to their mere military function, and the 'eastern' origin of this weapon. It may suffice to mention that the introduction of the horse and the chariot into Egypt took place during the late 17th century BC, i.e. roughly 400 years before the first Ramesside king succeeded on the Egyptian throne, although foreigners remained quite prominent members among the chariot makers throughout the whole New Kingdom. However, the presence of chariot workshops is not at the same time a forcing proof for the presence of foreign craftsmen. Over the centuries the Egyptians were capable enough of building and repairing chariots themselves. But matching the foreign aspects of chariotry to Egypt in history as well as terminology, it is likewise not surprising that military compounds, like the one at Qantir sketched here very briefly, provide us with proofs for the presence of non-Egyptians, whether they were craftsmen, soldiers or officials. At Qantir, this international aspect is further evident from the presence of Hittite and possibly even Mycenaean troops and craftspeople at the court of the

Ramessides (Pusch 1996). Besides a considerable amount of Mycenaean pottery found throughout the Ramesside strata in all sites excavated so far at Qantir, there was also unearthed a complete scale of a Mycenaean boar's tusk helmet, being the only hint for this type of head armour found so far outside the Mycenaean world. It may thus be allowed to interpret these later workshops at a similar level and setting to the postulated Memphite military arsenal (Sauneron 1954, Herold forthcoming), and – although there is no definite proof for a palace context at Qantir yet – preliminarily also to Akenaten's 'palace workshops' (Petrie 1894) of el-Amarna, or those from Kamid el-Loz (Frisch *et al.* 1985), in terms of patronage, organisation and political importance.

To summarise, the Late Bronze Age industrial complex at Piramesses comprises primary production workshops for glass, faience and Egyptian blue, and secondary, though extensive, workshops for bronze. Of these materials, three were certainly manipulated on an 'industrial' scale, while the extent of the production of Egyptian blue still needs to be established. The foundry and the faience production probably served the immediate needs of the growing city, while glass and possibly Egyptian blue may have been exported also to other sites. It has to be mentioned though that there is good evidence for the small scale craft or domestic working of bronze and Egyptian blue as well as the working of a wide range of other inorganic and organic materials outside the high temperature field, such as bone, leather and semi-precious stone. How much of this was related to the regular maintenance of the military equipment and installations which followed the initial casting installations is difficult to judge in the absence of representative collections of items produced and worked at the various workshops. Once the stratigraphic relationship of the different groups and classes of materials to each other is better understood, this picture may reveal more details regarding the internal organisation and succession of crafts and industries at Piramesses. At present, it may suffice to bear in mind the focal orientation on the working and processing of copper in a wide range of applications.

THE ORGANISATION OF THE EGYPTIAN GLASS INDUSTRY AT LARGE

The investigation of Egyptian glass has a long tradition in archaeometry, although practically all of the early work focussed on the chemical characterisation of individual finds, and the identification of raw materials, pigments and opacifiers used in the production of this glass. First Petrie (1894; 1926) and later Turner (1954) also addressed technological aspects of Egyptian glass production, using the material excavated by Sir Flinders Petrie at Tell el-Amarna some one hundred years ago. Since then, even the very basic question whether there was Egyptian primary glass making at all, or whether all glass found and worked in Egypt was imported from Mesopotamia, remained unresolved. Nolte (1968), as mentioned briefly above, postulated the existence of various stylistically distinct glass factories in New Kingdom Egypt, and further archaeological evidence for such workshops in addition to el-Amarna was unearthed at sites such as Lisht, Malqata and others. Only recently Nicholson and co-workers at Tell el-Amarna and the present authors at Qantir started a fresh approach to answer the basic question of Egyptian glass production in the New Kingdom.

The working hypothesis developed by us is based on the assumption that the

Egyptian glass industry comprised primary as well as secondary workshops, the former producing raw glass from silica and plant ash, and turning them into coloured glass ingots, the latter working these into finished objects. These workshops were connected by a far-reaching, international, trade network. The existence of different secondary workshops within LBA Egypt is long known, both archaeologically and from stylistic studies. Here, it is argued in addition that a number of different primary workshops or glass making centres, both in Egypt and in Mesopotamia, produced raw glass from plant ash and quartz. This raw glass production may have been done in open, sand-lined hearths with no or only a limited superstructure, but details of this remain to be explored. The raw glass was probably crushed and hand sorted to obtain a good quality, homogeneous, glass. Colorants were added to this primary glass which was then remelted in cylindrical crucibles (Rehren 1997). The resulting disk-shaped ingots were removed from the crucibles and traded throughout the Levant (Pulak 1997) to supply the secondary workshops with glass of a range of colours. It appears that some of these workshops were restricted to, or specialised in, a certain colour only, while others may have produced a wider spectrum of colours (Rehren 2000a).

A crucial issue in this respect is to prove the existence of a number of such primary workshops, and to locate them. The well-known chemical homogeneity of Egyptian glass, except for the elements introduced as colorants and opacifiers, is not in favour of divers production sites, and little archaeological evidence exists to prove their existence. We argue, however, that it is possible to separate different glass making sites from each other, using chemical characteristics of the glass produced, and that archaeological evidence is at hand to locate such primary workshops.

THE CHEMICAL IDENTIFICATION OF DIFFERENT PRIMARY WORKSHOPS

It has been argued elsewhere that the precise chemical composition of ancient Egyptian glass was controlled only broadly by the initial batch composition, and that modifications by subsequent glass forming reactions in the batch and the melting behaviour of the relevant phase diagram SiO_2-Na_2O-CaO contributed significantly to the narrow compositional scatter of the final glass composition (Rehren 2000a; 2000b). The constant use of similar raw materials, *i.e.* ash from the same species of plants and crushed quartz pebbles, would have provided only a roughly defined chemical characterisation of the initial batch composition, limited due to the inherent chemical heterogeneity of these raw materials.

Thus, the major components of the glass are of limited value for a chemical characterisation of different workshops. It is assumed, however, that the plants used as alkali sources reflect to some extent the local soil chemistry of their growing region, and that the concentrations of minor and trace elements are less affected by the phase diagram constrains. The analysis and interpretation of those minor and trace elements introduced via the plant ash should therefore offer an approach to characterise different glass workshops. Significant differences in the soda to potash ratio in early *vs.* later Egyptian glasses have been tentatively interpreted as a coarse regional marker (Brill in Lilyquist and Brill 1993; Rehren 2000a). Recent work at the Research Laboratory for Archaeology and History of Art at Oxford (Shortland and Tite 2000) on the cobalt-blue glasses from el-Amarna, which are compositionally distinct from most other contemporary glasses, also indicate the existence of such regional, possibly workshop-specific characteristics, as discussed by Rehren (2001). It is planned to study the plant ash chemistry of relevant species in more detail, and to expand the analytical programme of Egyptian glass by a number of trace elements identified as potentially diagnostic.

THE ARCHAEOLOGICAL IDENTIFICATION OF PRIMARY WORKSHOPS

There is some new information that can be given here beyond that already published by us elsewhere (most recently Rehren and Pusch 1999). The best indicator of primary glass production still seems to be the occurrence of significant numbers of cylindrical crucibles. Finds of layered black and white slag and glassy slag seem also to indicate primary glass production, although a comprehensive explanation and interpretation of these finds is still lacking. Nicholson and Jackson (1998) argue that a certain cylindrical type of furnace, recently excavated on behalf of the EES at Tell el-Amarna (Nicholson 1995), is related to the production of glass. A plausible explanation how this furnace would have operated, however, is still missing, and the experimental work done at el-Amarna is a first step at best (see Rehren 2000a for a more detailed discussion of these experiments). The two major problems in the proper identification of archaeological evidence for Egyptian glass production are the intimate mixing of remains from various activities and processes in the archaeological record, as described earlier in this paper for Qantir, and the fact that we still have a very limited knowledge about the exact nature of Egyptian glass making processes, and hence what to expect as diagnostic finds. Both problems can only overcome through the thorough and 'holistic' investigation of as many workshop assemblages as possible.

The occurrence of glass rods, however, at an archaeological site certainly indicates the former activity of a secondary workshop, and not the primary production of glass. Bearing in mind the textual and archaeological evidence of long distance trade in coloured glass ingots, one has to assume that these secondary workshops had access to a number of different sources of coloured glass. This does not exclude the possibility of a co-existence of primary and secondary workshops at any one site; to the contrary, most glass working sites known so far in Egypt seem to have hosted both. At present, only Piramesses appears as a predominantly primary site. This is further emphasised by the most recent discovery of a complete crucible, still filled with a powdery white calcium silicate material, probably badly corroded primary glass. The crucible has been excavated in autumn 2000 in site Q V, situated some hundred metres south of the royal horse stud in site Q IV. The context of this find seems to indicate a small workshop for the production of calcite inlays for faience tiles. However, further studies on the newly discovered material have to be awaited for more details and clarification.

CONCLUSION

The Egyptian glass industry of the New Kingdom was highly organised, boasting a number of different primary workshops producing glass ingots, and secondary workshops where glass of a range of colours was worked into finished objects. This industry was thoroughly integrated in an international trade network in ingots and objects, linking Egypt with Mesopotamia and the Eastern Mediterranean. The number of primary and secondary glass workshops active at any one time must have been small in comparison with metal workshops, and the relationship among contemporary glass workshops would have been rather close. This is indicated by the apparent frequent proximity of primary and secondary glass workshops in Egypt. We can not yet identify a pattern of a few glass making 'centres' integrated into a network of smaller, subsidiary, workshops. At present, each site has to be studied individually, and it will take much more chronological, contextual and compositional data before an explicit structure of this perceived network can be postulated.

These workshops, however, were by no means isolated entities, but formed parts of larger and more divers industrial complexes comprising installations for the production and / or working of other materials in close proximity to glass. Examples of such settings are the copper-centred industrial complex at Piramesses, and the palace workshops at Tell el-Amarna. The excavations at sites Q I and Q IV at Qantir-Piramesses so far indicate a dominating specialisation on working copper-related compounds such as bronze, red glass, Egyptian blue and faience, while pottery production obviously took place outside the excavated area. At el-Amarna, the evidence is for a somewhat broader range of glass production, including at least copper blue and cobalt blue, and secondary workshops working glass of all colours known in ancient Egypt. The processing of a number of other cobalt-coloured materials (Shortland and Tite 1998), including blue-coloured pottery, appears to be concentrated predominantly, though not exclusively, at el-Amarna. This colorant-specific pattern indicates a certain degree of material-oriented specialisation based on the availability of cobalt, and the know how necessary to use it, as specific aspects of the workshops there.

This complex and often well-integrated nature of the Late Bronze Age industries offers exciting research possibilities. On the other hand, it is also responsible for the thorough mixing of remains and debris in the archaeological record. Any attempt to reading, understanding and interpreting the material evidence therefore requires a detailed examination of all the materials present, in order to first identify the various operations and processes undertaken, and then allocate finds and processes properly. This is often an iterative process, with the results of tentative interpretations feeding back to the initial allocations of finds on which such interpretations were based. Based on our work so far, we are confident that the initial allocations may reliably be based on technologically reasoned assumptions, and that non-technical constrains played a minor role only in the internal organisation of these workshops. The workshops' overall emphasis, however, was controlled by the particular requirements and demands that led to their setup, and changed as the demands changed. An example for this is the development of the workshops at Piramesses, with the emphasis shifting from the large-scale production of architectural material to the production and maintenance of military equipment. Such changes obviously occurred within a few months or years, and may add to the difficulties in untangling the organisational web as it worked at any one time in the past.

REFERENCES

Bass, G., 1997, Prolegomena to a study of maritime traffic in raw materials to the Aegean during the fourteenth and thirteenth centuries B.C., in Laffinauer, R. and Betancourt, Ph. (eds.), *Techne, Craftsmen, Craftswomen and Craftsmanship in the Aegean Bronze Age*, 153–170.

Costin, C., 1991, Craft Specialization: Issues in defining, documenting, and explaining the organization of production, in Schiffer, M., (ed.), *Archaeological Method and Theory*, vol. 3, 1–55.

Davey, C., 1985, Crucibles in the Petrie Collection and hieroglyphic ideograms for metal, *Journal of Egyptian Archaeology*, 71, 142–8.

Drenkhahn, R., 1984, Ramses III, in Helck, W. and Westendorf, W. (eds.), Lexikon der Ägyptologie V, 114–119.

Frisch, B., Mansfeld, G. and Thiele, W. R., 1985, *Kamid el-Loz 6. Die Werkstätten der spätbronzezeitlichen Paläste*. Habelt, Bonn.

Hamza, M., 1930, Excavations of the Department of Antiquities at Qantir (Faqus District), *Annales du Service des Antiquités de l'Egypte*, 30, 31–68.

Herold, A., 1998, Gegossen – Geschmiedet – Tordiert. Zur Herstellung von bronzenen Trensen in der Ramses-Stadt Piramesse/Ägypten, *Metalla*, 5, 3–21.

Herold, A., 1999, *Streitwagentechnologie in der Ramses-Stadt. Bronze an Pferd und Wagen*, Forschungen in der Ramses-Stadt, 2, Zabern, Mainz.

Herold, A., forthcoming, *Streitwagentechnologie in der Ramses-Stadt. Knäufe, Knöpfe und Scheiben aus Stein*, Dissertation Hamburg 1999, accepted for publication in *Forschungen in der Ramses-Stadt*.

Lilyquist, C. and Brill, R., 1993, *Studies in Early Egyptian Glass*, The Metropolitan Museum of Art, New York.

Nicholson, P., 1995, Glassmaking and glassworking at Amarna: Some new work. *Journal of Glass Studies*, 37, 11–19.

Nicholson, P. and Jackson, C., 1998, 'Kind of Blue': glass of the Amarna period replicated, in McCray, P. (ed.), *The Prehistory and History of Glassmaking Technology*, The American Ceramics Society, USA, 105–120.

Nolte, B., 1968, *Die Glasgefäße im alten Ägypten*, Münchner Ägyptologische Studien 14, Berlin.

Ogden, J., 2000, 6. Metals, in Nicholson, P. and Shaw, I. (eds), *Ancient Egyptian Materials and Technology*, CUP, Cambridge, 148–176.

Petrie, W. M. F., 1894, *Tell el-Amarna*, London.

Petrie, W. M. F., 1926, Glass in the early ages, *Journal of the Society of Glass Technology*, 10, 229–234.

Pulak, C., 1997, The Uluburun shipwreck, in Swiny, S., Hohlfelder R. and Swiny, H. (eds.), *Res Maritimae, Cyprus and the eastern Mediterranean from Prehistory to Late Antiquity*, 233–262.

Pulak, C., 2000, The copper and tin ingots from the Late Bronze Age shipwreck at Uluburun, in Yalcin, Ü. (ed), *Anatolian Metal I*, 137–157.

Pusch, E., 1990, Metallverarbeitende Werkstätten der frühen Ramessidenzeit in Qantir-Piramesse/Nord. Ein Zwischenbericht, *Egypt and the Levant*, 1, 75–113.

Pusch, E., 1994, Divergierende Verfahren der Metallverarbeitung in Theben und Qantir? Bemerkungen zu Konstruktion und Technik, *Egypt and the Levant*, 4, 145–170.

Pusch, E., 1996, Pi-Ramesses-Beloved-of-Amun, Headquarters of thy Chariotry – Egyptians and Hittites in the Delta Residence of the Ramessides, in Eggebrecht, A. (ed.), *Pelizaeus-Museum Hildesheim Guidebook – The Egyptian Collection*, Mainz, 126–144.

Pusch, E., "000, Doors – statues – musical instruments? Large scale bronze production and casting at eh delte residence of Ramesses, Egypt, in Hickmann, E., Laufs, I. and Eichmann, R. (eds.), *Studien zur Musikarchaeologie II*, 219-232.

Rehren, Th., 1997, Ramesside glass colouring crucibles, *Archaeometry*, 39, 355–368.

Rehren, Th., 2000a, New Aspects of Ancient Egyptian Glassmaking. *Journal of Glass Studies*, 42, 13–24.

Rehren, Th., 2000b, Rationales in Old World Glass Making. *Journal of Archaeological Science*, 27, 1225–1234.

Rehren, Th., 2001, Aspects of the production of cobalt-blue glass in Egypt. Accepted for publication in *Archaeometry*.

Rehren, Th. and Pusch, E., 1997, New Kingdom glass melting crucibles from Qantir-Piramesses, *Journal of Egyptian Archaeology*, 83, 127–142.

Rehren, Th. and Pusch, E., 1999, Glass and glass making at Qantir-Piramesses and beyond, *Egypt and the Levant*, 9, 171–179.

Rehren, Th., Pusch, E. and Herold, A., 1998, Glass colouring works within a copper-centred industrial complex, in McCray, P. (ed.), *The Prehistory and History of Glassmaking Technology*, The American Ceramics Society, USA, 227–250.

Rothenberg, B., 1988, *The Egyptian Mining Temple at Timna*, iams, London.

Rothenberg, B., 1990, *The Ancient Metallurgy of Copper*, iams, London.

Sauneron, S., 1954, La manufacture d'armes de Memphis, *Bulletin de l'institute français d'archéologie orientale*, 54, 7–12.

Scheel, B., 1989, *Egyptian Metalworking and Tools*, Shire Egyptology, Aylesbury.

Shortland, A. and Tite, M., 1998, The interdependence of glass and vitreous faience production at Amarna, in McCray, P. (ed.), *The Prehistory and History of Glassmaking Technology*, The American Ceramics Society, USA, 251–265.

Shortland, A. and Tite, M., 2000, Raw materials of glass from Amarna and implications for the origins of Egyptian glass, *Archaeometry*, 42, 141–151.

Stern, M. and Schlick-Nolte, B., 1994, *Frühes Glas der alten Welt*, Hatje, Stuttgart.

Stos-Gale, Z., Gale, N. and Houghton, J., 1995, The origin of Egyptian copper. Lead isotope analysis of metals from el-Amarna, in Davies, V. and Schofield, L. (eds.), *Egypt, the Aegean and the Levant*, 127–135.

Thomas, S., 2000, *Aspects of Technology and Trade in Egypt and the Eastern Mediterranean during the Late Bronze Age*, Unpublished PhD thesis, University of Liverpool.

Tilley, C., 1991, *Material culture and text: the art of ambiguity*, Routledge, London.

Tite, M. S., Bimson, M. and Cowell, M., 1987, The technology of Egyptian blue, in Bimson, M. and Freestone, I. C. (eds.), *Early vitreous materials*, British Museum Special Publications, London, 69-78.

Turner, W.E.S., 1954, Studies in ancient glass and glassmaking processes. Part I. Crucibles and melting temperatures employed in ancient Egypt at about 1370 B.C., *Journal of the Society of Glass Technology*, 38, 436–444T.

Chapter 13

The Evolution of Glazing Technologies in the Ancient Near East and Egypt

Sarah Paynter and Michael Tite

Abstract

In the ancient Near East, a range of glassy materials were produced each making use of the same glass forming process involving firing alkalis, silica and colourants together. The resulting products also have a similar appearance (lustre and colour). These materials included: glazed stones and faience, which were produced from the 5th millennium B.C. in both Mesopotamia and Egypt; glass, which was produced from around 1500 B.C. in both Mesopotamia and Egypt; and glazed clay objects, which were produced from around 1500 B.C. but only in Mesopotamia. This paper describes how the examination of artefacts, and experimental replication, have been used to reconstruct the processes involved in the production of these different glassy materials. This information was used to address the following questions:

1. Why were glazed clay objects not produced until ~1500 B.C.?
2. Is it significant that glazed clay objects are first produced at the same time as glass vessels are first produced in Mesopotamia?
3. Why are alkali-glazed clay objects not produced in Egypt?

This study emphasises the influence of ancient cultures on the evolution of these technologies as well as the cultural significance of the technological processes and the materials that were produced. The ancient and replicated materials were examined using SEM, EDX, and WDS techniques. By viewing the production of glassy materials as part of a web of related technologies, all of which are embedded in a continually developing cultural setting, a fresh perspective on the evolution of glazing technology has been obtained.

INTRODUCTION

Over 6000 years ago, the first glazed materials were produced in the Near East and Egypt (Vandiver 1983; Moorey 1994). Blue glazed objects were made of faience, quartz and steatite stones. By about 1500 B.C., vessels and other objects were being made from glass across the Near East and Egypt (Lilyquist and Brill 1995). In Mesopotamia, glazed clay objects were also being produced (Moorey 1994). However it was not until around the 1st century B.C. that glazed clay objects made their first appearance in Egypt (Kleinmann 1986). Can the way that the technology evolved, to make this range of different materials, and products, be explained? This paper describes how scientific examination and experimental replication have been used to reconstruct the processes involved in the production of these different glassy material artefacts. These processes are then interpreted in the context of the ancient cultures, with the aid of preserved textual evidence and the archaeological record. By viewing the production of glassy materials as part of a web of related technologies, all of which are embedded in a continually developing cultural setting, a fresh perspective on the evolution of glazing technology has been obtained.

GLASSES AND GLAZES

A glassy material is a homogeneous, but amorphous, solid. A glaze is simply a thin coating of glass bonded to a substrate. The glassy materials manufactured in the ancient Near East were produced using the glass-former silica, the sources of which were quartz pebbles or sand. The melting temperature of silica is very high, around 1700°C. However certain oxides, 'fluxes', will react with silica to produce a melt at temperatures lower than 1000°C and when the melt cools a glass is formed. In antiquity, alkali fluxes were used for making glassy materials. The alkalis were obtained from the ashes of certain desert plants and were mainly sodium oxide with a small amount of potassium oxide, magnesium oxide and some other compounds in addition (Oppenheim *et al* 1970; Turner 1956). In Egypt, there was another source of sodium oxide in the form of natron, an evaporitic deposit found at the Wadi Natrun. The colourant used in the earliest glassy materials was a copper ore. Copper oxide dissolves in glassy materials and, depending on the composition of the glass, will colour it from blue to green. Only small amounts, around 2%, are required to produce a bright colour.

The properties of a glassy material are dependent on which compounds have been added to the silica and in what relative quantities. The way in which the material is then processed also plays a part. For example it is possible to vary the melting temperature, the refractive index, the opacity, the colour, the resistance to chemical attack and the thermal expansion coefficient of the glass, depending on what components and processing techniques have been used to produce it. As each oxide added to the silica has a different influence on the properties of the resulting glass, a combination of many oxides is usually required to produce a glass with optimum characteristics.

This description of glasses and glazes is based on current understanding but the perception of glassy materials in antiquity was very different. Fortunately some texts survive, written in the Sumerian and Akkadian languages in the cuneiform script on clay tablets, that describe the attitude of these ancient societies towards the glassy materials that they produced (Oppenheim *et al* 1970; Reiner 1995). These texts can supplement the information gathered through scientific examination of artefacts. The reconstructed chain of technological steps can then be viewed in the context of the culture at the time and place of production.

ARTEFACTS MADE FROM GLASSY MATERIALS

The first objects to incorporate glassy materials were glazed stones, mainly quartz and steatite, and faience. The main constituent of steatite is the mineral talc, a hydrated magnesium silicate. The composition of fired steatite is approximately 66wt% SiO_2 and 34wt% MgO. Faience is a material made from finely ground quartz. In the end product, the quartz particles are sintered together by small areas of glass and the surface of the object is glazed. Both quartz and steatite stones contain a large amount of silica, so these glazing processes are related in that they utilised silica-rich materials. This is an important factor to recognise in order to understand the subsequent development of glazing technology.

No glazed quartz objects from this period have been examined using modern techniques (Beck 1934 and 1935). However Predynastic Egyptian glazed steatite beads have been examined using scanning electron microscopy (Tite and Bimson 1989; Vandiver 1983) and the microstructure and composition of the glaze provide information on the way in which the glaze was produced. The artefact glazes contain large quantities of forsterite crystals (Mg_2SiO_4) and the bulk of the glaze has a composition of approximately 27wt% MgO, 55wt% SiO_2, 11wt% CuO and 7wt% Na_2O. For the glaze to contain such high concentrations of magnesium it must have been produced by the reaction of a mixture of alkalis and copper ore with the stone itself during firing. The silica in the steatite has formed the basis of the glaze. The copper oxide dissolves in the glaze and colours it blue. Quartz stones and faience could be glazed in the same manner as they are both made up entirely of silica. An advantage of faience is that it was easier to produce objects with fine details, or elaborate shapes, from a paste formed from the ground quartz than by carving hard quartz stone.

Different methods, alone or in combination, could be used to expose the surface of the stones or faience to the glazing mixture. Exclusive to faience is the efflorescence process. This involves mixing the quartz, alkalis and copper ore with water to form a paste that could be moulded to shape in the same way as damp clay. As the object dries out and hardens, water migrates to the surface where it evaporates. The alkalis that were dissolved in the water are then deposited in a thin layer on the surface of the object. When the object is fired, the alkalis react with the quartz and copper ore to form a blue glaze over the surface. The cementation process involves burying the object in a glaze mixture followed by firing and this can be used on the stones and faience. The application process involves simply applying the glazing materials to the surface of the object, as either a powder or a slurry, and this too can be used on the stones as well as faience. If the applied glazing mixture contains only alkalis and copper ore then, as with the other glazing methods, the success of the process is dependent on the object being glazed containing sufficient silica with which the alkalis can react to form a glaze. The lustrous, bright colours of these glazed materials were desirable and potent attributes in the ancient Near East and Egypt. Materials with these properties, including semi-precious stones, such as turquoise and lapis lazuli, had considerable appeal and cultural significance.

There have been rare finds of glass beads and lumps of glass that date from the early 2nd millennium B.C.. However it is not until after 1500 B.C. that glass objects are found

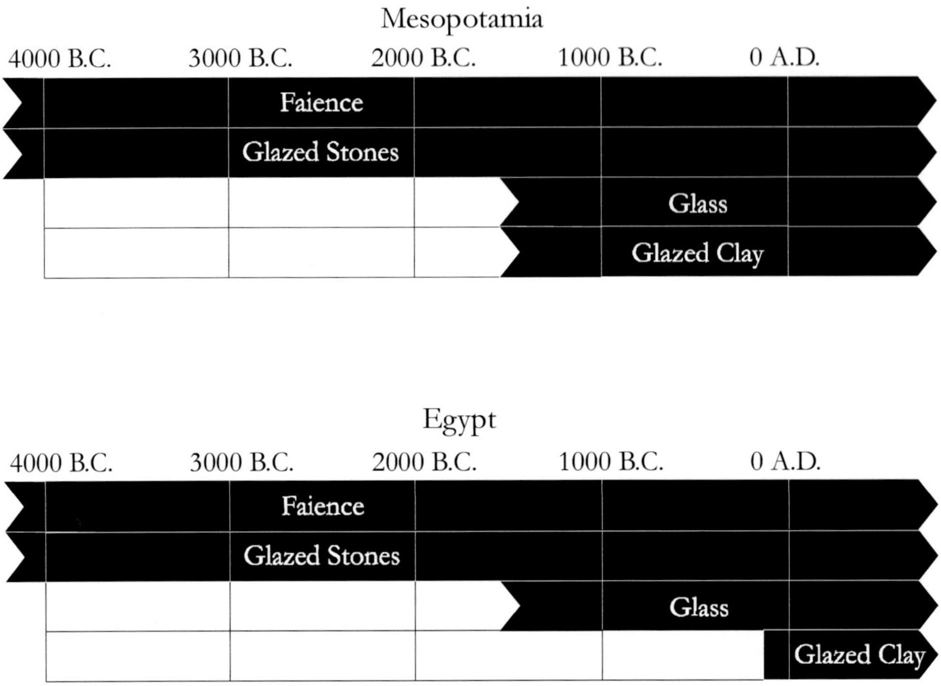

Figure 13.1: Diagram summarising the chronology of technologies for producing glassy materials in ancient Mesopotamia and Egypt.

in significant quantities across Mesopotamia and Egypt. From then on a range of glass artefacts are produced, including vessels, plaques and inlays. In Mesopotamia, glazed clay objects are introduced at about the same time. The earliest clay glazes known, from the 15th century B.C., are copper blue / green. Examples of simple polychrome glazes have been found dating to the 14th / 13th century B.C. and glazes came to be produced in a wide range of colours. In Egypt, glazed clay vessels are not produced until around the 1st century B.C., and then only rarely. These vessels are glazed with lead fluxed glazes, a different technology to the alkali fluxed glazes and glasses manufactured in the Near East until that time. Lead-glazing technology is believed to have originated elsewhere, probably in the Mediterranean region (Kleinmann 1986).

The temporal and spatial distribution of glassy material technologies in the ancient Near East is summarised in Fig. 13.1. This chronology gives rise to three questions which will be discussed in the following text.

QUESTION 1: WHY WERE GLAZED CLAY OBJECTS NOT PRODUCED UNTIL ~1500 B.C.?

Before 1500 B.C., certain types of stone and faience objects were glazed, but only after 1500 B.C. are glazed clay objects produced in Mesopotamia. Clay was a heavily utilised material in Mesopotamia, and many methods were devised to decorate clay surfaces using plaster and paint, coloured slips and small pieces of inlaid stone (Moorey 1994). Since glazed stones and faience were being produced long before 1500 B.C., why were clay objects not glazed until afterwards? One possible prohibitive factor that has been highlighted in the past was the high thermal expansion of alkali glazes, like those produced in antiquity (Hedges 1982). Whilst quartz and steatite stones have high thermal expansion coefficients, comparable to those of alkali-glazes, in contrast most clays have much lower thermal expansion coefficients. Therefore as a clay object, glazed with an alkali-glaze, cools after firing the glaze is stretched and eventually cracks. The fine network of cracks that forms in a glaze as a result of this contraction mismatch is known as crazing. It has further been suggested that, with the advent of glass manufacture, artisans might have come to understand how the changes in the composition of the glaze would affect the thermal expansion coefficient. They could then tailor the glaze to fit the clay and the mismatch between the clay and the alkali-glazes could be overcome or at least minimised. This would explain the co-emergence of the glass manufacturing and clay glazing technologies. However there is a more fundamental reason for glazing being restricted to silica-rich stones and faience during the early stages of glazing technology development and this is outlined in the following text.

Replicating the Glazing Technology

Quartz and steatite stones were glazed in the laboratory using mixtures of ground alkali and copper oxides and firing to 1000°C. Different glazing methods, such as application and cementation, were attempted. A blue glaze was produced on the quartz stones and a greenish glaze on the steatite stones. Figs. 13.2A and 13.2B show scanning electron microscope images of examples of the glazes produced on these stones. Attempts to produce a glaze using the same method on a sample of calcareous clay, of similar composition to Mesopotamian clay, failed. Figure 13.2C shows a scanning electron microscope image of this failed glaze. The clay contains about 50wt% silica, which can react with the alkali to form regions of glaze, but the remaining clay components cause this glaze to be very viscous. It cannot flow to produce a smooth surface and bubbles remain trapped underneath, producing blisters. Many crystalline phases form which cause the glaze to be opaque and matt, and the copper oxide does not go into solution, remaining as a metallic black layer on the surface. However if the clay is modified by adding fine quartz powder, such that the clay composition was in excess of 65wt% SiO_2, the glazing process is successful. The Egyptian Nile silt is also a very silica rich clay type and this clay can be successfully glazed by the reaction of alkalis and copper ore with the silica in the clay itself. However the resulting glaze has an unattractive appearance due to severe crazing and the dark clay colour corrupting the colour of the glaze.

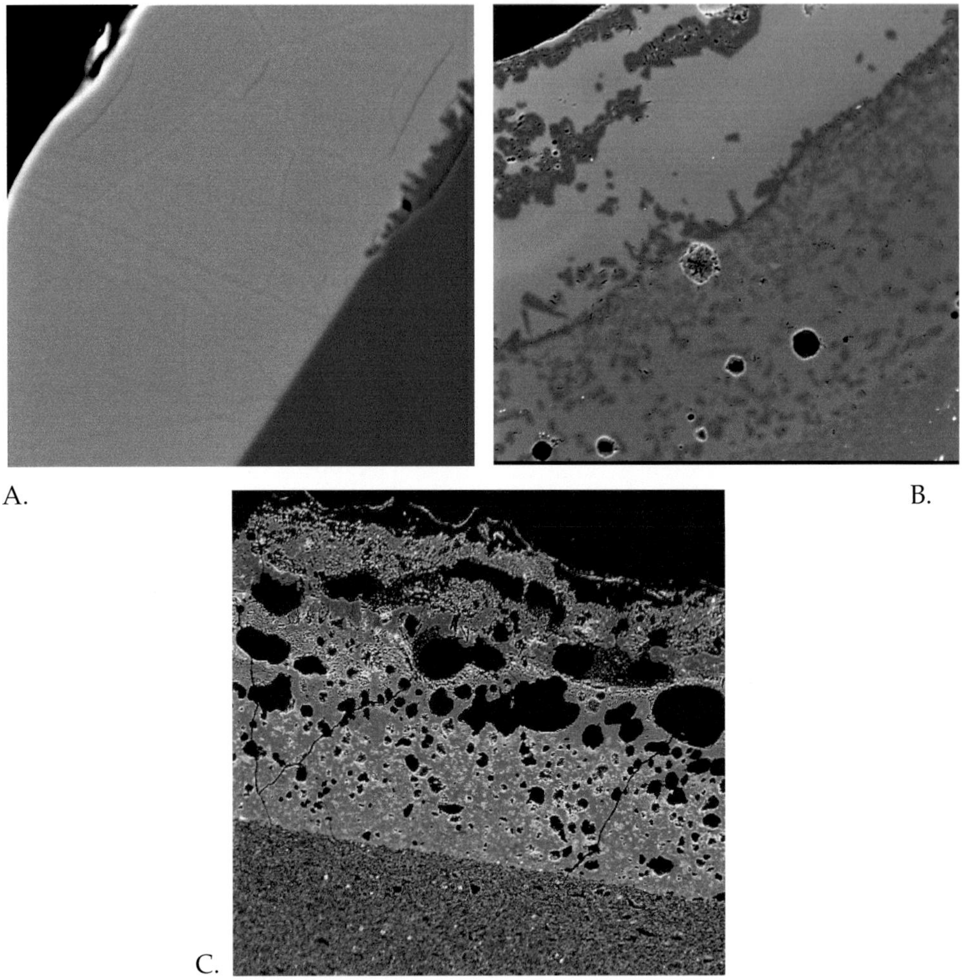

A. B. C.

Figure 13.2: Scanning electron microscope back-scattered electron image of glazed vitreous materials: A. Quartz (dark grey) with clear glassy glaze (light grey), x1000; B. Steatite (dark grey, bottom right) with mainly glassy glaze (light grey) containing aboundant forsterite crystals (dark grey), x200; C. Clay (dark grey stippled, bottom) with a partly crystalline, vesicular reaction layer that cannot really be called a glaze, x100.

Therefore, a glazing technology based on the application of a mixture containing only alkalis and a copper compound is limited to silica-rich stones (and silica-rich clay), because the object being glazed must contain enough silica with which the applied alkalis and copper oxide can react to form a blue glaze. The method is not successful on the Mesopotamian alluvial clays and therefore the development of a different technology, although based on the same raw materials, was necessary before clay objects could be successfully glazed and this is discussed in the next section of this paper.

QUESTION 2: IS IT SIGNIFICANT THAT GLAZED CLAY OBJECTS ARE FIRST PRODUCED AT THE SAME TIME AS GLASS VESSELS ARE FIRST PRODUCED IN MESOPOTAMIA?

As discussed above, glazed clay objects cannot be readily produced from Mesopotamian alluvial clay using a glazing mixture containing only alkali and a copper compound. Instead, the glazing mixture must also contain quartz. The alkali, quartz and copper compound could have been merely ground and mixed together, or have been first fritted and then reground. However, because of the solubility of alkali in water, the recognized ethnographic practice for achieving an even and successful glaze on a clay body is first to frit the glazing mixture and then apply the ground frit in a water suspension (Rye and Evans 1978).

Unfortunately, only limited information is available on the methods of glazing used on Egyptian faience and steatite from the Predynastic through to New Kingdom period, and virtually none for the equivalent period in Mesopotamia. It is likely that there were many developments in glazing technology prior to 1500 B.C.. However, in the absence of archaeological evidence, it is not possible to estimate, with any confidence, the extent to which applied glazes containing additional quartz were used on steatite and faience prior to their first use on clay around 1500 B.C.. The most comprehensive survey of glazing methods has been undertaken on faience from Egypt by Vandiver (1983, 1998) using macroscopic evidence (e.g. variations in glaze thickness, drips, firing marks) obtained by visual examination and low power optical microscopy. She established that, prior to the New Kingdom period, the great majority of faience was glazed by the efflorescence or cementation methods. However, there is evidence that applied glazes containing additional quartz were used during the Second Intermediate Period in glazing tiles from Kerma (Vandiver 1983:A93, Vandiver 1998:133), although it is not clear whether or not the mixture had been prefritted.

Nevertheless, the fact that the glazing of clay objects coincides with the production of the first glass vessels in Mesopotamia around 1500 B.C. indicates a potential link between the two technologies that needs to be investigated. In Table 13.1 the compositions of an average glass and an average glaze on a clay object from Mesopotamia are compared, each based on over twenty analyses by the authors and from the literature. Both glasses and glazes were made from plant ashes, silica and copper oxide colorant, although the copper oxide concentration has not been included in the table. (If natron had been used as the source of alkali there would be characteristically lower concentrations of magnesium and potassium oxides). From this table it can be seen that the compositions of the glazes and glasses are extremely

Table 13.1: Average compositions of glazes and glasses from Mesopotamia

	SiO_2	Na_2O	K_2O	CaO	MgO	Fe_2O_3	Al_2O_3
Blue glass	68.8	15.0	2.8	6.9	4.4	0.9	1.2
Blue glaze	63.9	15.4	4.5	6.7	3.9	2.8	2.8

similar. If ground up glass was applied to the surface of a clay object, and fired to 1000°C, a glaze would form.

The slightly larger quantities of alumina Al_2O_3, potash K_2O and iron oxide Fe_2O_3 in the glaze, relative to the glass, may be due to the addition of illitic clay to the glazing mixture. If the glaze was applied as a water suspension, or 'slop', the addition of a little fine clay to the slop would help to keep the glassy material in suspension and would improve the adhesion of the glaze mixture to the object until it could be fired. However since some glazes (and more rarely glasses) contain more potash than magnesia but little aluminium oxide, other potential causes must also be considered.

One way for the glaze to have become enriched in alumina, potash and iron oxide is for these components to have diffused from the clay itself into the glaze during firing. However experiments with glazes and clays of different compositions, using different firing conditions, have shown that it is unlikely that all of these components could have diffused throughout the thick Mesopotamian glazes (0.5mm in some cases) to such an extent that a uniform composition could be obtained. In particular alumina shows a low level of mobility. Also other diagnostic features would be present if a large amount of diffusion had occurred, for example greater concentrations of calcium rich crystalline material in the glaze as a result of increased calcium diffusion from the clay.

It seems probable, therefore, that potters made use of a different variety, or quality, of plant ash or that differences in the fritting and preparation procedures used by potters as compared to glassmakers resulted in a different ratio of potash to magnesia in the final product. This indicates that, although potters made use of similar raw materials to glassmakers, they did not use glass cullet for glazing, instead preparing their own glazing frit. In this context, it is worth emphasising that New Kingdom Egyptian faience glazes differ in composition from both Egyptian and Mesopotamian glass, in having higher potash contents but significantly lower lime and magnesia contents (Shortland 2000). Thus, it would appear that yet different raw materials were also being used by the faience makers.

Once glass was being produced, artisans had ready access to a technology and materials for successfully glazing clay objects which had not been available before, explaining the emergence of clay glazing technology at the same time as glass making became geographically widespread. It is interesting to also note that early glazed clay vessels are of the same shape as the early glass vessels (Peltenburg 1987). In addition, even though there is scattered evidence of some experimentation with glass for producing small beads just prior to 1500 B.C., glazed clay objects are not produced until glass vessels are being made, after 1500 B.C.. Therefore it is likely that glass technology provided not only the means, but also the inspiration, for the glazing of clay objects. The technology for glass production had widespread repercussions on many related technologies. A change in the microstructure and composition of glazes on steatite stones produced in Egypt at this time has also been noted, indicating that the method used to glaze them was probably modified (Tite and Bimson 1989). There is also a large increase in the number of colours in which faience objects are produced.

In Mesopotamia there was clearly a demand for glazed clay objects because once the necessary glazing technology became available, it appears to have rapidly become widespread geographically. Further, the popularity of glazed clay does not seem to have been adversely affected by the fact that the alkali glazes used will have exhibited a slight tendency to craze.

QUESTION 3: WHY ARE ALKALI-GLAZED CLAY OBJECTS NOT PRODUCED IN EGYPT?

Although Egyptian artisans began to produce glass vessels at approximately the same time as those in Mesopotamia, glazed clay vessels were not produced in Egypt until circa the 1st century B.C.. If glass provided the technical means to produce glazes for clay objects, why is there a delay between the introduction of glass making and the adoption of clay glazing technology in Egypt? One possibility that has been raised is that the clays available in Egypt were intrinsically less suitable for glazing than those in use in Mesopotamia. Another theory is that the Mesopotamian potters may have developed a technique that facilitated the glazing of their clays, of which the Egyptian artisans were unaware. In order to evaluate these theories, it is first necessary to characterise the clays used by Mesopotamian potters to produce glazed clay ware and compare them to the clays available to Egyptian potters. There are three characteristics that must be considered in assessing the suitability of these clays for glazing with alkali glazes. These are the thermal expansion coefficient of the clay, the extent of the reaction between the clay and the glaze and the colour of the fired clay.

The thermal expansion coefficient of the clay

When an alkali glaze is applied to a clay object and fired, the glaze becomes molten and flows over the object. As the object cools after firing the glaze re-solidifies at around 550°C. From this point on, the glaze can no longer accommodate the different thermal expansion coefficient of the clay. If the glaze shrinks much more than the clay, the increasing stresses on the glaze cause it to crack or 'craze'. Therefore it is the shrinkage of the clay, as it cools from 550°C to room temperature, that is critical in determining how severe crazing will be. A clay thermal expansion coefficient that is high leads to a decrease in crazing and vice versa. The way in which the clay is processed and the type, and distribution, of the minerals that the clay contains, determines the thermal expansion coefficient.

Mesopotamian pottery is extremely uniform because standardised methods of production were adopted from around the 3rd millennium B.C. (Mynors 1986) using well mixed alluvial clays. The clays have a high calcium oxide content that, along with other factors, contributes to the high thermal expansion coefficient of the clay (Fig. 13.3). In contrast in Egypt a wide range of clays were utilised by the potters in antiquity. Each of these clays contains different minerals in varying proportions dependent on where the clay was obtained and how it was prepared. The Nile silt is probably the best-known Egyptian clay. It has a very low calcium oxide content and this contributes to the low thermal expansion coefficient of this type of clay (Fig. 13.3).

The extent of the reaction between the clay and the glaze

When a glazed clay object is fired the glaze and the clay will react. Some of the clay

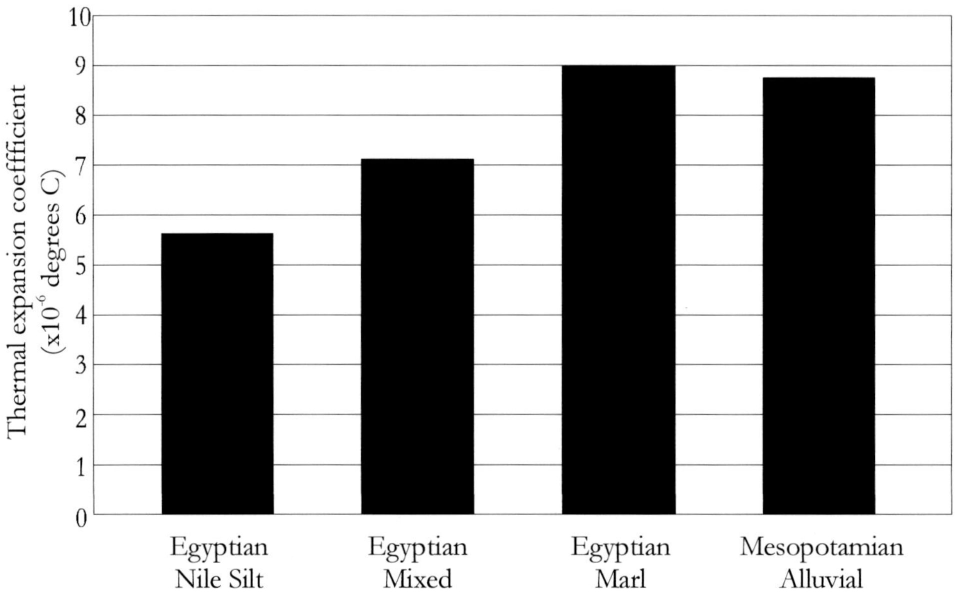

Figure 13.3: Comparing the thermal expansion coefficients (on cooling from 550°C to room temperature) of Egyptian and Mesopotamian clays, sampled from artefacts.

dissolves in the glaze and crystals form at the interface between the glaze and the clay. This region of interaction can act as a type of buffer between the glaze and the clay because it has properties that are intermediate between the two. The stresses on the glaze are thereby reduced and the glaze is less likely to craze. Therefore if the reaction between the glaze and the clay is increased, the crazing of the glaze will be reduced. The extent of this reaction is influenced by variables such as the duration and temperature of firing, whether the clay has been fired before glazing, the composition of the clay and the composition of the glaze. Clays that contain high concentrations of calcium and magnesium oxides (alkaline-earths) react with alkali glazes to a greater extent than other clays, all other factors being equal. Such clays are better suited to glazing with alkali glazes than other clays, as the glazes will craze less. A scanning electron microscope image of the interaction layer formed between an alkali glaze and an alkaline-earth rich clay is shown in Fig. 13.4.

Mesopotamian clays have high calcium and magnesium oxide contents (Fig. 13.5). This facilitates the formation of the interaction layer and helps reduce crazing. In contrast the Egyptian Nile silt contains very little calcium or magnesium oxides and so there is little interaction between the glaze and the clay (Fig. 13.5).

The colour of the fired clay

The composition of the clay, amongst other factors, also contributes to the colour that

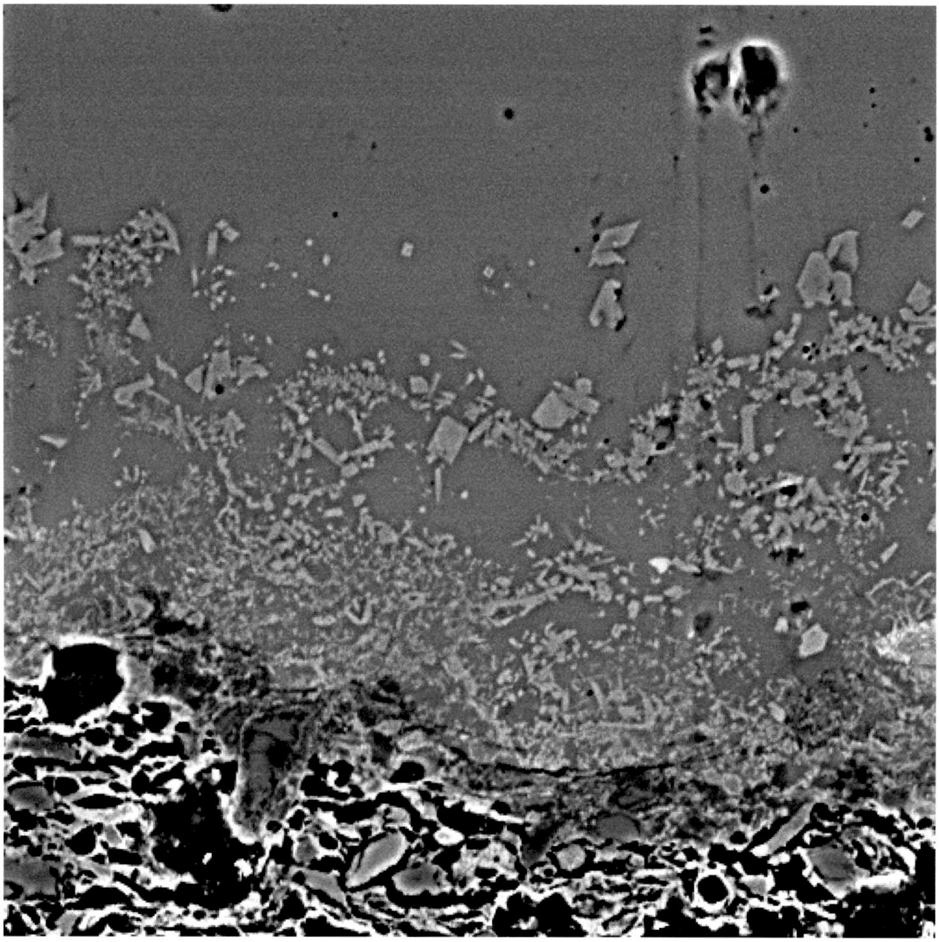

Figure 13.4: Scanning electron microscope back-scattered electron image of the interaction layer formed between an alkali glaze (top) and an alkaline-earth clay (bottom). Angular diopside crystals are visible in the interaction layer (x600).

develops when the clay is fired. Under the same firing conditions, clays that contain high concentrations of iron oxide will fire to a red colour. However, clays that contain a high concentration of calcium oxide will fire to a pale buff colour due to the formation of crystalline silicate phases, even if the clays also have high iron oxide contents. A glaze applied to a clay with this light colour will appear brighter and less discoloured than if it were applied to a clay with a dark colour.

The high calcium oxide of Mesopotamian clays causes them to have a pale buff colour when they are fired, very suitable for displaying a coloured glaze. In contrast, since the calcium oxide content of the Egyptian Nile silt is low but its iron content is high, it fires to a deep red colour under the same firing conditions. A blue coloured glaze appears almost black against this surface.

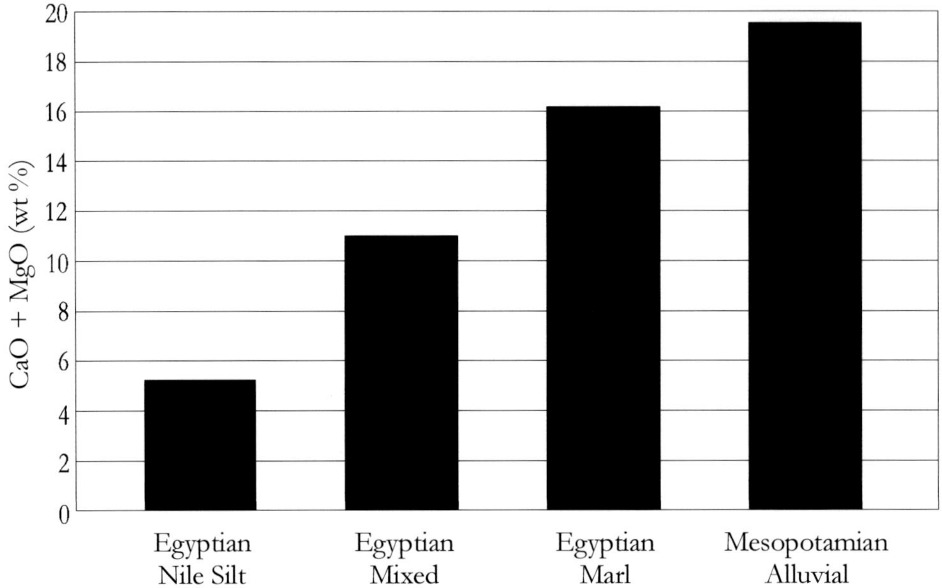

Figure 13.5: Comparing the alkaline-earth contents of Egyptian and Mesopotamian clays, sampled from artefacts.

Thus alkali glazes applied to Mesopotamian clays generally craze little and have a bright blue colour whereas alkali glazes applied to Egyptian Nile silts craze severely and have a dark colour. However, as stated previously, Egyptian potters made use of many different types of clay in antiquity including Egyptian marl clays. In contrast to the Nile silts, Egyptian marl clays contain high levels of calcium oxide (Fig. 13.5) and so, like Mesopotamian clays, have properties suited to glazing with alkali glazes (Fig. 13.3). There is also a range of other Egyptian clays with intermediate compositions and properties, one of which is illustrated (Fig. 13.3 and Fig. 13.5).

In summary it has been demonstrated that it was possible to obtain, or prepare, clay suitable for glazing with alkali glazes in ancient Egypt. It has also been established, through examination of glazed clay artefacts, that Mesopotamian artisans were not using any additional techniques, unknown to Egyptian potters, to facilitate glazing. These findings substantiate the argument that cultural factors are largely responsible for the technology for glazing clay not being adopted in Egypt.

Cultural Influences

In Mesopotamia stone was a rare commodity and was used very selectively only for certain applications. Therefore clay was used to build structures, even monumental ones, and for statues and figurines. The most widely available stone was limestone, which was used from an early date throughout Mesopotamia. However even this is usually found in a subsidiary role to clay, hidden in foundation courses or behind plaster (Moorey 1994). Prior to 1500 B.C. these clay objects and structures were decorated using plaster, paint and other media. For example the walls of a late 4th millennium B.C. temple at Uruk were covered in cones of clay that had been dipped in pigment at one end and then driven into the wall to form geometric patterns. Many mud brick walls were probably white-washed and then painted with murals but, unfortunately, few examples of these survive in the archaeological record. Glazes which could be applied to clay provided a means of giving figurines and architectural details, such as nails, wall plaques and bricks, a shiny, hard wearing and brightly coloured appearance.

In contrast, stone was employed on a large scale in Egypt, for many applications including monumental buildings and statuettes. The types of stone used included limestone, sandstone, granite, alabaster (calcite), quartzite, basalt, breccia, diorite and marble, and the stone was similarly decorated using paint. In these circumstances, clay glazing may not have been such a directly applicable or desirable technology. Where a glassy surface was required on an object, such as a vessel, a small figurine or a tile, faience was used instead (Lucas and Harris 1962). The Egyptian artisans were extremely skilled with faience and it was probably seen as being the only appropriate material for certain applications, not just because of its material properties but also because of its cultural associations (Friedman 1998). Gradual modifications of the faience making tradition occurred over time, such as additions to the palette of colours available and the range of glazing methods used, and are particularly noticeable in the period when glass begins to be produced. By 1500 B.C. the technique of inlaying faience became so developed that great detail could be represented in very bright colours on different coloured backgrounds and so, in many instances, the desired appearance could be achieved best in faience.

In Mesopotamia, faience was used to produce objects such as amulets, pin-heads, beads, vessels and occasionally for inlay, but after 1500 B.C. glazed clay, glass and frit were used as well for many of these applications and were also used extensively for architectural decoration. (Frit is a heterogeneous material formed from partially reacted silica, alkali and colourants but unlike faience, frit does not have a glaze layer on the surface). However, it is testimony to the aesthetic appeal of faience that, a thousand years later, artisans reverted to a faience type material instead of clay to produce glazed bricks for decorating the palace of Darius I. Later still, Islamic potters returned to faience type materials to produce a white base for the coloured designs on some of their glazed wares (Mason and Tite 1994).

It is likely that the technology for glazing clay objects was not adopted in Egypt at the same time as in Mesopotamia for a combination of reasons. The embeddedness of the faience and stone working traditions in Egypt, and less dependence on clay for use in monumental structures and symbolic objects, could be one influential factor. The appearance of faience, with its uncrazed glaze and vivid colours, was probably preferred over that of glazed clay. This factor is particularly important as the power of materials such as stone, glass and faience in Mesopotamian and Egyptian cultures was specifically associated with their colour, lustre and patterning. The cultural beliefs that led to faience being used for many applications associated with the gods and funerary practices would also have reinforced the preferences of Egyptian society for this traditional craft.

This period in the ancient Near East and Egypt was a time during which the armies of Tuthmosis I campaigned far outside the traditional borders of Egypt, invading the Levant and appointing local rulers loyal to Egypt. The archaeological record and documentary evidence, such as the Amarna letters, give an insight in to the extent of the tribute and trade that passed between the various powers in the Near East and Egypt. Items exchanged included glass, semi-precious stones, metals, faience and frit objects. The fact that there is no archaeological evidence that glazed clay objects were imported in to Egypt would indicate that these objects were simply not sought after.

CONCLUSIONS

It is apparent that glazing technology did not develop in isolation but was influenced by developments in other crafts. Close contact between artisans involved in a range of different technologies is more generally a crucial factor in the development of new technologies (Tite *et al* 2001). Thus, Peltenburg (1987) has suggested that, since the production of glass vessels involves the manipulation of hot, viscous fluids, the development of this new technology required input from metal workers. The working practices or political situation at this particular time may have facilitated this inter-craft exchange of technological experience. For example, artisans were expected to be skilled with many materials and might work in close proximity to other artisans in royal workshops. Also the formation of new and powerful states in Mesopotamia and Egypt during the period around 1500 B.C. may have led to renewed sponsorship of artisans (Moorey 1989).

Since, in the scientific study of ancient technologies, the different materials (e.g. pottery, glass, metals, organics) are often considered separately, the evidence of overlap and commonality between technologies, which might indicate the transfer of a processing technique or raw material from one to the other, is harder to detect. Thus an interdisciplinary approach has great potential in the study of other technologies in other periods and geographical areas.

The development of technology is often entirely attributed to restrictive technological factors such as the firing temperatures attainable by the kilns, the raw materials available and the technical expertise of the crafts-person. As a result, other important factors, such as the social, political and ideological significance or associations of a particular process or product in an ancient culture, are sometimes overlooked. The processes for producing an artefact must be interpreted in the context of the culture in which the artefact was created, and not in the context of the culture in which it is being studied. For example, a glass might be considered suitable for a particular application today on the basis of price, thermal expansion coefficient, refractive index and density. However, in the ancient Near East it is more likely to have been selected for its colour, opacity, and lustre; attributes which were considered to imbue it with certain medicinal powers (Reiner 1995). Finally, the excerpt of ancient text given below, which is thought to pertain to glass making in ancient Mesopotamia (Oppenheim *et al* 1970), illustrates the interrelationships of technological processes and the culture in which they develop.

'When you set up the foundation of a kiln to make glass, you first search in a favourable month for a propitious day, and then only you set up the foundation of the kiln. As soon as you have completely finished you go and place there kubu images, no outsider or stranger should thereafter enter the building, an unclean person must not even pass in front of the images. You regularly perform libation offerings before the kubu images. On the day when you place the "metal" in the kiln you make a sheep sacrifice before the kubu images. You place juniper incense on the censer; you pour out a libation of honey and ghee, and then you make a fire in the hearth of the kiln and place the "metal" in the kiln. The persons who you allow to come near the kiln must be ritually clean so that you can allow them to come down to the kiln. The wood you burn in the hearth of the kiln should be thick, peeled poplar wood: logs which have no knots, bound together with leather straps and cut in the month of Abu. Only this wood should go in to the kiln'

ACKNOWLEDGEMENTS

SP would like to thank NERC for a Research Studentship that provided funding whilst undertaking this research. We would also like to thank Marius Vendrell and his colleagues at the University of Barcelona for completing the dilatometry of the clay samples. The considerable expertise of Eleanor Robson and Roger Moorey, and the invaluable advice of the team at the RLAHA, including Chris Doherty and Andrew Shortland, are also greatly appreciated.

REFERENCES

Beck, H. C., 1934, Notes on glazed stone – Part I Glazed steatite, *Ancient Egypt and the East*, 69–88.

Beck, H. C., 1935, Notes on glazed stone – Part II Glazed quartz, *Ancient Egypt and the East*, 19–37.

Friedman, F. D., (ed.), 1998, *Gifts of the Nile-Ancient Egyptian faience*, Thames and Hudson, London.

Hedges, R. E. M., 1982, Early glazed pottery and faience in Mesopotamia, in Wertime, T. A. and Wertime, S. E. (eds.), *Early pyrotechnology*, 93–103, Smithsonian Institution Press, Washington D.C.

Kleinmann, B., 1986, History and development of early Islamic pottery glazes, in Olin, J. S. and Blackman, M. J. (eds.)*Proceedings of the 24th Archaeometry Symposium*, 73–84, Smithsonian Institution Press, Washington D.C.

Lilyquist, C. and Brill, R. H., 1993, *Studies in early Egyptian glass*, Metropolitan Museum of Art, New York.

Lucas, A. and Harris, J. R., 1962, *Ancient Egyptian materials and industries*, Arnold, London.

Mason, R. B. and Tite, M. S., 1994, The beginnings of Islamic stonepaste technology, *Archaeometry*, 39, 41–58.

Moorey, P. R. S., 1989, The Hurrians, the Mittani and technological innovation, in de Moyer, L. and Haerinck, E. (eds.), *Archaeologica Iranica et Orientalis: miscellanea in honorem Louis vanden Berghe*, Ghent, 273–286.

Moorey, P. R. S., 1994, *Ancient Mesopotamian materials and industries*, Oxford University Press, Oxford.

Mynors, H. S., 1986, *Mesopotamian ceramics of the third millennium B.C.: with analysis of pottery from Abu Salabikh, Kish and Ur*. Unpublished PhD thesis, University of Southampton.

Oppenheim, A. L., Brill, R. H., Barag, D. and von Saldern, A., 1970, *Glass and glass making in ancient Mesopotamia,* Corning Museum of Glass, New York.
Peltenburg, E. J., 1987, Early faience: recent studies, origins, and relations with glass, in Bimson, M. and Freestone, I. C. (eds.), *Early Vitreous Materials,* British Museum Occasional Papers 56, London, 5–29.
Reiner, E., 1995, *Astral magic in Babylonia,* American Philosophical Society, Philadelphia, 119–132.
Rye, O. S., and Evans, C., 1978, *Traditional pottery techniques of Pakistan,* Smithsonian Institution Press, Washington D.C.
Shortland, A. J., 2000, *Vitreous materials at Amarna,* BAR International Series S827, Oxford.
Tite, M. S. and Bimson, M., 1989, Glazed steatite: an investigation of the methods of glazing used in ancient Egypt, *World Archaeology,* 21, 87–100.
Turner, W. E. S., 1956, Studies in ancient glasses and glass making processes. Part III The chronology of glass making constituents, *J Society Glass Technology,* 40, 39–52.
Tite, M. S., Shortland, A. J. and Paynter, S, 2001 (in press), The beginnings of glass and glazed pottery production in ancient Mesopotamia and Egypt, in Reedy, C. and Hansen, E. (eds.), *Theory building in studies of technological innovation and style,* Cotsen Institute of Archaeology, Los Angeles.
Vandiver, P. B., 1983, Egyptian faience technology, Appendix A in Kaczmarczyk, A. and Hedges, R. E. M., *Ancient Egyptian faience,* Aris and Philips, Warminster.
Vandiver, P. B., 1998, A review and proposal of new criteria for production technologies of Egyptian faience, in Colinart, S. and Menu, M. (eds.), *La couleur dans la peinture et l'emaillage de l'Egypte ancienne,* Edipuglia, Bari, 121–139.

Chapter 14

Problematising the Transition from Bronze to Iron

Peter Haarer

Abstract

In Greece and Cyprus the twelfth century is usually seen as the beginning of the 'Transition' from bronze to iron – a decisive phase of one or two hundred years during which iron became cheaply and widely available for the first time, displacing bronze as the main metal for tools and implements. This paper attempts to quantify how the economic value of iron in fact changed in relation to silver and to bronze. The scheme proposed suggests that a different interpretation is possible, in which the Transition occurs with less speed, magnitude and impact than is traditionally assumed.

INTRODUCTION

From the end of the Bronze Age onwards the archaeological record in the eastern Mediterranean world shows an increase in the appearance of iron artefacts, especially in relation to bronze artefacts. This change, known as 'the transition from bronze to iron' (the 'Transition' hereinafter) implies the introduction of an innovation in ferrous technology.

The model generally employed for the description and explanation of the Transition is the 'Three Stage Scheme' of Snodgrass which is based mainly on the study of material from Cyprus and the Aegean (see especially Snodgrass 1971, chapter 5; 1980a; 1982; 1989; but also Snodgrass 1980b, 49–54; 1987, 170 & 176; and 1994). In Stage 1 of this scheme, iron is known, but used in small quantities and has a semi-precious value as a curiosity. In Stage 2, centuries of *'desultory experimentation'* end when necessity, in the form of a shortage of bronze, catalyses experimentation with iron in applications for which bronze had hitherto traditionally been used. Iron is found to be particularly suited to implements for cutting and piercing, and this leads to Stage 3, in which iron becomes widely available. The absolute chronology proposed by Snodgrass is shown in Table 14.1 (see esp. 1982 291–292 & fig. 2):

Table 14.1: Absolute chronology proposed by Snodgrass

	Stage 1	Stage 2	Stage 3
Cyprus	up to ca. 1150	ca. 1150 to ca. 1050	ca. 1050 onwards
Aegean*	up to *pp* ca. 1150	*pp* ca. 1150 to ca. 1000	ca. 1000 onwards

* metallurgically advanced areas only; *pp* - *paulo post.*

Snodgrass suggests that the Transition from bronze to iron had important economic and social effects which in turn affected the development of other technologies. He proposes that by Stage 3 iron had become a cheap metal, because of its common availability, and so tools and implements could be made from it which were affordable to the many, as opposed to those of bronze which had been expensive and therefore remained the preserve of the few. The arts and crafts were therefore 'democratised' by the advent of iron, which was therefore elemental during the Dark Ages to the reconstruction of society, Snodgrass here adapting a concept first proposed by Childe (Childe 1942, 183).

The Three Stage Scheme implies a number of assumptions, many of which tend to be repeated in variant interpretations of the Transition. It is presented as the single decisive phase in the development of iron during antiquity, which by implication is the result of a single major technical innovation. By extension it was brought about by a set of unique circumstances which forced society to discover the means for exploiting iron on a significant scale. In the Three Stage Scheme the catalyst is identified as an acute shortage of bronze (Snodgrass 1971, 237–239; 1980a, 348; 1980b, 50; 1982, 293; 1989, 29). This hypothesis has now been doubted (Maddin 1982, 303; Zaccagnini 1990, 496–500; E.S. Sherratt 1994, 61; Muhly 1996, 47–48) leading to the proposal of a hypothesis of fuel = tree shortage (Wertime 1982; Wertime 1983; Horne 1982a; Horne 1982b; Stager 1985, 44). In turn this has been thoroughly discredited (Åström *et al.* 1986, 28; Muhly *et al.* 1990, 162–163; Pickles & Peltenberg 1998, 81–84) and a further new explanation advanced by E.S. Sherratt (1994). This explains the Transition in terms of Cypriot entrepreneurs who needed something to sell to compensate for the loss of a market in ceramic products. Accordingly, they turned to iron, which they then cleverly marketed in the form of knives (E.S. Sherratt 1994).

A second set of assumptions, which may be observed in the Three Stage Scheme and in the rival explanations for the advent of the Transition, is that in antiquity iron was inherently undesirable. Therefore, the small quantities of iron available in the Bronze Age would have been regarded as oddities of mere incidental interest. During the Transition, we must interpret the increased number of iron artefacts as necessities, or as objects which people had to be persuaded to use. After the Transition, the abundance, by Bronze Age standards, of iron goods in the Iron Age must be viewed as purely utilitarian items which were mass produced once the mechanical potential of the metal had been appreciated.

A third assumption is that, when set against the broad backcloth of the great epochs, the Transition was a fast process completed within one to two centuries. A fourth assumption is that the value of iron plummeted to very low levels as a result of the Transition, becoming very cheap especially relative to bronze.

This paper tests this final economic assumption using evidence mainly from the Aegean area, but with reference also to Ancient Near Eastern material. In so doing it presents evidence which raises difficulties with the model presented by Snodgrass and proposes that a different interpretation of the Transition may be possible.

PRICES OF IRON IN FOURTH-CENTURY GREECE

What was the value of iron in Early Iron Age Greece? It is useful to begin by discarding the notion commonly cited that by the eighth /seventh centuries, two thousand measures of iron had the same purchasing power as one measure of silver. This rate of exchange does not appear in documentary evidence, but was inferred by Courbin (1959) from an ancient tradition that in Archaic Greece iron roasting spits known as *obeloi*, surviving examples of which are typically rods of metal about 75 to 180cm in length with a square cross-section of 0.5cm to 1.5cm, had functioned as the immediate precursors of a denomination of the first silver coins, also called *obeloi* (later *oboloi*, Eng. 'obols'). The one supposedly superseded the other implying that at the point of transition there was an equivalence in value between an iron spit and a lump of silver with the same weight as an obol coin. Accordingly, Courbin deduces the exchange rate of iron for silver by reconstructing the weight of what he believes to be a 'standard' iron spit of monetary function and comparing this weight with that of an early silver obol.

The result is unconvincing for several reasons. The spits chosen by Courbin for his calculation come from a massive bundle of dozens of such artefacts discovered at the Argive Heraion. Their manufactured length is critical to his argument, yet this cannot be ascertained with meaningful accuracy because the only known measurement of the length of the bundle from before it was moved from its context and every spit within it fragmented beyond reconstitution (Svoronos 1906, 196; Furtwängler 1980, 93 & n58; Courbin 1983, 152) is the excavator's estimate of '*about five feet*' (Waldstein 1902, 77), which translates into metric as ca. 153cm +/-15cm. By joining a number of fragments together Courbin is able to establish that the lower limit must be at least 142cm (Courbin 1959, 210, 215) but it is clear that he has no sound material foundation to proceed to his conclusion that the spits originally measured almost exactly 150cm (Courbin 1959, 215; Courbin 1983, 151 & n19; Haarer 2000, 1.83–86). Further, Courbin assumes, as have many others, that the bundle relates to a story about the minting of the first coinage by the Argive tyrant Pheidon, which is preserved in the fifth-century AD *Etymologicum* of Orion of Thebes (Sturz 1820 column 118).

The study of spits from sanctuary contexts throughout Greece shows that the bundle from the Argive Heraion was not unique, but belongs to a class of votive gift dedicated like many other metal objects, such as bronze tripod cauldrons, as an item of value, though without any function as a specialised and favoured form of money (Haarer 2000, 1.107–108, 115, 120). Moreover, historiographic analysis of the story about

Pheidon indicates that it was invented in the fourth century to flatter the Macedonian Royal Family (Brown 1950). We may add here that the traditions of the use of spits as pre-coinage money attached to Sparta and Byzantion seem equally unreliable (Haarer 2000, 1.161–171, 176–178). Finally, epigraphic evidence from the fifth century onwards indicates that iron spits were bought and sold for one or more drachmas each, considerably more than one silver obol (*IG* I^3.421 column II fragment c lines 93–94 dated ca. 414; *IG* XI.2.148 dated 297; *IG* XI.2.287 dated 250; *ID* 2.403 dated 189), thereby throwing further doubt on the possibility that one spit had the same value as one silver obol (Haarer 2000, 1.157–158). When set against all the above, methodological and computational flaws in the convoluted route by which Courbin arrived at his exchange ratio of 2000 : 1 hardly help its credibility.

A more solid foundation from which to determine how the value of iron changed during the Early Iron Age in Greece is offered by actual prices of iron recorded in inscriptions, the earliest of which belongs to the mid fourth century. A text of this date from Delphi gives maximum and minimum prices for the purchase of one mina of iron as follows (*FD* 3.5 document 27 column IIA, commentary *FD* 3.5, 142):

maximum: 1 silver drachma
minimum: 1 silver obol, 3 *chalkoi*

Assuming that the sums of money and weights referred to in this inscription follow the Attic-Euboic 'standard' of eight *chalkoi* = one obol = 0.72g, six obols = one drachma = 4.31g and one hundred drachmas = one mina = 431g, then exchange ratios may be extrapolated as follows:

maximum price = 431g iron : 4.31g silver or 100 : 1
minimum price = 431g iron : 0.99g silver or 435 : 1.

To this may be added two Eleusinian building accounts, also from the fourth century. The first of these concerns the year 330/29, and records an expenditure for iron of 8½ obols per stater (*IG* II^2 1672.205–206), while the second concerns the year 327/26 and records an expenditure of 4 obols per stater (*IG* II^2 1673.53). The staters in which the iron is measured in these inscriptions may be assumed to be 'Solonian' weight staters, equivalent in weight to one hundred and five silver 'didrachmas', that is 4.31g x 2 x 105 = 905.1g (for an explanation of the relationship between the weight of a silver drachma and the weights by which commodities were bought and sold in Athens in the fourth century see Lang & Crosby 1964, 2–4 & 18–21. Their value of 4.36g per drachma is substituted here with the more commonly cited value of 4.31g). Therefore the exchange ratios implied for iron : silver may be calculated as 148 : 1 and 314 : 1. Further to these documents, an account from Epidauros dated 365 to 335 gives a price of 14 drachmas 2 obols per talent of iron (*IG* IV^2 103), which equates to a ratio of iron : silver of 419 : 1. (Sixty minas = one talent = 25860g. Much of the material presented here may be found in Treister 1996, 250).

The inscriptions listed above therefore indicate that in fourth-century Greece the price of iron relative to silver varied between 100 and 435 measures of iron to one measure of silver. Before proceeding it will be helpful to consider why iron retailed within such a wide range of prices. A similar range is found in the Ancient Near East in Neo-Babylonian documents of the first half of the sixth century which give ratios for the exchange of iron for silver ranging between 240 and 831 measures of iron to one of silver. Powell (1990) comments extensively on these ratios and explores a variety of explanations for the range. It is conceivable that it simply reflects fluctuations in the value of each metal, the lowest ratio being produced by high iron and low silver prices, and *vice-versa*, but Powell argues that the purity of silver when used as money was fairly consistent, and that its value was relatively stable during the Neo-Babylonian period (Powell 1990, 80). He bases his argument on the lack of qualifiers for the purity of silver in the texts, the fact that silver was used as the measure for the value of gold, and the relative stability of silver equivalences for a range of other objects, though he concedes that there is at present no scientific evidence with which to test his *'methodological postulate'*.

The range of ratios might also be explained by the mechanics of procurement and supply, by which iron may have been obtained in bulk by metal stockists at the lower price, and sold on in much smaller quantities to end consumers at the higher price. Powell argues convincingly that this factor lies behind a range of prices for Old Assyrian tin, where the silver : tin ratios all fall within 15% of two levels of prices, one of which is twice the other (Powell 1990,85). The Neo-Babylonian iron prices do not, however, conform to this pattern, but lie in a wide scatter of different values.

Powell proposes two further possible explanations for the variation in iron prices, comprising fluctuations in supply and demand, and what he terms *'unrecorded factors of a personal social or economic nature'*, though due to the nebulous state of the evidence for such forces it is impossible to gauge their effects (Powell 1990, 78, 99). Powell suggests, nevertheless that the range of prices attested for iron in the Neo-Babylonian period are far too great to be accounted for entirely by any of these factors, and concludes that they relate primarily to the sale of iron of different grades and specifications (Powell 1990, 78). This scenario seems most likely given the nature of pre-industrial ferrous metallurgy which may be inferred. As illustrated by the archaeological research by experiment undertaken by Peter Crew coupled with the analytical work of Salter on the iron industry of pre-Roman Britain, considerable variations of qualities and properties would result from the choice of raw materials, the processes employed, and the degree to which the metal was forged (see esp. Crew 1991 & 1994; Salter 1989, esp. 255–258). In the Ancient Near East, as in other historical societies, lexical evidence for this phenomenon may have been masked by the tendency to translate with the same modern word a variety of ancient terms which probably refer to different types of metal (Powell 1990, 83). The range of prices in fourth-century Greece may perhaps be explained in the same way. If so, this level of complexity has been largely absent from the discussion of iron during the Early Iron Age.

Where do the prices place iron in terms of affordability in the fourth century? At the rate of 100 measures of iron to one of silver, one Attic silver drachma would buy 431g iron. At a rough guess, this would provide more than sufficient metal for a dagger blade or spear-head, or enough for a good sized axe, almost enough for a reasonable sword, and just under half the quantity required for a roasting spit. At the rate of 435 measures of iron to one of silver, the same drachma would buy 1875g iron. This would provide enough metal for about half a dozen dagger blades or spear heads, four or so good sized axes, at least two swords, and one large roasting spit. However, in most of these applications, the quality of the metal (by which I mean its consistency, ability to be forged, and physical properties) would be critical, either to the forming of the artefact, or its performance, if not both (Crew 1991; Salter 1989, 255–258). For many objects, therefore, expensive iron would not merely be preferred, but would be essential.

What was the Attic silver drachma worth in real terms? Markle (1985) has studied rates of pay at Athens for jury service and attendance of the assembly, and has considered whether these would have offered adequate compensation for the loss of earnings from regular employment to those who had to work full-time to support households. As part of these calculations Markle estimates that during the fourth century a 'moderately active man' with a wife and two children would need to spend about two and a half obols per day on subsistence. Therefore jury pay or assembly pay at the rate of three obols per day would leave him with half an obol spare (Markle 1985, 277–280).

This margin would be increased if the man had an occupation for which he was remunerated at the rate of one drachma per day worked. Markle makes an educated guess that the average Athenian citizen would work on two hundred and fifty days of each year and so if he brought home a drachma on each of these days his annual earnings would total two hundred and fifty drachmas. This equates to just over four obols for each day of the calendar year (Markle 1985, 297).

However, in the fourth century, Eleusinian building accounts from the 320s suggest that an unskilled labourer was paid one and a half drachmas per day worked, while a semi-skilled labourer was paid two drachmas per day worked and a skilled labourer two and a half drachmas per day worked (*IG* II^2.1672; Markle 1985, 293). Following the same formula, these sums would give three hundred and seventy five drachmas per annum, or about one drachma per calendar day, five hundred drachmas per annum, or about one drachma two obols per calendar day, and six hundred and twenty five drachmas, or about one drachma and four obols per calendar day.

How far would income after subsistence stretch? Some information about the cost of various goods has been assembled by Pritchett (1956) commenting on the *Hermokopidai* stelai of 414/3 (*IG* I^3.421) which record the sale by auction of property confiscated by the Athenian state from a number of citizens who had been convicted of serious religious crimes and a small selection of this material may be presented here for the purposes of illustration. Amongst items of clothing, the price of a *himation*, a loose outer garment like a cloak, varies between five and fifty drachmas between the fifth and third centuries (Pritchett 1956, 206). Amongst items of furniture, beds are sold for six to eight drachmas in the stelai, but are worth twelve to twenty drachmas in third century documents (Pritchett 1956, 227–228). Amongst livestock, a *bous*, denoting any sort of bovid, seems typically to have had a worth of about fifty drachmas per animal in the fifth and fourth centuries (Pritchett 1956, 255). Slaves tended to be more expensive. In the stelai they average 174 drachmas per person, but a passage of Xenophon states that the price of a slave could vary between fifty and one thousand drachmas per person (Xenophon *Memorabilia* 2.5.2), while a passage from a fourth-century speech attached to Demosthenes ([Demosthenes] 59.29) claims that the finest courtesan might change hands for three thousand drachmas. (Pritchett 1956, 276–278).

Using the information presented above as a rough guide, it seems likely that iron tools were economically accessible to most Greeks by the fourth century, though iron was not perhaps that 'cheap'. The purchase of a good quality iron tool would represent a small investment, and its breakage or loss would be lamented.

VALUES OF IRON BEFORE THE FOURTH CENTURY

How do the fourth-century prices of iron relate to the economic value of the metal in earlier periods? There is no direct evidence from Greece with which to answer this question but documentary sources from elsewhere give some indication about other eastern Mediterranean regions. As mentioned briefly above, Ancient Near Eastern texts of the Neo-Babylonian period (612–547) indicate that in Babylon silver was exchanged for iron at ratios of 1:831, 1:573, and even as much as 1:229. In the Syro-Palestinian region the rate was 1:361, and in a place probably to be identified as Cyprus it stood at 1:240 (Powell 1990, 78). A much earlier legal text from the Nippur region dated to 1033 gives the price of an iron dagger blade as two silver shekels (Brinkman 1988, 157 n63). If the weight of the dagger blade is reckoned as 200g to 300g, and that of a silver shekel as 8.3g, then the exchange rate which may be inferred is approximately 12 to 18 measures of iron to one of silver. Finally, from just before the end of the Bronze Age, the well-known 'iron letter' from the Hittite King Hattusilis III (ca. 1289–1265) probably to the Assyrian ruler Shalmaneser I (ca. 1273–1244), and dated to soon after the latter's succession (see Goetze (1940) for a text and commentary), indicates that the quantity of iron in a single dagger blade was of sufficient value to be worthy of exchange between two of the most powerful rulers in the Ancient Near East (Craddock 1995, 257). Therefore it seems likely that the price of the blade at that time would have been considerably more than two shekels.

The number of texts cited here is small, and where documents stand in isolation from comparative texts of similar geographical and/ or chronological context the reliability of any conclusions drawn can be guaranteed to a lesser extent than would be desirable. This observation seems especially pertinent to the lone eleventh-century document from the Nippur region. With this *caveat* in mind, and in the absence of fuller sources, we may note that the documentary evidence from the Ancient Near East seems to imply that there was a decline in the value of iron between the Late Bronze Age and the sixth century. This decline can be quantified tentatively as a drop of between thirteen-fold and seventy-fold between the eleventh and sixth centuries.

How do these figures relate to Greece and Cyprus? Is there evidence for a similar decline in the value of iron during the Early Iron Age, and if so, at what rate over what period? Is the decline as dramatic and rapid as that implied in the Three Stage Scheme of Snodgrass?

One possible indicator of how the value of iron changes over time is provided by the way in which types of artefact feature in the archaeological record, and the iron roasting spit provides a good subject for such a study for a number of reasons. One of the main attractions of the type for this purpose is that its form is relatively simple, comprising as it does a simple iron bar with few complex or ornamental features besides a handle at one end. (Illustrations of good examples of spits may be found in Courbin 1957, 369 figure 52 (from Argos South Cemetery Area Tomb 1) and Kunze & Schleif 1944 plate 71d (from Olympia). The use of such artefacts in animal sacrifice is represented in many scenes painted on vases, many of which are illustrated in van Straten 1995). The material value of a spit in the Early Iron Age in Greece therefore seems likely to have been determined more by the metal from which it was made, than by artistic enhancements. Moreover, we may be sure that spits were made from good quality or in other words expensive iron, because inferior grades cannot be drawn down into a bar with the long length and slender thickness of a spit. As such, the roasting spit is in effect not far removed from a billet of high grade iron (and the type does in fact eventually adopt a role as such when it reaches north-western Europe and Britain at a much later date: see Crew 1994 and 1995). One further advantage is that compared to other iron types spits have been well published in Greece thanks to the interest generated in them by the ancient tradition that they were once used as units of pre-coinage money.

Iron spits first appear in the eleventh/ tenth centuries in funerary contexts on Cyprus at Lapithos (*Kastros* Tomb 417 = Cypro-Geometric IA (ca. 1050–1000)), Palaepaphos (*Skales* Tomb 67 = Cypro-Geometric IA or IB /II (ca. 1050–1000 or ca. 950)), and Kaloriziki (Tomb 6b = Cypro-Geometric I-II (ca. 950)). The context of the find at Kaloriziki remains largely unpublished, but the wealth of information about the tombs at Lapithos and Palaepaphos allows us to conclude that iron spits were placed only in the richest burials at these cemeteries, which are among the richest known anywhere on Cyprus at this time (Coldstream 1989, 329 and 333 table 37.3). The exclusive association between iron spits and rich burials seems to be repeated in Krete from the tenth century onwards at Knossos (the North Cemetery and Fortetsa = Early Proto-Geometric onwards (ca. 970 onwards)) and at Eleftherna (Area K) from the Geometric onwards (ninth century onwards) (Haarer 2000, 1.24–29). The association is also repeated slightly later in the heart of the Aegean world on Euboia at Lefkandi (Toumba cemetery Pyre 13 = Sub Proto-Geometric II (ca. 875–850)) (Haarer 2000, 1.30–31). The inference which may be drawn from these finds is that between the eleventh and ninth centuries iron spits were prestigious items which only the few had the means to obtain. If, as suggested above, the value of the simple iron spit was largely governed by the material from which it was made, then we could conclude that during this period iron still had a high value, similar perhaps to that which it seems to have had in the eleventh century in the Ancient Near East. We may note in this context that the iron spit had a wooden counterpart. Therefore, although spits need not have been purely items of material wealth, and probably also had non-economic values, the iron version was akin to 'best china'.

In the eighth century in Greece there are signs that iron spits are losing value. At sites such as Argos, iron spits begin to be included amongst the grave goods of individuals further down the social scale (South Cemetery Area Tomb 14/3 and the grave from the Theodoropoulou Plot = Late Geometric I (ca. 765–730): Haarer 2000.1, 34–35). There are also signs that the form of iron spits is changing under the influence of inflationary pressures. Firstly, they tend to be longer. The biggest intact example dated securely to the preceding period measures 75cm, whereas that from the eighth century measures 162cm (Haarer 2000, 2.55–58). Secondly, the number of spits in the largest individual deposits increases from one or two in the eleventh and tenth centuries, to up to twelve in eighth-century Greece and up to eighteen in Cyprus (Haarer 2000, 1.202–203).

By the sixth century, there are signs that the value of iron spits has decreased still further. The largest single deposits are now found at sanctuaries and comprise massive bundles of dozens of spits, such as that from the Argive Heraion (see above). By the fifth century, the deposition of spits in graves and sanctuaries has all but ceased, and the demise of the massive bundle perhaps indicates that the number of spits needed to make an impression had become too high to assemble. Spits also begin to be recycled as scrap (at Olympia and the Portico of the sanctuary of Athena Chalkioikos at Sparta: Haarer 2000; 203, 2.27 and 35). This evidence correlates well with a passage from Herodotos (2.135), written in the late fifth century, which implies that perhaps the only massive bundle of iron spits deposited at a sanctuary still visible by that time, at Delphi, could not be interpreted as evidence that its dedicator, supposedly one Rhodopis, had '*great wealth*' (Haarer 2000, 1.203).

The patterns outlined above indicate that the economic value of iron spits declined considerably between the eleventh and fifth centuries, but did so gradually throughout this period. How closely does this scheme relate to the value of iron? Parts of the Homeric epics have sometimes been cited as evidence for the value of iron in the pre-Classical period, but these seem too chronologically ambiguous to prove helpful, because it is possible to argue with equal strength that references from the *Iliad* or *Odyssey* in which iron features in a context indicative of value, such as a ransom, belong to either the Late Bronze Age setting or the Early Iron Age context of composition (see esp. Pleiner 1969, 9 for the former view and I. Morris 1986, 12 for the latter). This problem does not apply to Hesiod, but though he offers much evidence on the possible symbolic and emotive value of iron (see esp. E. S. Sherratt 1994, 80–81), he offers no direct evidence about its economic value relative to other metals.

Developments in the handles of iron knives have been used with more conviction to show changes in the value of the metal during the Early Iron Age. These show a progression from ivory to wood via bone, while the rivets which held them together change from bronze to iron over the course of the eleventh century in Cyprus and tenth century Aegean Greece. These changes have been taken to imply a drop in the value of the iron from which the knives were made to minimal levels (E.S. Sherratt 1994 71–72 & 74–75). However, other evidence indicates that, as suggested by the spits, iron retained much of its Late Bronze Age value throughout the Early Iron Age. Åström *et al.* analyse the assemblages of metal grave goods from Palaepaphos-*Skales*, Lapithos and Amathous, and show that during the early Cypro-Geometric period iron is usually found in rich tombs and with gold (Åström *et al.* 1986, 37–40). Similarly, Rupp has observed that the quantity of iron deposited in Cypro-Geometric graves is similar to that of gold (Rupp 1989, 353). There also seems to be a correlation at Lapithos between burial with iron and burial with a sacrificed slave, the latter being a marker of the high prestige of the primary burial (Gjerstad *et al.* 1934, 228, 234–235, 265; Haarer 2000, 1.17,

19). Much later in time, there is evidence that iron still had significant value in Aegean Greece as late as the eighth to sixth centuries in the form of hoards at sanctuary sites (Pleiner 1969, 17). Excluding spits these comprise billets of Ancient Near Eastern bi-pyramidal type found in the sanctuary of Athena Pronaia at Delphi (Perdrizet 1908, 213), 'hooks' found at the sanctuary of Hera at Perachora (Payne 1940, 189), a mass of iron found in a votive context at Kalapodi (Felsch *et al.* 1980, 96f), and perhaps a mass of spearheads found at Dodona (Carapanos 1878, plates 57 and 58, etc).

THE VALUE OF IRON AND BRONZE

As outlined above in the introduction, one of the crucial questions issues when trying to understand what happened to iron during the Early Iron Age is how its value compared with that of bronze. In Greece, a late fifth-century inscription records the purchase of 150 talents of tin for 34852 drachmas and of 37 talents of copper for 1301 drachmas 6 obols (*IG* I^3 472 column 4c lines 139–141 and 142–145 dated ca. 421/20 - 418/7; Price 1968, 103; Treister 1996, 248). Per talent these prices equate to 233 drachmas for tin and 35 drachmas one obol for copper, or exchange ratios of 1 : 26 for silver : tin and 1 : 171 for silver : copper. The content of tin in bronze tended to be varied according to the physical properties required for the application to which the metal would be put (Northover 1988; Pickles & Peltenberg 1998, 75–77), but if we choose a mid-range value of 10% then we may use the ratios given in the inscription to estimate an exchange ratio for silver : bronze of 1 : 157. Given the pattern of progressive decline, it seems likely that iron would have had a higher value during the fifth century than it has in the fourth-century documents listed above. Therefore, during the fifth century it seems likely that while the value of the cheapest and probably poorest quality iron was lower than that of bronze, that of the most expensive and probably best quality iron was higher.

For prices before this time it is again necessary to appeal to documentary evidence from the Ancient Near East. A text from Uruk shows that during the Neo-Babylonian Period (612–547) iron was less valuable than tin, copper and therefore bronze, the exchange ratio of iron to silver given being 240 to 361 : 1, compared with 6 to 9 : 1 for tin and 180 : 1 for copper (Oppenheim 1967; Powell 1990, 85). However, in texts from the Old Assyrian period (ca. 2000–1600) iron is much more valuable than tin, copper, and therefore bronze. Iron is exchanged for up to ninety times the quantity of silver and up to ten times that of gold (see Powell 1990, 94 n82, 95 n83 & n84 and 96 n88 for references and bibliography to the relevant texts, also Muhly 1980, 35–36 and Moorey 1994, 287), while tin and copper trail far behind. The ratio of tin : silver is about 14 +/ -15% : 1 for the buying price and 7 +/-15% : 1 for the selling price (Powell 1990, 85), while that of copper : silver is 46 to 180 : 1 (Muhly 1980, 39).

These documents indicate that iron lost value relative to bronze between the Old Assyrian period, when it was more valuable, and the Neo-Babylonian period, when it was less valuable. It follows that at some point in time the metals must have achieved parity, but was this before or after the beginning of the Iron Age? If the prices of tin and copper remained more or less consistent within broad limits between the Old Assyrian and Neo-Babylonian periods (Powell 1990, 85), then the exchange ratio for iron of 12 to 18 measures for one of silver, which was inferred from the legal text of 1033 from the Nippur region, would imply that good iron had about the same value as tin at this time, and so considerably more value than bronze. We must repeat here the caveat that this text stands in relative isolation and therefore the extent to which it is representative is unclear. Provided that the ratio inferred gives a reasonable guide, if the value of iron from the late eleventh century onwards moved only slowly to its Neo-Babylonian levels it seems likely that it did not provide a cheap alternative to bronze in the Ancient Near East until long after the start of the Iron Age. If Greece followed a similar pattern of changes in the value of iron relative to silver it is possible that it also followed a similar pattern in changes relative to bronze up to the fifth century.

It seems, therefore, that in Greece iron declined slowly in value throughout the first half of the first millennium, rather than suddenly over a period of one to two centuries. Furthermore, the decline was smaller than is usually implied, and the better quality grades of iron did not become cheap substitutes for bronze. What are the implications of these conclusions? At a practical archaeological level, the interpretation of deliberate deposits of significant quantities of iron must perhaps be revised. For example, our understanding of the social status of the male individual buried in ca. 900 in the area of the later Agora at Athens in the well-known Grave 27 (Blegen 1952) shifts markedly depending on whether we think that the iron tools with which he was interred were available to the many or the restricted few. At a more general level, although iron would have become ever more economically accessible during the Early Iron Age, it seems unlikely that it would have had the 'democratising' effect with the rapidity or intensity envisaged by Snodgrass. One further point to stress is that the high value of iron inferred here for the Early Iron Age suggests that iron was a desirable material before, during and after the Transition. This factor, rather than necessity or purely functional considerations was perhaps the driving force behind the increased exploitation of the metal (cf. B.G. Scott 1990, 31, 34 who proposes a similar theory to account for the adoption of iron in prehistoric Ireland).

GREECE AND THE ANCIENT NEAR EAST

The material presented above has further ramifications for the interpretation of the Transition. The disparity between the prices for iron in Greece and those in the Ancient Near East has possible implications for the claim by Snodgrass that the more forward looking areas of the Aegean were second only to Cyprus in pioneering and developing ferrous metallurgy during the Early Iron Age. Snodgrass has parts of the Aegean world enter his Stage 2 in ca. 1150 along with Cyprus, Palestine, Syria and Iran. Those areas which lag far behind include Mesopotamia, which enters Stage 2 only in ca. 900. Cyprus is the first region to move into Stage 3, followed by parts of Greece in ca. 1050, with Palestine and Iran trailing behind in ca. 950, and Syria in ca. 900. Mesopotamia enters Stage 3 only in ca. 750 (Snodgrass 1982, 291 fig. 2).

In terms of economic accessibility, if iron in the Ancient Near East was consistently cheaper than iron in Greece, as may be inferred from the evidence presented above, then it seems difficult to justify the assertion that Greece was among the first areas to enter Stage 3. This is all the more apparent if, as Snodgrass states, his Stage 3 was primarily an economic change by which iron became significantly cheaper than before (Snodgrass 1980a, 368). Furthermore, there is evidence that the differences in the price of iron between the two regions reflects a difference in the quantities of iron available in each. This in turn seems to imply a difference in the scale with which the metal was being exploited.

As a case study we may turn to Assyria from which relatively good and well studied archaeological and textual evidence is available. It seems clear that by the ninth century the iron industry of Assyria was well developed. The Assyrians were at the height of a period of imperialist expansion which had been achieved by a succession of strong rulers who had led large armies to victory. These mighty armed forces had been equipped with weapons and tools of iron which must have created a requirement for the production of this metal on an unprecedented scale. A vivid illustration of the use of iron for military applications at this time is offered by Hasanlu, a city which lay in the border territories between Urartu and Assyria but shows stronger material links with the latter (Pigott 1989, 74–78). About two thousand iron artefacts were recovered from late ninth century destruction levels at this site, amongst which were about seven hundred arrowheads, five hundred spearheads and seventy swords (Pigott 1989, 73). From the end of the eighth century, an illustration of the quantities of iron which could be amassed comes from Khorsabad, where a hoard of 140 to 150 tonnes of trade iron was found in a store room (Place 1867, 1.84–88; Haarer 2000, 2.160–164). Other finds from the ninth, eighth and seventh centuries show that virtually all tools and weapons of this time were made of iron (Curtis *et al.* 1979, 382–384). The archaeological evidence is mirrored by textual evidence, especially in lists of tribute and booty. These documents suffer from serious biases: the lists contain formulaic phrases, and some rulers exaggerated their achievements by inflating the quantities of goods obtained or including some not obtained (for example, the records of the rulers between the reigns of Shalmaneser III and Tiglath-pileser III appear to be somewhat fantastical). Iron, however, does not seem to suffer this fate, and the lists may be used to plot a steady increase in the quantity of iron in use (Pleiner & Bjorkman 1974, 292f; Moorey 1994, 289f). They indicate that iron had become so plentiful and unremarkable by the reign of Sargon II (ca. 722–705) that it is scarcely mentioned, and even then, only as finished artefacts which were 'numberless'.

The evidence from ninth and eighth-century Greece presents a very different picture. There are no large single assemblages of iron artefacts to compare with those of ninth-century Hasanlu or late eighth-century Khorsabad, and while it is true that this is in part the product of archaeological formation processes and the emphasis of excavation away from settlement contexts, it also seems true that where iron is found in Greece during the comparable period it appears in much smaller quantities to which considerably more importance is attached. What little evidence there is from settlement contexts seems to imply that most tools and implements in Greece were being made of wood, bone and stone, indicating that the circulation of iron (and for that matter bronze) for everyday practical purposes was restricted (I. Morris 1986, 9–11). Iron is concentrated in deliberate deposits left by members of the elite, such as the assemblage from the Panoply Tomb from Late Geometric Argos which Coldstream has described as *'the richest haul of Geometric offerings ever found in the Argolid'* in which *'wealth is expressed in iron'* (Coldstream 1977, 146). The context certainly suggests that the set of twelve iron spits and the associated pair of iron warship-shaped firedogs and pair of iron axes found here would have been the envy of the leading citizens of the city (Haarer 2000, 1.31–37). It seems unlikely that the same goods would have evoked the same response among the elite of Hasanlu or Khorsabad.

Therefore, at least as far as the end of the Geometric Period, the scale in which iron was produced and circulated in Aegean Greece seems to have been less than that in Assyria. That this is should be so should cause no surprise because of the difference between the socio-economic conditions operating in each region. Assyria was politically unified at the start of the first millennium in a way which Greece was not, and was therefore able to embark on collective endeavours such as military conquest, or the centralised procurement of iron weapons for this task on a scale which almost certainly did not become possible in Aegean Greece until much later.

CYPRUS AND THE ORIGIN OF IRON WORKING

What of Cyprus? As seen above, the one price which we have for iron from there of 240 measures of iron to one of silver lies within the range of prices known from elsewhere in the Ancient Near East during the Neo-Babylonian period (612–547). However, we have no idea whether this price would have been regarded as relatively high or low in Cyprus. Nevertheless, the behaviour of iron roasting spits in the archaeology of Cyprus is similar to that of Greece and this may indicate that the values of iron in both areas, and therefore perhaps their availability, were also similar. If so, then Cyprus would also follow rather than lead the Ancient Near East in terms of the development of iron during the Iron Age. Such a conclusion would add to the doubts expressed by others that Cyprus was the cradle in which the breakthrough bringing about the greater abundance of iron after the Bronze Age was discovered (Waldbaum 1982).

There are further grounds for scepticism. In particular, in certain areas of the ancient Near East, archaeological evidence, supplemented where possible with documentary sources, attests to tremendous and progressive development in the exploitation of iron in terms of its geographical distribution, and in the quantity, range of types and applications in which it appears from the seventh to fifth millennia to the end of the Bronze Age. These may be traced in the excellent summary provided by Waldbaum (1980), and include a gradual widening of the areas from which iron objects are attested, an increase in the number of artefacts per area and an increase in the repertoire of objects. By the end of the Bronze Age, textual evidence from Hittite Anatolia suggests that progress up to that point amounted to considerably more than what Snodgrass dismisses as *'inconsequential experimentation'*. The relevant documents list hundreds of iron objects from a repertoire which is far greater than that attested archaeologically, and which includes the use of iron in applications such as the fronts of headbands, statuettes of men and animals, an altar, a throne, a bath of 90 minas in weight, and vessels including a spouted pitcher, and rhyton (Košak 1986, 131 1i, 129–130 1g, 128 1d, 133 2c, 128 1d, 131 1j, 130 1h2, 129 1f). There are also many utilitarian items such as daggers, knives, spears, hammers, axes, maces, pegs /bolts and staffs /poles (Košak

1986, 126–128 1c, 133 2b), and the list of articles given here is far from exhaustive of Košak's catalogue.

The archaeology of Cyprus has produced little evidence to match this iron ancestry, and certainly not enough to indicate that preparation for a ferrous revolution was under way at the end of the Bronze Age. The claim that the exploitation of iron was pioneered on Cyprus is instead based largely on the supposition that the wealth of experience of Cypriot smiths in non-ferrous metal production would have encouraged them to experiment with the production of iron, and that thereafter, the new-won skills could have been passed on through extensive and well-established trading networks (Waldbaum 1982, 325).

In addition to the difficulty that the use of iron in Cyprus during the Bronze Age is much less advanced than that in other areas, the iron types found in Cyprus during the twelfth and eleventh centuries are limited in range, and it does not necessarily follow that they were manufactured there, especially since no early ironworking sites have yet been identified on the island. Moreover, although there is better evidence during the Early Iron Age for the carburisation, quenching and tempering of iron in Cyprus, this is mostly a product of the distribution of metallographic studies and the significance of such evidence is hard to assess (Waldbaum 1982, 326).

A better candidate for the development of the ferrous technology of the Iron Age may be found if we consider what happened at the end of the Bronze Age to those who had been employed elsewhere in the production of iron goods, and especially to those working in Hittite Anatolia with its relatively well developed iron industry. Cook proposes that as Anatolia became engulfed in chaos at the end of the Late Bronze Age, smiths there moved to the relative security of Cyprus taking their skills them (Cook 1988, 31). However, such a movement could have taken place before this time, especially if Cyprus came under some sort of Hittite control at the end of the Bronze Age, as intimated by textual evidence (Zaccagnini 1990, 495). No arrival of such new technology is indicated by archaeology, but it is possible that the evidence has been obscured. The symptom of a change in ferrous technology during the Early Iron Age is an increase in the quantity of iron being made. Therefore, the introduction near the end of the Bronze Age of the innovations through which this was achieved would perhaps go undetected, because collective output would almost certainly have been diminished by the chaos into which the eastern Mediterranean world descended. If so, then in metallurgical terms the beginning of the Early Iron Age far from being a progressive phase was perhaps a recessive period, during which the development and spread of an important innovation was arrested.

IS THE CONCEPT OF A TRANSITION USEFUL?

Our final point concerns the concept of the Transition from bronze to iron. If it is characterised by changes in the quantities of iron artefacts in circulation and by a fall in the value of iron from a higher value to a lower, and was the only significant period of development in ferrous technology during antiquity, then it should be readily identifiable as an unstable phase sandwiched between two stable phases. However, the decline in the value of iron from the beginning of the Early Iron Age onwards seems to have been continuous, with no sign in the evidence examined above that it was heading towards a definable conclusion. Moreover, although the scale of the decrease in value could be described as dramatic if viewed over a five hundred year period, there is evidence that the value of iron had decreased to an equal or greater extent during previous periods. As mentioned above, in the Old Assyrian period (ca. 2000–1600), texts indicate that iron was up to ninety times more valuable than silver.

Changes in the quantities of iron attested across the eastern Mediterranean world offer a similar story. The inception of the Iron Age is marked by a substantial increase in the quantity of iron in circulation. However, this was not the only time at which such a phenomenon occurred, as again becomes evident from the survey of pre-Iron Age iron by Waldbaum (1980). For example, only fourteen small objects are listed in her category of finds dating to before 3000. In contrast, her entry for the Late Bronze Age includes more than seventy finds of a wide variety of types ranging from jewellery to tools and implements while, as referred to above, Hittite documents indicate that by this same period hundreds of iron objects were in circulation (Siegelová 1984, Košak 1986).

Set in this broad chronological and geographical context, the phenomenon known as the Transition from bronze to iron may only be described as one period of more rapid development punctuating a long continuity stretching over millennia. To single it out as the only phase of real consequence does not seem justified, and is not helpful when trying to understand the history of iron.

CONCLUSIONS

There are many major issues which have not been addressed in this paper. To what extent does the price of iron reflect the availability of iron rather than other factors? How compatible are the types of evidence used in conjunction, whether of different types or from different cultural contexts? How representative of the whole is the small sample of evidence used here? In the absence of the relevant discussions, the conclusions offered serve more to problematise the Transition from bronze to iron and the model for its interpretation presented by Snodgrass, than to offer a better alternative. With this in mind, the possibilities raised may be summarised as follows:

1) At the end of the Late Bronze Age an important development in ferrous technology took place resulting in iron becoming available in unprecedented quantities. We may label this phase 'the Transition from bronze to iron', but it was perhaps not the single definitive moment in the development of iron.
2) The social impact of this developmental phase has been exaggerated in terms of time-scale and magnitude. Iron did not become available to the masses as a cheap alternative to bronze and so democratise the arts and crafts within one or two centuries. Rather, for a long time during the Early Iron Age, at least in Greece, it remained the expensive preserve of the few. As such it is likely to have been perceived as a desirable metal rather than a practical necessity.
3) The innovation which brought about the greater abundance of iron apparent in the archaeological record in the Early Iron Age was probably not developed in Cyprus, and may have been developed initially towards the end of the Late Bronze Age. If so, then we must envisage a rather different social context for this technological change.

ACKNOWLEDGEMENTS

I would like to thank Andrew Shortland for giving me the opportunity to present this paper at the conference. I would also like to thank Chris Salter and Peter Crew for their advice on technical matters, and Robin Osborne, Roger Moorey, Sue Sherratt and Eva Baboula for comments on earlier drafts of the material presented here. The views expressed and all inaccuracies remain my own.

Dedicated to my father
JIM HAARER
28.XII.1924 – 24.IV.2001

REFERENCES

Non-standard abbreviations

FD III *Fouilles de Delphes* vol. III., Boccard, Paris.

ID *Inscriptions de Délos*, F. Dürrbach (ed.), Académie des Inscriptions et Belles-Lettres, Paris, 1923–1927.

IG *Inscriptiones Graecae*, Deutsche Akademie der Wissenschaften zu Berlin, George Reimer, Berlin 1873–.

- *IG* I³ *Inscriptiones Atticae Euclidis anno anteriores*, 3ed.
- *IG* II² *Inscriptiones Atticae Euclidis anno posteriores*
- *IG* IV².1 *Inscriptiones Epidauri.*
- *IG* XI.2 *Inscriptiones Deli.*

Åström, P., Maddin, R., Muhly, J. D., Stech, T., 1986, Iron Artifacts from Swedish Excavations in Cyprus, *Opuscula Atheniensia*, 16.3, 27–41.

Blegen, C. W., 1952, Two Athenian Grave Groups of about 900 B.C., *Hesperia*, 21, 279–294.

Brinkman, J. A., 1988, Textual Evidence for Bronze in Babylonia in the Early Iron Age, 1000–539 B.C., in Curtis, J. E. (ed), *Bronzeworking Centres of Western Asia c. 1000–539 B.C.*, Kegan Paul International Ltd., London and New York, in Association with the British Museum, London.

Brown, W. L., 1950, Pheidon's Alleged Aeginetan Coinage, *Numismatic Chronicle,* 6th Series Volume 10, 177–204.

Carapanos, C., 1878, *Dodone et ses Ruines*, Librairie Hachette et Cie, Paris.

Childe, V. G., 1942, *What Happened in History*, Penguin Books Ltd., Harmondsworth.

Coldstream, J. N., 1977, *Geometric Greece*, Ernst Benn Ltd., London.

Coldstream, J. N., 1989, Status symbols in Cyprus in the eleventh century BC, in Peltenberg, E. (ed), *Early Society in Cyprus*, Edinburgh University Press in association with the National Museums of Scotland and A.G. Leventis Foundation, Redwood Burn Ltd., Trowbridge, 325–335.

Cook, V., 1988, Cyprus and the Outside World During the Transition from the Bronze Age to the Iron Age, *Opuscula Atheniensia*, 17.2, 13–32.

Courbin, P., 1959, Valeur Comparée du Fer et de L'Argent lors de l'introduction du monnayage, *Annales Économies Sociétés Civilisations*, 14(2), 209–233.

Courbin, P., 1983, Obéloi d'Argolide et d'ailleurs, in Hägg, R. (ed), *The Greek Renaissance of the Eighth Century BC: Tradition and Innovation*, Proceedings of the second international

Symposium at the Swedish Institute in Athens, 1981, Swedish Institute at Athens/ Paul Åströms Förlag, Stockholm, 1983, 149–156.

Craddock, P. T., 1995, *Early Metal Mining and Production*, Edinburgh University Press, Edinburgh.

Crew, P., 1991, The Experimental Production of Prehistoric Bar Iron, *Historical Metallurgy*, 15.1, 21–36.

Crew, P., 1994, Currency Bars in Great Britain, Typology and Function, in Mangin, M. (ed.), 1994, *La sidérurgie ancienne de l'Est de la France dans son contexte européen, Archéologie et archéométrie*, Actes du Colloque de Besancon, 10–13 November 1993, Annales littéraires de l'Université de Besancon, 536, Diffusion des Belles Lettres, Paris, 345–350.

Crew, P., 1995, Aspects of the Iron Supply, in Cunliffe, B. W., 1995, *Danebury: an Iron Age hillfort in Hampshire, Volume 6 a hillfort community in perspective*, CBA Research Report 102, Council for British Archaeology, York, 276–284.

Curtis, J. E., Wheeler, T. S., Muhly, J. D., Maddin, R., 1979, Neo-Assyrian Ironworking Technology, *Proceedings of the American Philosophical Society*, 123, 369–390.

Felsch, R. C. S., Kienast, H. J., Schuler, H., 1980, Apollon und Artemis oder Artemis und Apollon? Bericht von den Grabungen im neu entdeckten Heiligtum bei Kalapodi 1973–1977, *Archäologischer Anzeiger*, 1980, 38–123.

Furtwängler, A. E., 1980, Zur Deutung der Obeloi im Lichte samischer Neufunde, *in* Hampe, R. (ed), *Tainia, Festschrift für Roland Hampe*, Verlag Philipp von Zabern, Mainz, 81–98.

Gjerstad, E., *et al.*, 1934, *The Swedish Cyprus Expedition Volume I, Finds and Results of the Excavations in Cyprus 1927–1931*, Stockholm.

Goetze, A., 1940, *Kizzuwatna and the Problem of Hittite Geography*, Yale Oriental Series, Researches Volume XXI, Yale University Press, Yale.

Haarer, P. S., 2000, *Obeloi and Iron in Archaic Greece*, unpublished D.Phil. thesis, Oxford.

Horne L., 1982a, Fuel for the Ancient Metalworker, the Role of Charcoal and Charcoal Production in Ancient Metallurgy, *Expedition*, 25.1, 6–13.

Horne L., 1982b, The Demand for Fuel: Ecological Implications of Socio-Economic Change, in Spooner, B. & Mann, H. S. (eds), *Desertification and Development, Dryland Ecology in Social Perspective*, Academic Press, London, 201–205.

Košak, S., 1986, The Gospel of Iron, in Hoffner, H. & Beckman, G. (eds), *Kaniššuwar, a tribute to Hans G Güterbock*, the Oriental Institute of the University of Chicago, Assyriological Studies 23, Chicago, 125–135.

Kunze, E. & Schleif, H., 1944, *Olympische Forschungen I*, Archäologisches Institut des Deutschen Reichs, Verlag Walter de Gruyter & Co., Berlin.

Lang, M. & Crosby, M., 1964, *The Athenian Agora Volume X Weights, Measures and Tokens*, The American School of Classical Studies at Athens, Princeton, New Jersey, Printed in Germany at J.J. Augustin, Glückstadt.

Maddin, R., 1982, Early Iron Technology in Cyprus, in Muhly, J. D., Maddin, R., Karageorghis, V. (eds), *Early Metallurgy in Cyprus 4000–500BC*, ACTA of the International Archaeological Symposium, Pierides Foundation, Larnaca, Cyprus, 303–314.

Markle, M. M., 1985, Jury Pay and Assembly Pay at Athens, in Cartledge, P. A. & Harvey, F. D. (eds), *CRUX. Essays Presented to G.E.M. de Ste. Croix on his 75th Birthday*, Imprint Academic, Sidmouth Printing Ltd., Sidmouth, 265–297.

Moorey, P. R. S., 1994, *Ancient Mesopotamian Materials and industries, The Archaeological Evidence*, Clarendon Press, Oxford.

Morris, I., 1986, Gift and Commodity in Archaic Greece, *MAN (N.S.)* 21, 1–17.

Muhly, J. D., 1980, The Bronze Age Setting, in Wertime, T.A. & Muhly, J.D. (eds), *The Coming of the Age of Iron*, Yale University Press, New Haven and London, 25–76.

Muhly J. D., Maddin R., Stech T., 1990, The Metal Artefacts, in Fritz, V., *Kinnert, Ergebnisse der Ausgrabungen auf dem Tell el - 'Oreme am See Gennesaret 1982–1985*, Otto Harrasowitz, Wiesbaden, 159–175 (=Ch. 13).

Northover, J. P, 1988, Alloy Design in the Bronze Age, in Jones, J.E., *Aspects of Ancient Mining and Metallurgy. Acta of a British School at Athens Centenary Conference, Bangor 1986*, British School at Athens, Bangor, 44–56.

Oppenheim, A. L., 1967, Essay on Overland Trade in the First Millennium B.C., *Journal of Cuneiform Studies*, 21, 236–254.

Payne, H., 1940, *Perachora I, the Sanctuaries of Hera Akraia and Limenaia; Excavations of the British School of Archaeology at Athens, 1930–1933*, Clarendon Press, Oxford.

Perdrizet, P., 1908, *Fouilles de Delphes V, Monuments Figurés, Petits Bronzes, Terres-Cuites, Antiquités Diverses*, École Française D'Athènes, Paris.

Pickles, S. & Peltenberg, E., 1998, Metallurgy, Society and the Bronze /Iron Transition in the East Mediterranean and the Near East, *Report of the Department of Antiquities, Cyprus*, 1998, 67–100.

Pigott, V.C., 1989, The Emergence of Iron Use at Hasanlu, *Expedition*, 31 Nos. 2–3, 67–79.

Place, V., 1867, *Ninive et L'Assyrie*, Imprimerie Impériale, Paris.

Pleiner, R., 1969, *Iron Working in Ancient Greece*, National Technical Museum, Praha.

Pleiner, R. & Bjorkman, J.K., 1974, The Assyrian Iron Age, The History of Iron in the Assyrian Civilization, *Proceedings of the American Philosophical Society*, 118, 283–313.

Powell, M.A., 1990, Identification and Interpretation of long Term Price Fluctuations in Babylonia: More on the History of Money in Mesopotamia, *Altorientalischer Forschungen*, 17, 77–99.

Price, M. J., 1968, Early Greek Bronze Coinage, in Kraay, C. M. & Jenkins, G. K. (eds), 1968, *Essays in Greek Coinage Presented to Stanley Robinson*, Clarendon Press, Oxford, 90–104.

Pritchett, W. K., 1956, The Attic Stelai Part II, *Hesperia* 25, 178–328.

Rupp, D. W., 1989, Puttin' on the Ritz: manifestations of high status in Iron Age Cyprus', in Peltenberg, E. (ed), *Early Society in Cyprus*, Edinburgh University Press in association with the National Museums of Scotland and A.G. Leventis Foundation, Redwood Burn Ltd., Trowbridge, 336–362.

Salter, C. J., 1989, The Scientific Investigation of the Iron Industry in Iron Age Britain, in Henderson, J. (ed), *Scientific Analysis in Archaeology and its Interpretation*, Oxford University Committee for Archaeology, Monograph No. 19, UCLA Institute of Archaeology Research Tools 5, Oxbow, Oxford, 250–273.

Scott, B. G., 1990, *Early Irish Ironworking*, Ulster Museum.

Sherratt, E. S., 1994, Commerce, Iron, and Ideology: Metallurgical Innovation in 12th-11th century Cyprus, in Karageorghis, V. (ed), *Cyprus in the 11th Century B.C., Proceedings of the International Symposium Organised by The Archaeological Research Unit of the University of Cyprus and the A.G. Leventis Foundation, Nicosia 30–31 October 1993*, Nicosia, 59–106.

Siegelová, J., 1984, Gewinnung und Verarbeitung von Eisen im Hethitischen Reich im 2. Jahrtausend v.u.Z, *Annals of the Naprstek Museum (Prague)*, 12, 71–168.

Snodgrass, A. M., 1971, *The Dark Age of Greece: an archaeological survey of the eleventh to the eighth centuries BC*, Edinburgh University Press, Edinburgh.

Snodgrass, A. M., 1980a, Iron and Early Metallurgy in the Mediterranean, in Wertime, T.A. & Muhly, J.D. (eds), *The Coming of the Age of Iron*, Yale University Press, New Haven and London, 335–374.

Snodgrass, A. M., 1980b, *Archaic Greece, The Age of Experiment*, Dent & Sons, London.

Snodgrass, A. M., 1982, Cyprus and the Beginnings of Iron Technology in the Eastern Mediterranean, in Muhly, J.D., Maddin, R., Karageorghis, V. (eds), *Early Metallurgy in Cyprus*

4000–500BC, ACTA of the International Archaeological Symposium, Pierides Foundation, Larnaca, Cyprus, 285–295.

Snodgrass, A. M., 1987, *An Archaeology of Greece, the Present State and Future Scope of a Discipline*, Sather Lectures Volume 53, University of California Press, Berkeley, Los Angeles, Oxford.

Snodgrass, A. M., 1989, The Coming of the Iron Age in Greece: Europe's Earliest Bronze/Iron Transition', in Sørensen, M. L. S. & Thomas, R. (eds), *The Bronze Age - Iron Age Transition in Europe, Aspects of Continuity and Change in European Societies c.1200 to 500 B.C.*, BAR International Series 483(i), Oxford, 22–35.

Snodgrass, A. M., 1994, Gains, Losses and Survivals: What we infer for the 11th century BC, in Karageorghis, V. (ed), *Cyprus in the 11th Century B.C., Proceedings of the International Symposium Organised by The Archaeological Research Unit of the University of Cyprus and the A.G. Leventis Foundation, Nicosia 30–31 October 1993*, Nicosia, 167–174.

Stager, L. E., 1985, The Archaeology of the Family in Ancient Israel, *Bulletin of the American Schools of Oriental Research*, 260, 1–35.

Sturz, F. G., 1820, *Orionis Thebani Etymologicon, ex Museo Frid. Aug. Wolfii*, Ioa Aug. Gottl. Wiegel, Leipzig.

Svoronos, I. N., 1906, Obeloi sideroi, *Journal International d'Archéologie Numismatique*, 9, 192–202.

Treister, M. Y., 1996, *The Role of Metals in Ancient Greek History*, Mnemosyne Biblotheca Classica Batava, Supplementum Centesimum Quinquagesimum Sextum, E.J. Brill, Leiden, New York, Köln.

van Straten, F. T., 1995, *Hiera Kala. Images of Animal Sacrifice in Archaic and Classical Greece*, Religions in the Graeco-Roman World Vol. 27, E. J. Brill, Leiden.

Waldbaum, J. C., 1980, The First Archaeological Appearance of Iron and the Transition to the Iron Age, in Wertime, T. A. & Muhly, J. D. (eds), *The Coming of the Age of Iron*, Yale University Press, New Haven and London, 69–98.

Waldbaum, J. C., 1982, Bimetallic Objects from the Eastern Mediterranean and the Question of the Dissemination of Iron, in Muhly, J. D., Maddin, R., Karageorghis, V. (eds), *Early Metallurgy in Cyprus 4000–500BC*, ACTA of the International Archaeological Symposium, Pierides Foundation, Larnaca, Cyprus, 325–349.

Waldstein, C., 1902, *The Argive Heraion, Vol. I: General Introduction, Geology, Architecture, Marble Statuary and Inscriptions*, Houghton Mifflin & Co., Boston and New York, the Riverside Press, Cambridge.

Wertime, T. A., 1982, Cypriot Metallurgy against the Backdrop of Mediterranean Pyrotechnology: Energy Reconsidered, in Muhly, J. D., Maddin, R., Karageorghis, V. (eds), *Early Metallurgy in Cyprus 4000–500BC*, ACTA of the International Archaeological Symposium, Pierides Foundation, Larnaca, Cyprus, 351–361.

Wertime, T. A., 1983, The Furnace Versus the Goat: the Pyrotechnologic Industries and Mediterranean Deforestation in Antiquity, *Journal of Field Archaeology*, 10, 445–452.

Zaccagnini, C., 1990, The Transition from Bronze to Iron in the Near East and in the Levant: Marginal Notes, *Journal of the American Oriental Society*, 110.3, 493–502.